Laboratory Manual
for Human Anatomy
and Physiology:
Fetal Pig Version

Laboratory Manual for Human Anatomy and Physiology:
Fetal Pig Version

281-930059

Eileen Walsh   Kathryn E. Malone   Jane M. Schneider

Westchester Community College

West Publishing Company

St. Paul   New York   Los Angeles   San Francisco

West's Commitment to the Environment

In 1906, West Publishing Company began recycling materials left over from the production of books. This began a tradition of efficient and responsible use of resources. Today, up to 95 percent of our legal books and 70 percent of our college texts are printed on recycled, acid-free stock. West also recycles nearly 22 million pounds of scrap paper annually—the equivalent of 181,717 trees. Since the 1960s, West has devised ways to capture and recycle waste inks, solvents, oils, and vapors created in the printing process. We also recycle plastics of all kinds, wood, glass, corrugated cardboard, and batteries, and have eliminated the use of styrofoam book packaging. We at West are proud of the longevity and the scope of our commitment to our environment.

Production, Prepress and Binding by West Publishing Company.

The procedures and experiments in this book are designed to be conducted in a laboratory setting under the supervision of an instructor. You and your instructor are responsible for your safety when conducting these procedures and experiments. Neither the authors nor the publisher will be responsible for any injury or damage that may occur as a result of any procedure or experiment.

Dedication . . .

to our mothers with gratitude

Contents in Brief

Contents

Preface

*"It is clear that the body
of an animal cannot be simple."*
–Aristotle

This laboratory manual has been written, in terms of style, with the student in mind. It is, we believe, clear with respect to directions and guidelines and easy to follow. The organization and scope would permit the use of this manual with either a one or two semester course. It is not expected that all exercises would be done in any one course, but rather, instructors can select the exercises based upon the needs of their particular students and their equipment inventory.

 ## Features

There are many pedagogical features incorporated into this manual which should promote greater student understanding. These include:

1. A brief *overview* statement at the beginning of each unit that provides generalized information about the activities to be done in each unit;
2. An overall *background* that gives an introduction in terms of content to the material to be covered in the unit;
3. A specific *discussion* section for each major topic covered within a unit in which the key words are in boldface type and definitions are provided;
4. Clear *materials* list and *procedures* for each exercise except Unit 10 where the description of muscles and directions for dissection are combined into a single section;
5. Numerous *tables* and *illustrations* with added information that provides students with a substantial information base to effectively complete the exercise;

6. Short *Quick Quizzes* dispersed within the unit for students to test themselves and *Unit Tests* at the back of the manual for either instructor or student use;
7. An extensive number of *color illustrations including photomicrographs of the tissues and a photo gallery* near the back of the manual with color photos of preserved specimens that correspond to the line drawings within the text;
8. Units that have a *balanced approach* to the content of anatomy and physiology by including activities in gross anatomy, microscopic anatomy, and physiology;
9. A *safety precaution section* at the beginning of the manual that incorporates procedures that are based upon recommendations from the Centers for Disease Control; warnings that refer students to these safety precautions are found throughout the manual;
10. *Appendixes* that provide additional data, e.g., normal body fluid components, anatomical landmarks, etc.;
11. Thorough *index* for quick reference to content.

 ## Acknowledgments

The completion of this manual has been achieved through the cooperative efforts of many individuals. We wish to thank all of them and in particular express our appreciation to the staff at West Educational Publishing including: Ron Pullins for his encouragement and a special thanks to his assistant Denise Bayko for her patience and good humor, also to Deanna Quinn for her fine efforts in the production aspects of this work. Her suggestions were insightful and her affability was appreciated.

Additionally we would like to thank Richard Morel for the initial editing of the manuscript, and Rolin Graphics for the art program.

We are also grateful to Jeff Grosscup for the fine photos of the dissection specimens and to James DeLorenzo, Paul Hinchey and Katie Horan for their help in preparing the specimens.

We also wish to thank the following reviewers for their constructive comments during the writing of this manual:

Mary Schwanke, *University of Maine, Farmington*
Reinhold J. Hutz, *University of Wisconsin, Milwaukee*
James B. Larsen, *University of South Mississippi*
Cynthia Carey, *University of Colorado, Boulder*
Harvey Liftin, *Broward Community College*

Jean Cons, *College of San Mateo*
Donald W. Green, *New Mexico Junior College*
James E. Hall, *Central Piedmont Community College*
Louis C. Renaud, *Prince George's Community College*
Robert E. Nabors, *Tarrant County Junior College*
Linda L. MacGregor, *Bucks County Community College*
David Smith, *San Antonio College*
Dwayne H. Curtis, *California State University, Chico*
Phillip Cooper, *Suffolk County Community College*
Eugene R. Volz, *Sacramento City College*
William C. Matthai, *Tarrant County Junior College*
A. Kenneth Moore, *Seattle Pacific University*

Eileen Walsh
Kathryn Malone
Jane Schneider

Safety Precautions

Safety precautions and guidelines are outlined here for your protection in the laboratory. They should be read *before* embarking on your laboratory work. In addition, where they are referred to specifically in a unit, they should be reviewed before beginning work.

 ## A. General Guidelines

ALWAYS FOLLOW THE DIRECTIONS OF YOUR INSTRUCTOR.

1. Take time to read and become thoroughly familiar with the procedures and precautions for a specific laboratory exercise before beginning to work.
2. If you are unclear of any procedure, ask your laboratory instructor for clarification. **DO NOT** initiate experiments on your own and **DO NOT** work unsupervised.
3. Keep your lab work area and yourself free of encumbrances. Place unnecessary books, clothing, etc., away from your work area.
4. Dispose of all waste material according to your instructor's directions.
5. **DO NOT** eat, drink, or smoke in the laboratory and never place anything in your mouth while you are in the laboratory.
6. Become familiar with the location and use of equipment to be used in emergencies, e.g., fire extinguisher, fire blankets, first aid kits. Be aware of the nearest exit in the laboratory.
7. Always clean your work area and your hands before leaving the laboratory. Use a rapid acting antimicrobial skin cleanser (such as Hibiclens).

 ## B. Precautions Concerning Work With Body Fluids, Animal Samples, and Preserved Specimens

1. When appropriate, use only your own body samples.
2. Always wear protective gloves.
3. **DO NOT** pipet anything by mouth. Use mechanical devices.
4. Dispose of the following in a BIOHAZARD container:
   a. test tubes that have held human or animal samples
   b. slides that were used for sample fluid smears
   c. used capillary tubes
   d. used urine containers
   e. used mouthpieces for the spirometer
5. Clean your laboratory work area with a 10% Chlorox or Lysol solution at the beginning and end of the laboratory session.
6. Clean reusable instruments as described in number 5 above and wash with soap and water.
7. Always use care when working with instruments where splashing or aerosolizing (blending, centrifuging) may occur.

C. Precautions Concerning Reagents/Chemicals

1. Report any accidents (spills, broken glass, cuts, etc.) to your instructor.
2. Use care when handling chemicals and if they make contact with your eye flush the eye with running water for 5 minutes. Report to your lab instructor any contact of eyes or skin with chemicals and seek appropriate medical help.
3. Avoid breathing fumes from reagents or preserved specimens for a prolonged period. A large fan should be used if the lab's air circulation is sluggish.
4. If you are directed to note the odor of a chemical, waft the vapors from the end of the vessel containing the chemical toward your nose.
5. Avoid pointing the open end of a test tube toward anyone, including yourself.
6. When heating any container, use heat-resistant glassware and handle with appropriate caution.
7. Pour reagents at the supply area and **DO NOT** return unused materials to stock bottles.
8. Follow the direction of your instructor regarding the proper disposal of all chemicals.

 ## D. Precautions Concerning Dissections

1. Care should be used in handling dissection equipment both during the dissection and at clean-up.
2. Should any cuts or puncture wounds occur from the equipment or specimens, wash the wound with disinfectant soap, notify your instructor and seek appropriate medical attention.
3. Hold preserved specimens under the faucet and allow water to rinse the specimen for several minutes to wash away excess preservative.
4. Return **cleaned** dissecting equipment to its original place.
5. Discard particulate matter from preserved specimens according to your instructor's directions.

UNIT 1

Terminology and Body Regions

OVERVIEW

In this unit you will become familiar with the standard vocabulary used in anatomical descriptions. Specifically, you will learn the major directional terms, body planes and sections and the body regions and cavities.

OUTLINE

Terminology and Body Regions

Exercise 1   Identification of Directional Terms

Exercise 2   Body Planes and Sections
Exercise 3   Body Regions and Cavities

 Terminology and Body Regions

Background

To standardize the location of structures in the human body, the assumption is made that the body is in the **anatomical position.** This term defines the body as standing erect with feet together, the upper limbs hanging at the sides, with the palms facing forward. From this position we can move in a well-defined way to various areas of the body. Keep in mind as you look at the belly side of the body in this position that the **body's right side** is on **your left side.**

EXERCISE 1
Identification of Directional Terms

Discussion

Figure 1.1 locates the terms frequently encountered in anatomical description on the human body (a biped) and on the pig (a quadruped). Table 1.1 summarizes these terms.

Materials

A model of the human torso.

Procedure

Locate the following structures and record the appropriate directional terms in the Observations section.

a. The navel relative to the nipples.

b. The shoulder blade relative to the breastbone.

c. The testicle relative to the hips.

d. The abdominal muscles relative to the intestine.

e. The knee relative to the groin.

Observations

Use the appropriate terms to describe the directional relationships of the structures a through e in the Procedure section.

■ **FIGURE 1.1 Directional terms**

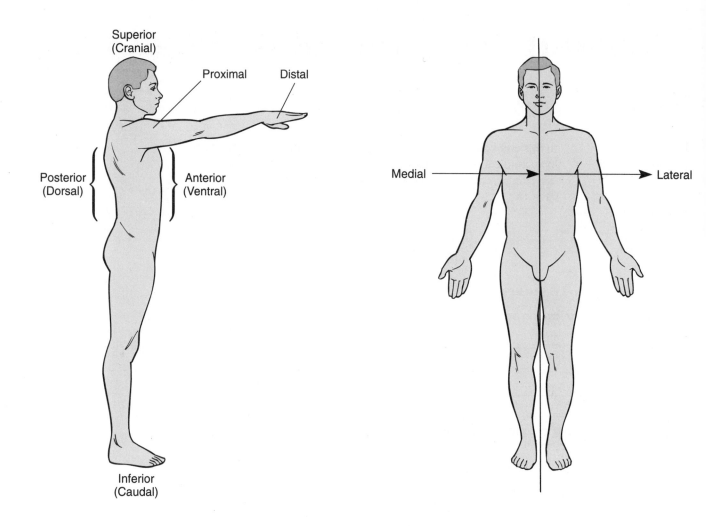

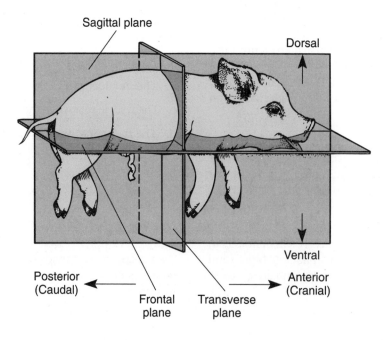

■ **TABLE 1.1 Directional Terms**

TERM	DEFINITION	EXAMPLE
Anterior (ventral)	Situated in front of the front of the body	The chest is on the anterior surface of the body
Posterior (dorsal)	Situated in back of the back of the body	The buttocks are on the posterior surface of the body
Superior (cranial)	Toward the head; relatively higher in position	The eyebrows are superior to the eyes
Inferior (caudal)	Away from the head; relatively lower in position	The mouth is inferior to the nose
Medial	Toward the midline of the body	The breast is medial to the armpit
Lateral	Away from the midline of the body	The hip is on the lateral surface of the body
Proximal	Closer to any point of reference, such as the attached end of a limb, the origin of a structure, or the center of the body	The arm is proximal to the forearm
Distal	Farther from any point of reference, such as the attached end of a limb, the origin of a structure, or the center of the body	The hand is distal to the wrist
Superficial (external)	Located close to or on the body surface	The skin is superficial to the muscles
Deep (internal)	Located further beneath the body surface than superficial structures	The muscles are deep to the skin

a. _____

_____

b. _____

_____

c. _____

_____

d. _____

_____

e. _____

_____

EXERCISE 2
Body Planes and Sections

Discussion

It is often convenient to examine a specific structure from various views. Therefore, a cut is made to allow for this alternative perspective. For example, it is important to observe certain bones of the skull from both a lateral and a medial aspect. To do this, a cut down the center of the skull from ventral to dorsal is necessary. Table 1.2 and Figure 1.2 illustrate the body planes and sections.

Procedure

1. Define the type of plane that divides the structures listed in Step. 2.

2. Record your answers in the Observations section. (More than one term may be used.)
   a. The plane that divides the kidney into front and back halves.
   b. The plane that separates the stomach from the urinary bladder.
   c. The plane that separates the two pairs of upper incisors.

Observations

Record the appropriate planes for those areas described in a through c in the Procedure section.

a. _____

b. _____

c. _____

■ **TABLE 1.2   Body Planes and Sections**

PLANE/SECTION	DESCRIPTION	EXAMPLE
Sagittal	A longitudinal section that divides a structure into right and left portions	A longitudinal cut of the head through the eye
Midsagittal	A sagittal cut which divides the body into *equal* left and right portions	A cut through the umbilicus
Parasagittal	A sagittal cut which divides the body into *unequal* portions	A cut through the chest to the left of the sternum
Frontal/Coronal	A longitudinal section made at a right angle to the sagittal sections. It divides the body into anterior and posterior portions	A section which separates the breastbone from the vertebrae
Transverse	A cut across the length of a part which divides it into superior and inferior portions	A guillotine makes a transverse cut

■ **FIGURE 1.2   Body planes and sections**

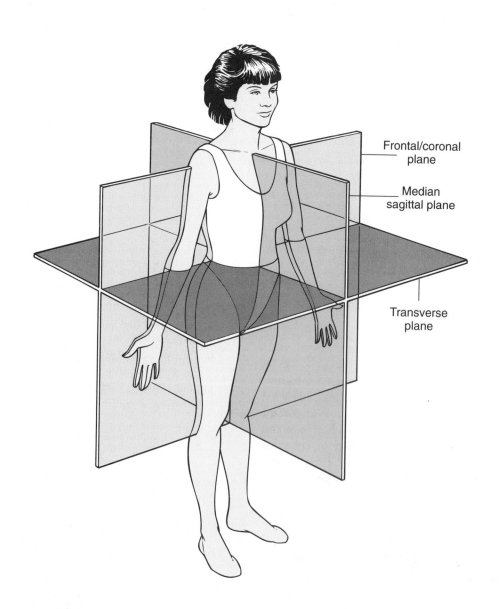

Frontal/coronal plane

Median sagittal plane

Transverse plane

EXERCISE 3
Body Regions and Cavities

Discussion

Table 1.3 summarizes some of the body regions commonly referred to, and Figure 1.3 illustrates some of them. In addition, there are two major body cavities: the **dorsal/anterior** (consisting of the cranial and spinal subdivisions) and the **ventral/posterior** (consisting of the thoracic and abdomino-pelvic cavities separated by the diaphragm). These are illustrated in Figure 1.4.

Finally, specific regions of the abdominal surface are illustrated in Figure 1.5 (page 8).

Materials

A model of the human torso.

Procedure

1. Locate the following structures as to region or cavity. Record your findings in the Observations section.
   a. In which region is the Adam's apple located?
   b. In which cavity is the brain located?
   c. In which cavity are the lungs located?
   d. In which region are the ovaries located?
   e. In which region of the abdomen is the navel located?

Observations

Use the proper terms to identify the regions or cavities described in a through e in the Procedure section.

a. _____

_____

b. _____

_____

c. _____

_____

d. _____

_____

e. _____

_____

■ **TABLE 1.3 Body Regions**

REGION	AREA	EXAMPLE
Cephalic/Cranial	Head region	The nose is in the **cephalic** region.
Cervical	Neck region	There are seven **cervical** vertebrae in humans.
Thoracic	Chest region	The heart is located in the **thoracic** region.
Axillary	Armpit region	The breast is medial to the **axillary** area.
Brachial	Upper arm region	The triceps muscle is located in the **brachial** region.
*Abdmonial	Lower ventral trunk region	The stomach is located in the **abdominal** region.
*Pubic/Pelvic	Genital region	The penis is located in the **genital** area.

*Sometimes called abdominopelvic since there is no physical border between the two.

■ **FIGURE 1.3   Body regions**

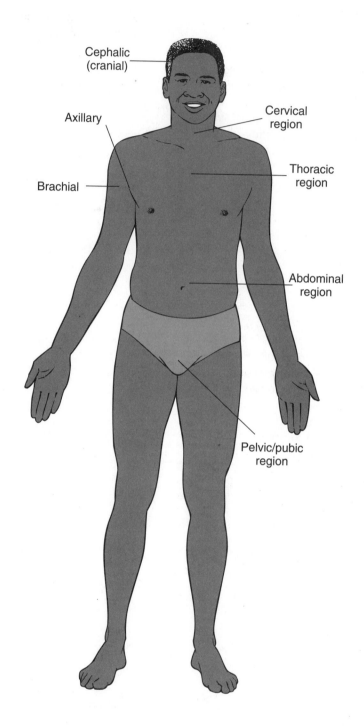

Cephalic (cranial)

Axillary

Brachial

Cervical region

Thoracic region

Abdominal region

Pelvic/pubic region

■ **FIGURE 1.4 Body cavities**

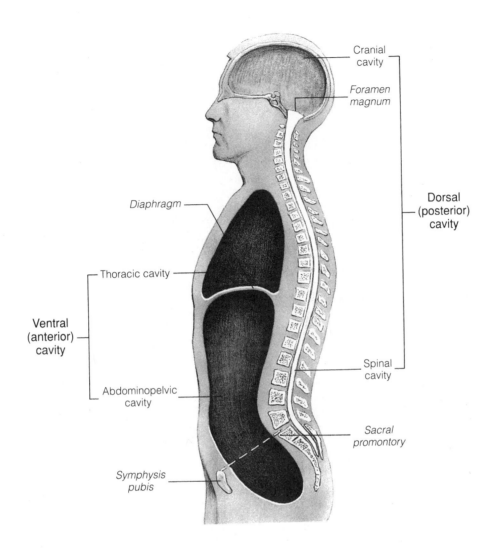

■ **FIGURE 1.5 Abdominal regions**

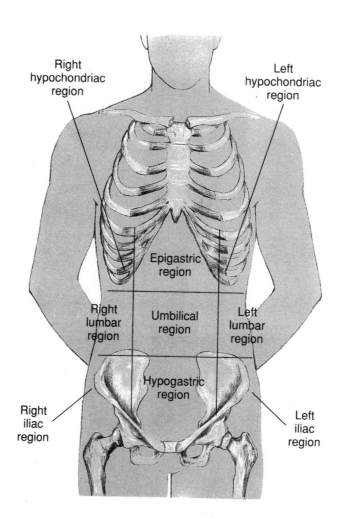

U N I T   2

Measurements and the Compound Microscope

O V E R V I E W

In these exercises you will work with everyday measurements, converting between English and metric systems, and within the metric system as well so that you learn the basis of the metric system. In addition, you will learn to use the compound microscope as a basic tool of histologic examination for the study of anatomy. This means that you need to become familiar with the parts of the compound microscope as well as their functions.

O U T L I N E

Measurements

Exercise 1   Conversions Within the Metric System

The Compound Microscope

Exercise 1   Parts and Functions of the Microscope
Exercise 2   Use of the Microscope
Exercise 3   Care of the Microscope
Exercise 4   Preparation of a Cheek Smear

Measurements

Background

While measurements in the United States are commonly made in units of the English system (pounds, quarts, inches, etc.), the metric system is becoming increasingly more important in our lives. Food cans give alternative weights in grams; car engine capacities are advertised in cubic centimeters (cc's), and wrench sets are stamped with "cm" and "mm" sizes. The scientific community has long used the metric system to record its measurements. The dimensions of the cells and structures you observe under the microscope, and indeed all measurements related to the human body, ultimately have meaning only in the metric system (see Appendix A, "Measurements").

EXERCISE 1
Conversions Within the Metric System

Discussion

The metric system is neatly set up on the basis of ten. Length measurements are used to illustrate the metric system, but comparisons to volume and weight should be noted (Appendix A). The **meter** is about a yard long and is used for large measurements like body height. When divided into hundredths, the **centimeter** is obtained. When divided into thousandths, the **millimeter** is obtained (milli = one thousandth), and now small structures such as eyelashes and freckles can be measured. Dividing the millimeter into a thousand equal parts gives the unit known as the **micrometer,** symbolized by μm. Body cells are usually measured in micrometers. For example, liver cells are 20–35 μm in diameter, and white blood cells are between 9–21 μm in diameter. Continued

9

subdivision of the micrometer gives us the **nanometer** (0.001 μm), and one further division by 10 produces the **angstrom** (Å). The two latter units are used when measuring cell organelles. Table 2.1 illustrates the hierarchy from meter (m) down to Angstrom (Å) and provides a "relative" measurement for some things with which you are familiar and some things with which you will become familiar during your study of anatomy and physiology. This table also provides information about size ranges: those which can be seen with the naked eye and those which require a microscope to be viewed.

Materials

A meter stick; a centimeter or millimeter ruler.

Procedure

1. Measure the height of your lab partner using the meter stick. Record your data in Table A, which follows.
2. Resting the centimeter ruler on the bridge of the nose, measure the width of your lab partner's eye, from the medial corner (near the nose) to the lateral corner (on the side). Record in Table A.

3. Measure the diameter of a button on your partner's shirt, or something of equivalent size. Record in Table A.
4. Practice making conversions within the metric system by converting the given numbers in Table 2.1 to the units on either side of each of the numbers.

Observations

■ TABLE A Measurements

DATA	m	cm	mm
Height			
Eye			
Button			

Conclusions

● What is the most appropriate metric unit for measuring adult height? _____

■ TABLE 2.1 Units of Length Measurement—Metric System

SPECIMEN[a]	METER	CENTIMETER (cm = 0.01 m)	MILLIMETER (mm = 0.001 m)	MICROMETER (μm = 0.001 mm)	NANOMETER (nm = 0.001 μm)	ANGSTROM (Å = 0.1 nm)	Limits
Nerve cell process	1	—	—				Naked eye
Ostrich egg yolk	—	10	—				Naked eye
Hen's egg yolk	—	—	10	—			Naked eye
Human egg			—	100	—		Naked eye
Red blood cell			—	10	—		Light microscope
Bacterium			—	1	—		Light microscope
Virus				—	100	—	Electron microscope
Protein molecule				—	10	—	Electron microscope
Amino acid molecule				—	1	—	Electron microscope

[a]The units given for each specimen are meant to convey "order" of size and not specific size. For example, the yolk of a hen's egg may measure 23 mm in diameter; therefore, the appropriate unit of measurement is tens of millimeters (i.e., 2.3 × 10 mm), not ones or hundreds (see Appendix A, note a).

- What is the most appropriate metric unit for measuring eye width? _____

- What is (are) the most appropriate metric unit(s) for measuring the button? _____

- If volume units are treated like length units, how does a millimeter (ml) relate to a liter?

  _____

- What is the value of determining the measurements in Table A?

  _____

  _____

  _____

❖ The Compound Microscope

Background

An important aspect of the study of anatomy is microscopic structure. Not only are observations and dissections of the digestive system, for example, carried out, but sections of the tissues that comprise the walls of the digestive tract also are examined. Only through a combination of both gross, **macroscopic anatomy** (which can be seen with the naked eye) and **microscopic anatomy** (which needs the aid of lenses to observe) will a really clear comprehension of the system emerge. Figure 2.1 illustrates the parts of the microscope and Table 2.2 illustrates their functions.

■ **FIGURE 2.1　The compound microscope**

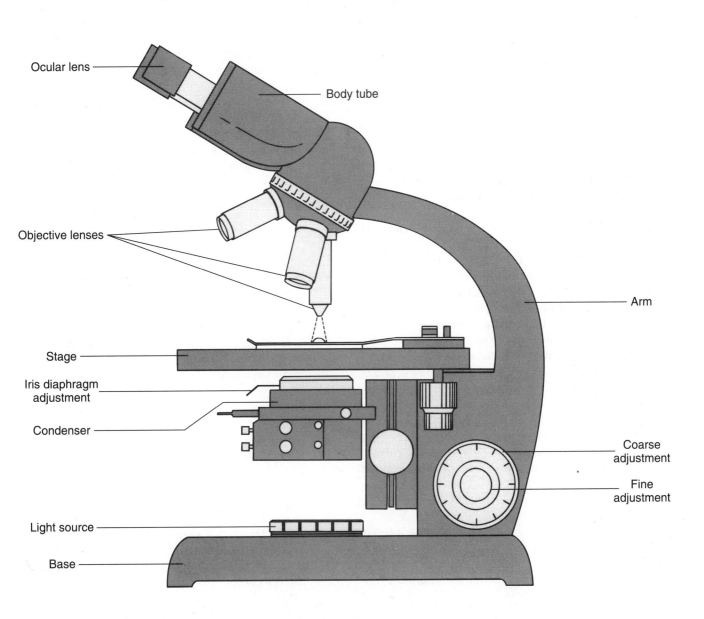

■ **TABLE 2.2 Functions of the Parts of the Compound Microscope**

PART	FUNCTION/DESCRIPTION
Light source	*Emits* light that passes through the specimen in varying degrees
Iris diaphragm	Controls the *amount* of light that passes upward toward the specimen
Condenser	*Concentrates* light on the specimen
Stage	Holds the specimen; may be mechanical or have clips
Objective lens	Lenses of different magnifying powers; located on the **nosepiece**, which allows the objective to snap into place in line with the ocular and the specimen; the magnifying power is written on the tube that holds the lens
Body tube	Maintains ocular lens and objective lens at proper distance from each other allowing for clear magnification
Ocular lens	The magnifying lens through which you look; microscopes may be **monocular** (you use one eye) or **binocular;** the magnifying power is written on the tube that holds the lens
Coarse adjustment	Allows for change in **working distance** for proper focusing; usually used for low power only
Fine adjustment	Used in high-power focusing, to make sensitive adjustments in working distance
Arm	Body tube support; microscope is always carried at arm and base
Base	Bears weight of microscope

The microscope is the instrument used to magnify tiny cells to a size that can be perceived. **Total magnification** of the light microscope is the product of two lenses used: the **ocular lens** and the **objective lens.** Thus

$$M_t = M_\propto \times M_{obj}$$

where

M_t = total magnification
$M_\propto$ = ocular lens magnification
M_{obj} = objective lens magnification

The magnifying power of each lens is written on the lens tube.

Just as the human eye has a limit to the size it can perceive, the ordinary light microscope has a useful magnification limit of about 1500X. A bacterium as small as 1 μm can be observed under the light microscope, but no detail can be discerned. To make out further detail, an instrument with greater resolving power, not greater magnifying power, is required. The resolving power of a microscope is the ability to clearly separate two points that are close together. **Resolving power** depends directly on the **wavelength of radiation** (λ) hitting the specimen, and indirectly on the **numerical aperture** (N.A.). The latter term is the product of the size of the cone of light entering the objective lens times the **refractive index** of the medium between the specimen and the objective lens. Because the electron microscope uses radiation whose wavelength is 10^{-5} (0.00001) that of the light microscope, it can discriminate much finer detail.

When the objective lens on the microscope labeled "oil" is employed, a drop of "immersion oil" is placed between the specimen and objective lens. This creates an optically homogeneous pathway from the slide to the objective lens. In this way a larger cone of light enters the objective lens, thus increasing the N.A., and the resolving power is greater. The oil lens is used when observing extremely small structures, such as cilia on a cell, or blood cells of various types.

There are many variations of the microscope available for use in the laboratory today. Many of these modifications make use of different types of light, or electromagnetic radiation as the "illuminating" source — for example, ultraviolet waves, and varying colors of light. These variations bring to view different cellular components.

EXERCISE 1
Parts and Functions of the Microscope

Discussion

Figure 2.1 diagrams a typical light microscope and its component parts. The functions of these parts are summarized in Table 2.2.

Materials

A compound microscope.

Procedure

1. Your instructor will demonstrate the location of microscope parts, which vary with different models.
2. Record the magnifications of all lenses in the Observations section.

Observations

Record the magnification of each of the following.

a. Ocular lens _____

b. Low-power objective lens _____

c. High-power objective lens _____

d. Oil-immersion objective lens _____

Conclusions

- What is the total magnification of your microscope under low power? _____

- What is the total magnification of your microscope when using the oil lens? _____

EXERCISE 2
Use of the Microscope

Discussion

One can think of the microscope as having three basic variables in these systems, each of which can be adjusted whenever the instrument is used. The variables in these systems are as follows:

a. Light source (turn on to proper intensity and refine with diaphragm and condenser).
b. Specimen (set into place securely).
c. Magnifying lenses (choose appropriate magnification and focus accordingly at that level of magnification).

The procedures that follow offer practice in controlling the variables in these systems.

Materials

A compound microscope; letter "e" slide;* transparent metric ruler.

Procedure

Because microscopes vary, your instructor will demonstrate appropriate sequences to use for the microscopes available in your lab. Remember the three variables in these systems: light source, specimen, and magnification (and focus).

_____

*Students may prepare these as wet mount, if desired.

1. Observe the letter "e" with the naked eye. Reproduce it in the Observations section below.
2. Observe the letter "e" with the microscope. Draw it in the space provided.
3. Move the slide to your left. Note how it moves when looking through the microscope. Record what you see.
4. Move the slide away from you. Note how it moves when looking through the microscope. Record.
5. Looking through the low-power objective of the microscope, lay your ruler on the stage and measure the size of the field. Record. Repeat for high-power objective and record.

Observations

Orientation of letter "e" with the naked eye. _____

_____

Orientation of letter "e" as observed under the microscope. _____

_____

Movement of the specimen under the microscope as you move the slide to your left. _____

Movement of the specimen under the microscope as you move the slide away from you. _____

Size of field:

Low power _____   Lens magnification _____

High power _____   Lens magnification _____

Conclusions

- What can you assume about the location of a subcellular structure that you observe to be in the upper right-hand corner of the cell through your microscope?

_____

- Which way would you move the specimen if you were attempting to move the cells at the lower left-hand corner of your field to the center of the field?

_____

- What happens to the working distance as you change from low-power to high-power magnification?

_____

_____

- Look up the definition of **parfocal**. Is your instrument parfocal?

  _____

- What is the proper positioning in the field of an object under low power that you wish to observe under high power?

  _____

  _____

- What "maneuver" would you perform with the microscope to see a structure that is located deeper within the cell?

  _____

  _____

- What does "depth of field" mean?

  _____

  _____

  _____

EXERCISE 3
Care of the Microscope

Discussion

A few rules are appropriate in caring for this instrument.

a. Always carry the microscope with two hands: one supporting the base, one holding the arm.
b. Always return the microscope to the appropriate storage compartment with the low-power objective in place, and the slide removed and returned to its box.
c. Avoid jarring the microscope, particularly when exchanging views with your lab partner; move yourself, not the microscope.
d. Use lens paper only to clean slides and lenses.
e. Always start your focusing process with low-power objective adjustments, working from the _smallest_ working distance and increasing it at the same time you are looking through the ocular lens.

EXERCISE 4
Preparation of a Cheek Smear

Discussion

By scraping cells from the outermost layer of cells of the mucous membrane lining your cheek, you acquire "fresh" material to observe. If you stain the specimen, certain cell components will appear dark and colored, since they form complexes with the organic material of the dye in the stain.

Materials

A wooden applicator stick; methylene blue stain (or iodine); distilled water; glass slides; microscope.

Be sure to review safety precautions before handling fresh or preserved specimens.

Procedure

1. Scrape the inside of your cheek lightly with the end of a wooden applicator stick.
2. Smear the scrapings on a glass slide. Let it air dry.
3. Slowly add a few drops of methylene blue (or iodine) stain on the smear and let it stand a few minutes.
4. Tap slide on paper towel and _blot_ off excess stain or wash it off _gently_ under running distilled water.

Note: If the stain is not buffered, let the stain and water sit together for a few minutes; then rinse off.

5. Air dry. Observe under the microscope. Make a sketch of your observations below and label the visible structures.

Observations

Cheek smear cells — low power

Conclusions

- What is the most appropriate unit of the Metric System to use in measuring the diameter of your cheek cell? (centi, milli, micrometer) _____

- What is the approximate diameter of your cheek cells?

  _____

UNIT 3

The Cell

OVERVIEW

In examining the structure of a typical animal cell you will come to better understand the basic unit of life. Then you will be ready to observe the various stages of cell division and examine the active and passive processes which operate as substances cross the cell membrane.

OUTLINE

Cell Structure and Cell Division

Exercise 1   General Cell Structure
Exercise 2   Cell Division: Mitosis and Cytokinesis

Quick Quiz 1

Transport Across the Cell Membrane

Exercise 1   Diffusion
Exercise 2   Factors That Affect Diffusion
Exercise 3   Osmosis

❖ Cell Structure and Cell Division

Background

Human cells are microscopic in size and are usually measured in micrometers (μm). Although they differ in size and shape, they have certain common characteristics: an outer limiting membrane, cytoplasm with organelles, and a nucleus.

EXERCISE 1
General Cell Structure

Discussion

The structure and function of the prominent parts of a typical cell are summarized in Table 3.1 and illustrated in Figure 3.1. The location and number of these structures vary from one type of cell to another, but generally all cells contain the structures listed and described in Table

3.1. Many of these structures are not visible with the light microscope, but their configurations have been determined by electron microscopic examination of cells. As indicated in Table 3.1, cells have a **cell membrane,** which is a lipid bilayer. The membrane forms a boundary for the cell and also regulates the movement of materials into and out of the cell.

Also present in all cells, at least in some form and at some point in the cell's life cycle, is the **nucleus.** This structure, set off from the rest of the cell by the nuclear membrane, contains the genetic material **DNA** (deoxyribonucleic acid) and the **nucleolus.** DNA is a nucleoprotein that "houses" the information for normal cell function and coils into chromosomes during cell division. The nucleolus contains the nucleoprotein RNA (ribonucleic acid) and apparently plays a role in protein synthesis.

The area between the nucleus and plasma membrane is called **cytoplasm.** This is a "colloidal" suspension of inorganic salts and organic molecules (including soluble enzymes) in water. It also contains the **organelles**—specialized structures that perform the functions of the cell. Endoplasmic reticulum, ribosomes, Golgi bodies, mitochondria, lysosomes, centrosome with centrioles,

15

■ **FIGURE 3.1 A typical animal cell**

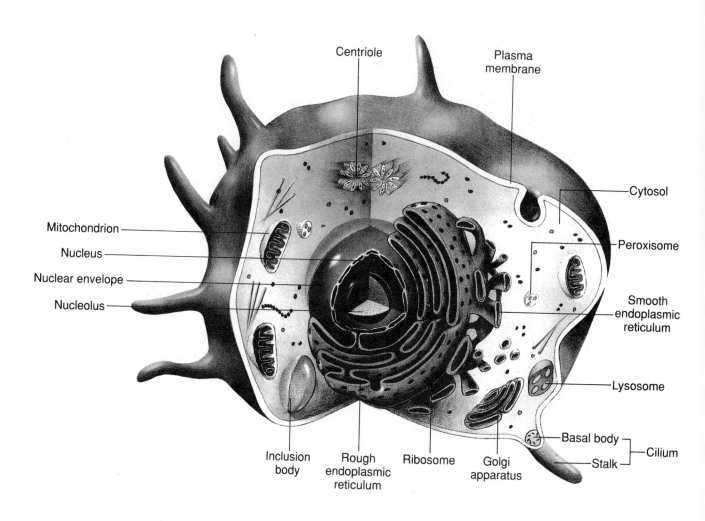

vacuoles, flagella, and cilia are organelles. Their functions are summarized in Table 3.1. The cytoplasm may also contain nonliving materials such as granules of various kinds, crystals, and pigments. These are called **inclusions.**

Materials

A microscope; prepared slides of respiratory epithelium, blood, and sperm.

■ **TABLE 3.1 Human Cell Components**

COMPONENT	STRUCTURAL ELEMENTS	REPRESENTATIVE FUNCTIONS
Nucleus	DNA and specialized proteins enclosed by a nuclear envelope	Control center of the cell containing genetic material
		Provision of specifications for synthesis of polypeptides and proteins that determine the specific nature of each cell
Plasma Membrane	Lipid bilayer containing proteins and carbohydrates	Selective barrier between cell contents and extracellular fluid
Cytoplasm Components		
Endoplasmic Reticulum	Network of interconnected, membrane-bounded tubules and flattened sacs	Production of lipids for new cell membranes; involved in manufacture of products for secretion
Golgi Apparatus	Flattened, membrane-bounded sacs	Processing, modifying, and sorting of polypeptides and proteins

■ **TABLE 3.1 (Continued)**

COMPONENT	STRUCTURAL ELEMENTS	REPRESENTATIVE FUNCTIONS
Lysosomes	Membrane-bounded vesicles containing digestive enzymes	Destruction of material such as foreign substances and cellular debris
Peroxisomes	Membrane-bounded sacs containing enzymes including catalase	Detoxification of potentially harmful substances
Mitochondria	Membrane-bounded structures surrounded by two membranes; inner membrane forms partitions called cristae	Major site of ATP production
Ribosomes	Particles composed of RNA and proteins	Assembly of amino acids into polypeptides and proteins
Secretory Vesicles	Membrane-bounded packages of secretory products	Storage of secretory products
Inclusion Bodies	Chemical substances in the form of particles or droplets	Storage forms of carbohydrates and triglycerides
Cytoskeleton	Microtubules, microfilaments, and intermediate filaments	Provide complex and dynamic structural network for the cell; involved in variety of movement processes

Procedure

1. Observe each of the prepared slides using low and high power and oil-immersion lenses. Observe the various types of cells for shape and size.

2. Record in Table A the parts of the cell you observe.

3. Also in the space provided, sketch the cells and label the nucleus, cytoplasm, cell membrane, and any observable organelles or inclusions.

(blank drawing area)

Observations

▪ TABLE A

CELL TYPE	RESPIRATORY EPITHELIUM	RED BLOOD CELL	WHITE BLOOD CELL	SPERM	OTHER
Cell Structure					
Membrane					
Nucleus					
Chromatin					
Nucleolus					
Organelles					
Vacuole					
Flagella					
Cilia					
Inclusions					
Other					

Conclusions

- From your observation, what can you determine about the size of most cytoplasmic organelles?

  _____

  _____

- Why do certain areas of the cell stain more readily than others?

  _____

  _____

  _____

  _____

EXERCISE 2

Cell Division: Mitosis and Cytokinesis

Discussion

You are probably aware that most cells have a limited life span and that during the course of a single day thousands of body cells are lost. However, many of the lost cells are continually replaced by similar cells through the process of **cell division**. During cell division, a parent cell undergoes a series of events in which duplicated genetic material is divided equally and the cytoplasm is split. The nuclear division is called **mitosis** and the division of cy-

toplasm is referred to as **cytokinesis**. The result of these events is the production of two **daughter cells**, each of which is the same as the **parent cell**. In between cell divisions, the cells are actively growing and metabolizing and this period is called **interphase**. The cell's life cycle of metabolism and growth and ultimate division is a continuous process that for the sake of convenience has been divided into phases. Although the classification is somewhat arbitrary, it provides scientists with a tool to describe the events that take place in the orderly division of a cell. Table 3.2 summarizes the major events during each of these phases, and Figure 3.2 illustrates these events.

Materials

A microscope; prepared slides of whitefish blastula and/or onion root tip; mitosis models and charts.

Procedure

1. Observe the different stages of mitosis using the diagrams, charts, and models available.
2. After you have familiarized yourself with these stages, observe the available slides.
3. Scan the slides first with low power.
4. Once you have located the area of dividing cells, use the high-power objective to find cells in each of the stages of mitosis. Note the characteristics of the stage and then in the space provided, sketch the cell, labeling all structures you are able to identify.

■ **TABLE 3.2 Major Events of Animal Cell Interphase and Cell Division**

PHASE	MAJOR EVENTS
Interphase	Basic life processes conducted; proteins and RNA being synthesized; DNA duplicates; centrioles are apparent in centrosome
Prophase	Centrioles move apart with projection of spindle fibers; chromosomes (consisting of paired sister chromatids) become visible and move toward equatorial plane; nuclear membrane and nucleolus disappear
Metaphase	Chromatid pairs line up on equatorial plane; centromeres of chromatid pairs attach to microtubules and split
Anaphase	Identical sets of chromatids (previously attached at the centromere) are moved along microtubules to opposite poles of the cell
Telophase	Chromosomes reach the centrioles and begin to assume the chromatin form; nucleoli reappear; nuclear membrane begins to reform around the two new nuclei; centrioles duplicate; cell membrane (which had begun to constrict during late anaphase) pinches inward and separates two daughter nuclei into two separate cells

Observations

Interphase	Prophase
Metaphase	Anaphase

Telophase

Mitosis

■ **FIGURE 3.2 Animal cell division**

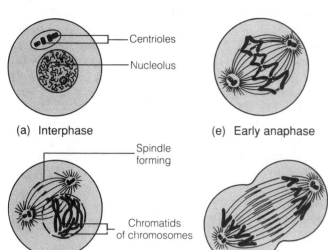

(a) Interphase

— Centrioles
— Nucleolus

(e) Early anaphase

(b) Prophase

Spindle forming

Chromatids of chromosomes

(f) Late anaphase

(c) Prometaphase

(g) Telophase

Daughter cells

(d) Metaphase

(h) Interphase

Conclusions

• What is the significance of interphase with respect to mitosis?

_____

_____

_____

_____

• What differences did you observe in cell division of animal and plant cells?

_____

_____

_____

_____

• What cells in the adult human body do not undergo mitosis?

_____

_____

_____

_____

❖ **Quick Quiz 1**

1. An efficient way to separate the aqueous inner environment of the cell from the aqueous external environment is by means of:
   (a) a water soluble protein layer
   (b) a lipid bilayer
   (c) a phospho-protein layer
   (d) none of these

2. The variety of specialized structures which perform various functions within the cytoplasm of the cell are:
   (a) cilia
   (b) inclusions
   (c) organelles
   (d) flagella

3. Which of the following would be visible with the use of a light microscope?
   (a) nucleus
   (b) centriole
   (c) mitochondria
   (d) ribosomes

4. Which of these structures has a microtubular support?
   (a) ribosome
   (b) Golgi bodies
   (c) nucleolus
   (d) flagella

5. Division of the cytoplasm of a cell is termed:
   (a) mitosis
   (b) cytokinesis
   (c) interphase
   (d) none of these

Transport Across the Cell Membrane

Background

The membrane that surrounds a cell not only forms a boundary between the cell and its environment, but also serves to regulate the movement of materials from the cell to the environment and vice versa. A membrane's permeability (the ease with which it permits substances to pass through it) is an important factor in the normal functioning of a cell. In general, cell membranes are characterized as being **selectively permeable**. This means that a membrane will allow some substances (e.g., water) to readily pass across the membrane, while other substances (e.g., protein) are prevented from doing so or pass across the membrane with some difficulty. The membrane is said to be permeable to molecules that find ready passage across it, and semipermeable or impermeable to those that have difficulty passing or cannot pass.

There are several processes by which materials can pass through a membrane. Some require no work on the part of the cell and are thus **passive mechanisms**. Example are diffusion, osmosis, and filtration. Other mechanisms, however, require the expenditure of energy by the cell. These are designated as **active mechanisms** and include the processes of active transport, phagocytosis, and pinocytosis. Table 3.3 briefly summarizes these mechanisms and Figure 3.3 illustrates some of them.

EXERCISE 1
Diffusion

Discussion

The process of diffusion refers to the **net movement** of solute molecules or ions from a region where they are greatly concentrated to a region of lesser concentration until equilibrium is reached. The movement of molecules is related to their kinetic energy and results in a random motion that causes the molecules to collide with each other. Whenever they can, the molecules become as distant from each other as possible. Thus molecules become evenly dispersed throughout a medium and establish an equilibrium with respect to the concentration. If you place a small sugar cube in a cup of hot tea and let it sit, the sugar molecules will begin to diffuse and ultimately disperse throughout the fluid. This is an example of diffusion, since the sugar molecules are moving from an area where they are highly concentrated (the cube) to an area where they are less concentrated (the tea). Diffusion can occur across a cell membrane as long as the cell membrane is permeable to the molecule, as for example with chloride ions. The two procedures that follow help to illustrate this principle.

A. Gas in a Gas

Materials

A large glass tube; two cotton plugs; ammonium hydroxide; concentrated hydrochloric acid; meter stick; ring stand and clamp.

Procedure

1. Secure the glass tube on the ringstand so that it is centered and horizontally level.
2. Saturate one cotton plug with ammonium hydroxide and the other with hydrochloric acid.
3. **Be careful handling the hydrochloric acid and ammonium hydroxide. Avoid contact with the skin and with the wet surfaces of cotton. Also, since these chemicals tend to vaporize readily, take care to avoid eye irritation, and breathing the vapors.**

■ **TABLE 3.3 Mechanisms of Transport across Membranes**

PASSIVE		ACTIVE	
NO CELLULAR ENERGY EXPENDITURE		**CELLULAR ENERGY EXPENDITURES**	
Diffusion	Solute molecules move from a region of higher to lower concentration until equilibrium is reached	Active transport	Molecules move from a region of lesser to higher concentration; involves carrier modules
Osmosis	Solvent (water) molecules move from a region of higher to lower solvent concentration (or lower to higher solute concentration)	Phagocytosis	Solid material is engulfed by cell through pseudopods and vacuole formation
Filtration	Molecules move from an area of higher pressure to lower pressure	Pinocytosis	Liquids and small solutes are taken into the cell through vesicle formation

■ **FIGURE 3.3 Some mechanisms of transport across membranes**

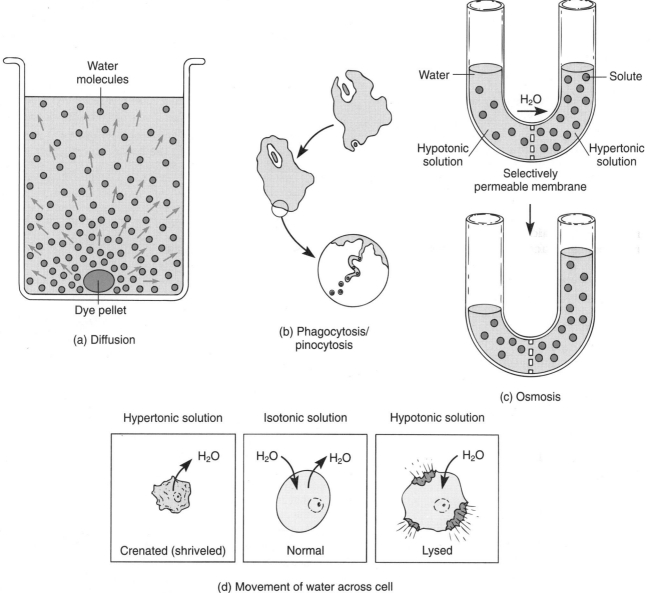

(a) Diffusion

(b) Phagocytosis/pinocytosis

(c) Osmosis

(d) Movement of water across cell

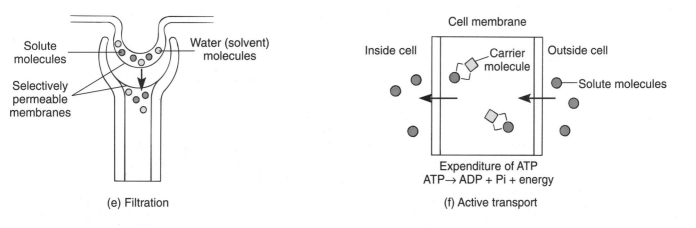

(e) Filtration

(f) Active transport

4. Insert the plugs simultaneously into opposite ends of the tube. Observe what happens. Measure the distance from either plug to the reaction point. Record below.

Observations

a. Distance ammonium hydroxide vapors traveled

_____

b. Distance hydrochloric acid vapors traveled

_____

Conclusions

• Based on your observations, do all molecules travel at the same speed? Explain.

_____

_____

_____

_____

_____

_____

_____

B. Solid in a Liquid

Materials

A small beaker; potassium permanganate crystals; forceps.

Procedure

1. Fill the beaker with water and gently place a crystal (about the size of a pea) of potassium permanganate on the bottom of the beaker.
2. Observe the results at 20-min intervals over the lab period and indicate what happens in the Observations section.

Observations

_____

_____

_____

_____

Conclusions

• If a crystal of greater molecular weight were used in this procedure, would the results be different? If so, how?

_____

_____

_____

EXERCISE 2
Factors That Affect Diffusion Rate

Discussion

The rate at which substances diffuse can be affected by such factors as concentration gradient, molecular weight, and temperature. In the next exercise you will observe the effect of the concentration gradient and temperature on diffusion.

When considering passage across the cell membrane by diffusion, another important factor comes into play, that is, lipid solubility. In this exercise you will observe the effect of molecular weight and lipid solubility on diffusion rates across the red blood cell membrane. Two generalizations can be made about molecular weight and lipid solubility: (1) the smaller the molecular weight, the greater the rate of diffusion; and (2) the greater the lipid solubility, the greater the rate of diffusion. If you think back for a moment about the structure of a cell membrane, the reasons for these generalizations should be apparent. See Figure 3.3a.

A. Molecular Weight

Materials

Whole blood (horse); four test tubes; 1-, 5-, and 10-ml pipettes; saline solution; test tube rack; 0.5 M solutions of ethylene glycol, glycerol, and glucose; stopwatch; wax pencil.

Be sure to review safety precautions before handling fresh or preserved specimens.

Procedure

1. Use whole blood or make a 3:1 suspension of red blood cells by placing 3 ml of blood in 9 ml of normal saline solution. This suspension can be made on a small scale for individual use or a larger volume can be made for the whole class. Gently mix the blood and saline by placing the thumb over the test tube and slowly inverting the tube a few times.
2. Mark three test tubes 1, 2, and 3. In tube 1 place 5 ml of the ethylene glycol solution; in tube 2 place 5 ml of glycerol solution; and in tube 3 place 5 ml of the glucose solution. Each of these substances is to be tested as indicated below, but be sure to *test one tube at a time*.
3. To the first tube add 3 drops of whole blood or 1 ml of the red blood cell suspension and mix the contents by gently inverting the tube as before. Start timing from the moment you add the cells to the ethylene glycol.
4. Hold the tube against the type of this book and record the time it takes for the print to become clear and readable (hemolysis time). Record the result in Table B.
5. Repeat this procedure for the remaining solutions (glycerol and glucose).

Note: Set aside remaining cell suspension for the next procedure.

Observations

■ TABLE B Hemolysis Time (Molecular Weight)

SOLUTION	TIME FOR PRINT TO BECOME CLEARLY VISIBLE (min)
Ethylene glycol	
Glycerol	
Glucose	

Conclusions

- What was happening to the red blood cells in each of the tubes? Why? What is this called?

_____

_____

_____

_____

_____

_____

- Which of these molecules has the greater molecular weight based on your observations? Why?

_____

_____

_____

_____

_____

- Calculate or look up the molecular weight of these substances. Did your observations confirm the generalization regarding the effect of molecular weight on diffusion rate? Are there any exceptions?

_____

_____

_____

_____

B. Lipid Solubility

Materials

Whole horse blood or 3:1 suspension of normal saline and red blood cells; 3 *M* solutions of methyl, ethyl, and propyl alcohol; three test tubes; test tube rack; stopwatch; wax pencil.

Procedure

1. Mark the test tubes 1, 2, and 3.
2. To tube 1 add 5 ml of methyl alcohol; to tube 2 add 5 ml of ethyl alcohol; and to tube 3 add 5 ml of propyl alcohol.
3. Test each tube separately. Add 1 ml of the red blood cell suspension to each tube, noting the time. Then record the time it takes to clearly read the print on this page (hemolysis time). Indicate in Table C which substance is most lipid soluble and which is least lipid soluble.

Observations

■ TABLE C Hemolysis Time (Lipid Solubility)

SOLUTION	HEMOLYSIS TIME (min)	LIPID SOLUBILITY
Methyl alcohol		
Ethyl alcohol		
Propyl alcohol		

Conclusions

• Increasing the length of the carbon chain of a molecule generally increases lipid solubility. If ethyl alcohol has a 2-carbon-chain backbone, would propyl alcohol have a longer or shorter carbon chain based upon your observations? Explain.

_____

_____

_____

• The formula of ethyl alcohol is C_2H_5OH. What is the formula for methyl alcohol based on your observations?

_____

_____

_____

EXERCISE 3
Osmosis

Discussion

Osmosis can be considered to be the net movement of solvent from an area of lower solute concentration to higher solute concentration through a **semipermeable** membrane. (In a salt-water solution, salt is the solute and water is the solvent.) If two solutions are separated by a semipermeable membrane (permeable to the solvent, impermeable to the solute), the solvent will move across the membrane in both directions. However, there will be a net movement of solvent into the compartment of greater solute concentration. Consider it as a "drawing force"—the more concentrated solution is drawing solvent away from the less concentrated solution in an attempt to establish an equilibrium. This drawing pressure is referred to as the **osmotic pressure**, and the more concentrated a solution is, the greater is its osmotic pressure. As solvent is drawn into the more concentrated solution, the increasing volume creates a "back-pressure," which is called **hydrostatic pressure**. When the hydrostatic pressure equals the osmotic pressure, the net movement of solvent ceases.

Water is the most common solvent in living systems, and the concentration of solutes and water on both sides of the cell membrane determines the direction and magnitude of osmosis. When cells are placed in a solution that has the same concentration of solutes as that inside the cell (normal saline), there is no net movement of water and the solution is considered to be **isotonic**. If cells are placed in a solution with a greater solute concentration and the membrane is impermeable to the solute, there will be a net movement of water out of the cell and the cell will shrivel up. This solution is called a **hypertonic** solution. Should the cells be placed in a solution with a lesser solute concentration, the net movement of water will be into the cell, which swells and eventually bursts. This solution is called **hypotonic**. These events are illustrated in Figure 3.3c and d. To illustrate the concept of osmosis, the first of the procedures that follow uses a substitute, nonliving membrane and the second employs the living membrane of red blood cells.

A. The Substitute Membrane

Materials

Dialysis tubing (2.5 cm width); string; solution of catalase (commercial or prepared from powder); commercial hydrogen peroxide solution (3%); beaker; distilled water.

Procedure

1. Cut about 3–4 inches of dialysis tubing, and secure one end by folding end of tubing back on itself and tying with string.
2. Fill the bag you have made from the dialysis tubing with the catalase solution, and tie the top end. The bag should be fairly soft.
3. Place the bag in a beaker which has been filled about half full with hydrogen peroxide. The beaker should be large enough for the bag to float untouched by the beaker.
4. Observe the size of the bag during the class hour, and make a qualitative comparison.

Observations

Has the bag changed in size during the course of 1½ to 2 hours? How?

_____

Conclusions

• Why did the bag change as it did? _____

_____

• What was moving across the surface of the bag?

_____

• What did NOT move across the surface of the bag?

_____

• How do you know what molecule was moving?

_____

• Write the reaction which was occurring in the bag.

_____

• By comparing the molecular movement of hydrogen and catalase, what can you say about their size?

_____

• How could dialysis tubing be used to cleanse the blood? _____

_____

B. The Living Membrane

Here are two exercises dealing with living membranes:

The Red Blood Cell

Materials

Glass slides and coverslips; a 10%, 0.9% saline solution, and distilled water in dropper bottles; paper towels; microscope.

Be sure to review safety precautions before handling fresh or preserved specimens.

Procedure

1. Place a drop of blood diluted 3:1 in isotonic saline on a slide and place a coverslip over the sample. Examine under the microscope and sketch below. During this and subsequent steps work as quickly as possible so as not to allow the samples to dry out.
2. Along the right edge of the coverslip place a drop of 10% saline solution, and "draw" it across under the coverslip by placing the torn edge of a paper towel alongside the left edge of the coverslip. Wait a short time and examine under the microscope. Sketch below.
3. Follow the procedure in step 2, replacing the 10% saline solution with distilled water. Examine under the microscope and sketch below.

Observations

Cells in isotonic saline	Cells in 10% saline	Cells in distilled water

Conclusions

- How did the cells look in each of the different solutions?

  _____

- Why was the hypertonic solution tested *before* the distilled water?

  _____

- What would you expect to see in a blood sample of a person who has drowned in salt water? Explain.

  _____

- Look up the definition for *crenation*. In which situation did this occur?

  _____

C. The Onion

Materials

Red/purple onion; 0.9% saline solution, 10% saline solution, and distilled water all in dropper bottles; microscope slides and coverslips; paper towels; microscope.

Procedure

1. Peel a finger-nail sized piece of the thin purple portion of the onion and place it on a microscope slide to which a drop of isotonic saline has been added. Cover with a coverslip. Observe under the microscope.
2. Place a drop of 10% saline at the right edge of the coverslip.
3. Tear a piece of the paper towel, and with the torn edge "draw" the hypertonic solution across under the coverslip from the right edge toward the towel at the left edge of the coverslip. Observe the onion cells under the microscope.
4. Place a drop of distilled water at the right edge of the coverslip. Proceed as in step 3. Observe the cells.

Conclusions

- What happened to the cells in 10% salt solution?

  _____

  Why? _____
- What happened in distilled water? _____

  _____

- Why did the cells respond this way? _____

  _____

- Why are some fresh vegetables soaked in tap water while waiting to be used? _____

  _____

Observations

Cells in isotonic saline	Cells in hypertonic saline	Cells in distilled water

Tissues

OVERVIEW

Now you are ready to examine groups of similar cells which operate "as one" in the body — the tissues. Every cell belongs to one of four categories of tissue. In addition to identifying these types of tissue you will become familiar with their locations in the human body and their functions.

OUTLINE

The Four Primary Tissue Types

Background

A tissue may be defined as a group of cells with a similar structure and embryonic origin working together to perform a particular function in the body. The four primary tissue types are epithelial, connective, muscle, and nervous tissue. Each of these types is specialized to perform specific functions, and the anatomical characteristics of each are adapted to effectively carry out these functions.

EXERCISE 1
Epithelial Tissue

Discussion

Epithelial tissue is generally found on what is termed "free surfaces" of the body. It is, therefore, found on the outer layers of skin, lining all cavities, and making up the walls of hollow organs and tubular structures such as those of the respiratory, digestive, excretory, reproductive, and cardiovascular systems.

The functions of this tissue are protection, absorption, filtration and secretion. Epithelial tissue can either be arranged in continuous sheets, (e.g., lining of bladder) or in secretory units called **glands**.

All epithelial tissues have certain common characteristics, including *(a)* closely packed cells with little or no intercellular substance; *(b)* absence of blood vessels; *(c)* location bordering on connective tissues from which they derive their nourishment; *(d)* a basement membrane, which "cements" the epithelial tissue to the underlying connective tissue; *(e)* high regenerative capacity.

Although all epithelial tissues exhibit these characteristics, still further differentiation and specialization provides the body with many varieties of epithelial tissues. These variations can be classified and named using the criteria of cell shape and number of layers in the tissue. There are three basic cell shapes: cuboidal, columnar, and squamous. The arrangement of layers is **simple** (one or two layers) or **stratified** (more than two layers). Tables 4.1 and 4.2 (pages 35 and 37) describe in more detail the various simple and stratified epithelial tissues, illustrated in Figures 4.1 to 4.8 (pages 35–38). Included also are two epithelial tissues that are not easily categorized using the criteria above. They are pseudostratified and transitional epithelial.

29

Materials

A microscope and prepared slides of: simple squamous, endothelium, or mesothelium; simple cuboidal or cross section (c.s.) of thyroid gland; simple columnar or c.s. of small intestine; stratified squamous or c.s. of skin; stratified columnar or c.s. of male urethra; stratified cuboidal or c.s. of sweat gland sac; transitional epithelium or c.s. of urinary bladder; pseudostratified epithelium or c.s. of trachea.

Procedure

1. Using low power, scan the slide to find the epithelial tissue, keeping in mind the tissue characteristics.
2. Once you have located the tissue, use high-power and/or oil-immersion lens to specifically identify the tissue type.
3. Sketch each of the tissues in the space provided in the Observations section.

Observations

Simple squamous	Simple cuboidal	Simple columnar
Stratified squamous	Stratified cuboidal	Stratified columnar
Pseudostratified epithelium		Transitional epithelium

Epithelial tissues

Conclusions

- What modifications of epithelial tissue do you see in the respiratory epithelium, and what purpose do they serve?

  _____

  _____

  _____

- Considering its location, why is a good regenerative capacity essential in epithelial tissues?

  _____

  _____

  _____

EXERCISE 2
Connective Tissue

Discussion

The most abundant tissue in the body is connective tissue. It is found, for example, connecting one organ to another, connecting organs to the body wall, embedding nerves and blood vessels, and as scar tissue. This type of tissue is referred to as **generalized connective tissue**. There are also several specialized connective tissues, including **adipose**, and **reticular tissues**. In addition, there are **supporting connective tissues, bone** and **cartilage**, and **fluid connective tissue, blood**. Connective tissues generally function in connection, support, and transport, and present the greatest diversification of the primary tissue types.

In contrast to epithelial tissue, connective tissue is more complex because there are components other than cells in the tissue. In addition to cells, connective tissue has protein fibers and a great deal of ground substance in between the cells. All these components play a role in the functioning of the tissue. Classification of connective tissue is difficult because of the great diversification and many variations. However, if the nature of the ground substance and functions are used as criteria, three major categories of connective tissue can be distinguished: **connective tissue proper, supporting tissue**, and **fluid tissue**.

The ground substance consists of carbohydrate-protein molecules in a suspending medium. Sometimes this combination can be gel-like as in the connective tissue beneath the skin. Sometimes it can be quite firm or even solid as in cartilage and bone. Sometimes this ground substance is liquid as in the plasma of blood. Figure 4.9 (page 39) provides a general scheme for the organization of connective tissue and Tables 4.3 and 4.4 (pages 39 and 40) describe the cell types and fibers found in general connective tissue. Tables 4.5, 4.6, and 4.7 (pages 40, 42, and 45) describe the different types of tissue found within each of the major categories of connective tissue. Figures 4.10–4.18 (pages 40–45) illustrate some of these tissues.

Materials

A microscope and prepared slides of areolar tissue; adipose tissue; reticular tissue or section of spleen; tendon; dermis of skin (irregular dense); hyaline cartilage or c.s. of trachea; elastic cartilage or c.s. of epiglottis; fibrocartilage or c.s. of intervertebral disc; ground bone—compact; blood smear.

Procedure

1. Examine the slides as you did for the epithelial tissue.
2. Sketch each of the tissues in the space provided in the Observations section.

Observations

Areolar

Adipose

Reticular

Connective tissue

Tendon

Dermis of skin

Fibrocartilage

Bone

Supporting tissue

Blood

Fluid tissue

Conclusions

- What characteristic does the presence of collagen fibers lend to a tissue?

_____

_____

_____

_____

- What characteristic does the presence of elastic fibers lend to a tissue?

_____

_____

_____

_____

- What is the function of fibroblasts, and why are they the most predominant cell in the generalized connective tissue?

_____

_____

_____

EXERCISE 3
Muscle Tissue

Discussion

Muscle tissue and nerve tissue are dealt with in this unit only very briefly. They will be discussed in greater detail in the specific units dealing with these tissues—Muscle (Unit 10) and The Nervous System (Unit 12).

The third primary tissue type is muscle that is characterized by its ability to contract. The main component of muscle tissue is an elongated cell, which upon stimulation shortens or contracts, thus providing the body with movement capabilities. The bulk of muscle tissue is associated with the skeletal system. However, most of the viscera (body organs) also contain muscle tissue, and the heart is also basically a mass of muscle. See Table 4.8 and Figures 4.19–4.21 (pages 46–47).

Classification of muscle can be done in several ways. A microscopic, anatomical method of classification is to categorize muscle according to the presence or absence of microscopic cross-striations within the cells. This type of classification provides two types of cells: striated and nonstriated. Another classification is based on the type of control over activity. Again two types can be identified: voluntary and involuntary. Using both criteria (i.e., striations and type of control), it is possible to distinguish three types of muscle: skeletal, smooth, and cardiac.

Materials

A microscope and prepared slides of skeletal [c.s. and l.s. (longitudinal section)] smooth and cardiac muscle.

Procedure

1. Examine the slide as you did for the epithelial and connective tissues.
2. Sketch each muscle type in the Observations section.

Observations

Skeletal

Smooth

Cardiac

Muscle tissue

- What is responsible for the cross-striations in skeletal and cardiac muscle? Why are there no striations in smooth muscle?

_____

_____

_____

_____

- What is the advantage of a syncytial arrangement in cardiac muscle?

_____

_____

_____

_____

- Which of the three types of muscle has more than one nucleus per cell?

_____

_____

_____

_____

_____

EXERCISE 4
Nerve Tissue

Discussion

Nerve tissue functions in conducting impulses through-out the body. It is composed of cells of two major types: **neurons**, which conduct impulses, and **neuroglia**, which are supportive and protective cells for neurons. Neuroglia do not conduct impulses. See Figures 4.22 and 4.23 (pages 47 and 48).

There are several different types of neurons, but all have certain characteristics in common. Typically, a neuron has a large cell body or **soma** (also called **perikaryon**) and long cell **processes** extending from the soma called **dendrites** and the **axon**.

Neuroglia, the supportive and protective cells of nerve tissue, make up the bulk of the tissue.

Materials

A microscope and prepared slides of giant multipolar neuron and myelinated nerve fiber (l.s. or c.s.)

Procedure

1. Examine the slides as before.
2. Make sketches in the space provided in the Observations section.

Observations

Multipolar neuron

Myelinated nerve fiber

Nerve tissue

Conclusions

• How does nerve tissue differ in function from the other primary tissue types?

_____

_____

_____

_____

• Why is an elongated arrangement advantageous in nerve tissue?

_____

_____

_____

_____

■ **TABLE 4.1 Simple Epithelial Tissue**

TISSUE TYPE	DESCRIPTION	LOCATION
Squamous	Broad, flat, irregularly shaped cells; arrangement is ideal for areas where diffusion and filtration are carried out	Lining of blood vessels and alveoli of lungs
Cuboidal	Cells cube shaped with relatively equal dimensions; a single centrally located nucleus	Surface of ovaries; lining of kidney tubules; follicles of thyroid
Columnar	Tall cells with a basal nucleus; modifications may include cilia, microvilli and mucus-secreting goblet cells	Lining of stomach, intestines, upper respiratory tract, uterine tubes
Pseudostratified	Columnar-shaped cells that *appear* to be in more than one layer; all cells, however, are in contact with basement membrane and some reach the free border; cilia and goblet cells may be present	Respiratory passages

■ **FIGURE 4.1 Simple squamous epithelium**

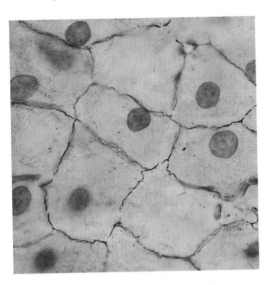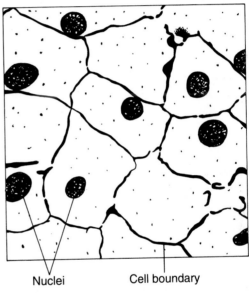

Nuclei   Cell boundary

■ **FIGURE 4.2 Simple cuboidal epithelium**

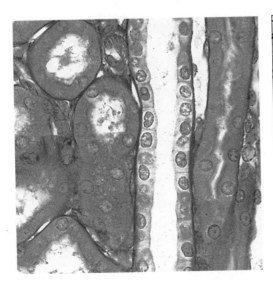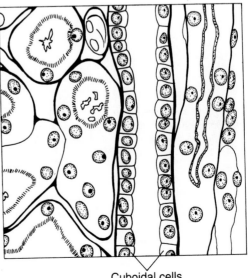

Cuboidal cells

■ **FIGURE 4.3 Simple columnar epithelium**

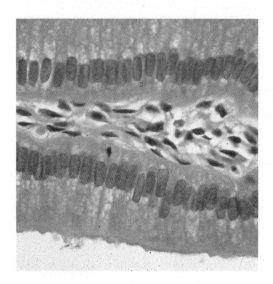

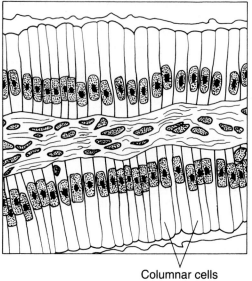

Columnar cells

■ **FIGURE 4.4 Pseudostratified columnar ciliated epithelium**

Goblet
cells

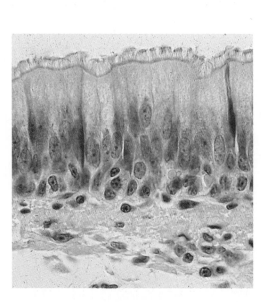

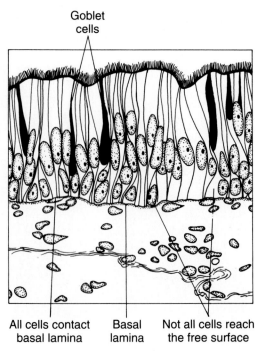

All cells contact   Basal   Not all cells reach
basal lamina   lamina   the free surface

■ **TABLE 4.2 Stratified Epithelial Tissue**

TISSUE TYPE	DESCRIPTION	LOCATION
Squamous	Upper layers typically flat with layers at base columnar or cuboidal in shape	Areas of wear and tear: skin; lining of mouth, esophagus, anus, vagina; may be keratinized or non-keratinized
Cuboidal	Has several rows of cuboidal cells; not a common tissue	Ducts of sweat glands
Columnar	Basal layer of polyhedral cells with superficial layer of columnar cells; not a common tissue	Ducts of mammary glands; larynx; male urethra
Transitional	Cells of the free border tend to round out when bladder is undistended; as bladder fills, upper layers stretch out and resemble squamous cells	Lining of urinary bladder

■ **FIGURE 4.5 Stratified squamous epithelium**

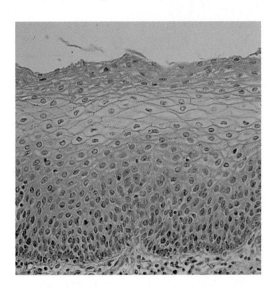
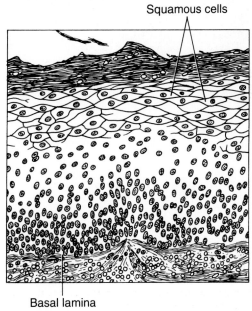

Squamous cells

Basal lamina

■ **FIGURE 4.6 Stratified cuboidal epithelium**

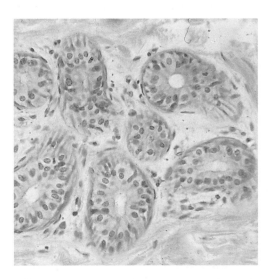
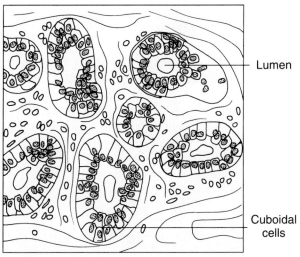

Lumen

Cuboidal cells

FIGURE 4.7   Stratified columnar epithelium

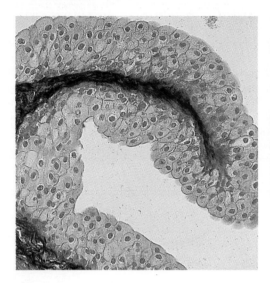
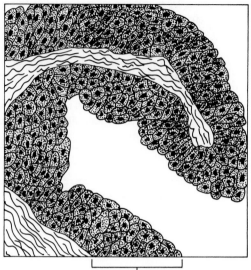

Multiple layers
of columnar cells

FIGURE 4.8   Transitional epithelium

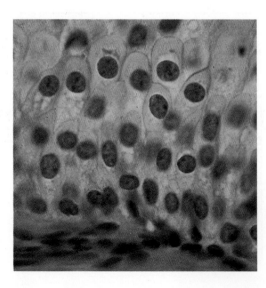
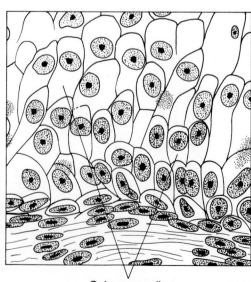

Columnar cells

■ **TABLE 4.3 General Connective Tissue Cells**

CELL TYPE	DESCRIPTION
Fibroblast	Small cell that functions to lay down ground substance and fibers
Histiocyte	Irregularly shaped cell with short branching processes; phagocytic; also called macrophage
Mast	Probably formed from the circulating basophil in blood, although larger; releases histamine, heparin, and serotonin
Plasma	May be a round or irregular small cell that develops from lymphocyte; involved in antibody production
Adipose	"Signet ring"-shaped cell with large vacuole for storage of triglycerides

■ **FIGURE 4.9 General organization of connective tissue**

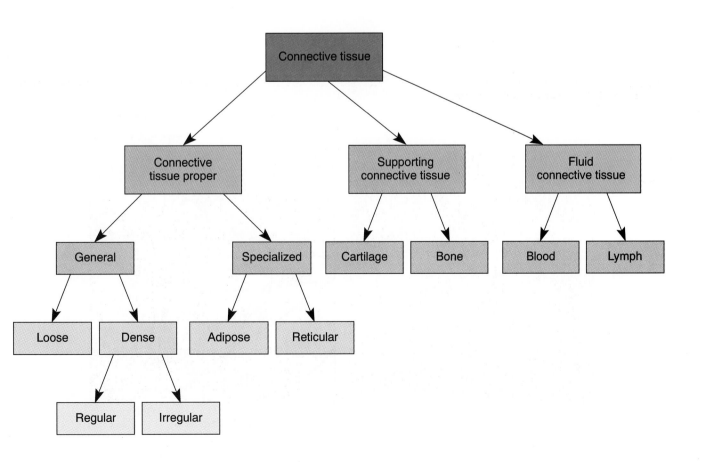

■ **TABLE 4.4 Connective Tissue Fibers**

FIBER TYPE	DESCRIPTION
Collagen	Made of the protein collagen; tough and strong; little elasticity
Elastic	Made of the protein elastin; characteristically branched and fused fibers; highly elastic
Reticular	Made of the protein reticulin (an antecedent of collagen); delicate branching fibers

■ **TABLE 4.5 Connective Tissue Proper**

TISSUE	DESCRIPTION	LOCATION
GENERAL	(Nonspecialized)	
Loose or areolar	No regular arrangement of cells; several different cell types; collagen and elastic fibers and reticular within the ground substance; cells present include fibroblasts, macrophages (histiocytes), mast, plasma, and adipose cells; predominant cell is fibroblast	Between organs and muscles; supporting blood vessels and nerves
DENSE	Tissue is packed with fibers; fibroblast is predominant cell type	
Regular	Fibers arranged in parallel fashion with cells in between	Tendon and ligament
Irregular	Fibers are randomly arranged within the tissue; usually seen in sheets that wrap around organs	Covering on bone and muscles
SPECIALIZED		
Adipose	Specialized areolar tissue in which adipose cells predominate	Subcutaneous fat
Reticular	A meshwork of cells and reticular fibers; phagocytic cells predominate	Framework for organs such as spleen, liver and lymph nodes

■ **FIGURE 4.10 Loose (areolar) connective tissue**

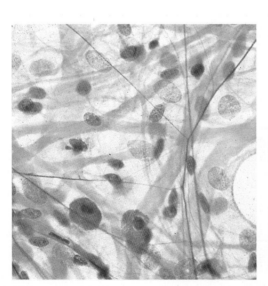
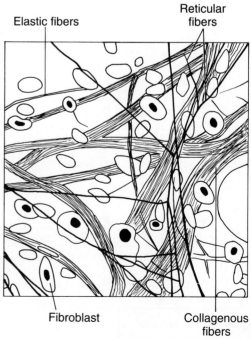

■ **FIGURE 4.11 Dense irregular connective tissue**

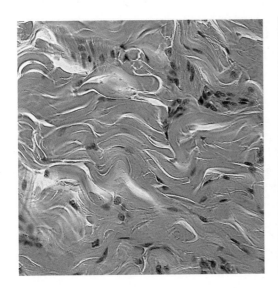
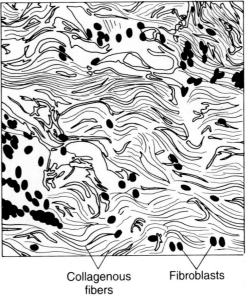

Collagenous
fibers   Fibroblasts

■ **FIGURE 4.12 Adipose tissue**

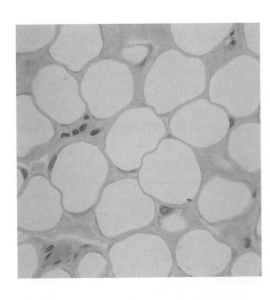
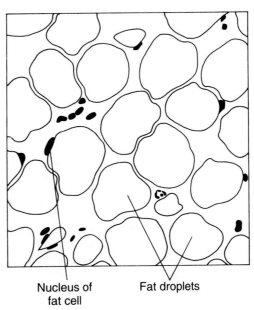

Nucleus of   Fat droplets
fat cell

■ **FIGURE 4.13 Reticular connective tissue**

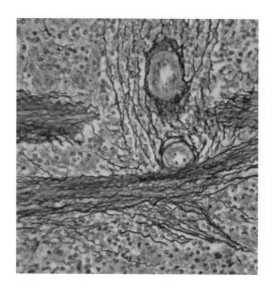 

Reticular
fibers

■ **TABLE 4.6 Supporting Connective Tissue**

TISSUE	DESCRIPTION	LOCATION
CARTILAGE	Cell type is chondrocyte and tissue types are distinguished on basis of predominant fiber and arrangement	
Hyaline	Few chondrocytes present in lacunae; an abundance of collagen fibers and ground substance (**Note:** Collagen fibers and ground substance have same refractive index.)	Articulating surfaces of long bones; embryonic skeleton
Elastic	Contains a greater abundance of elastic fibers than collagen fibers; heavy concentration of fibers forms a meshwork, with chondrocytes scattered in lacunae throughout the meshwork	External ear
Fibrocartilage	Bundles of dense connective tissue between which are regions of chondrocytes and matrix; lacks a perichondrium	Intervertebral disc
BONE	Distinguished on basis of arrangement of lamellae; cells are osteocytes	
Compact	Calcified matrix arranged in concentric lamellae (rings) enclosing the central Haversian canal; cells located in lacunae between lamellae; microscopic channels (canaliculi) radiate out from canal to cells and between cells; lamellae, Haversian canal, and cells collectively called Haversian system; systems closely packed	Shaft of long bone (diaphysis)
Cancellous	Lamellae arranged with macroscopic spaces; more porous and lighter bone of interlacing plates	End of long bone (epiphysis)

■ **FIGURE 4.14 Hyaline cartilage**

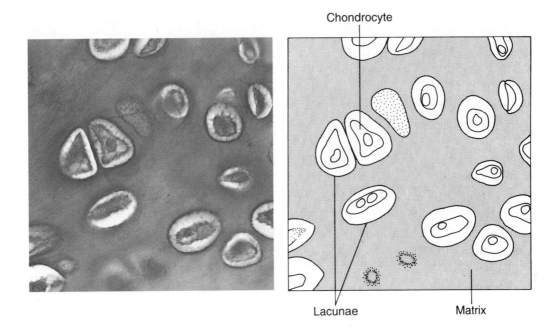

■ **FIGURE 4.15 Elastic cartilage**

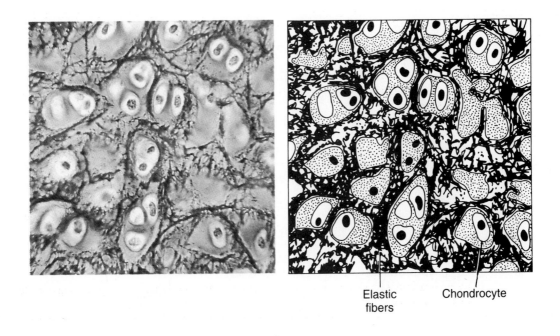

■ **FIGURE 4.16   Fibrocartilage**

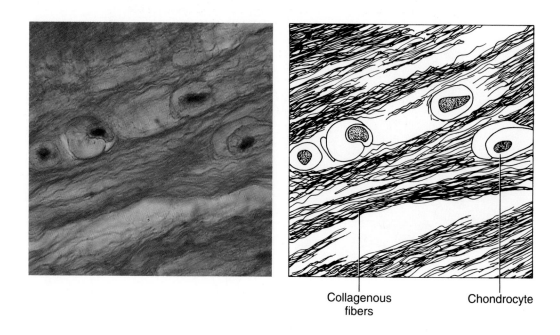

Collagenous
fibers

Chondrocyte

■ **FIGURE 4.17   Compact bone**

Matrix

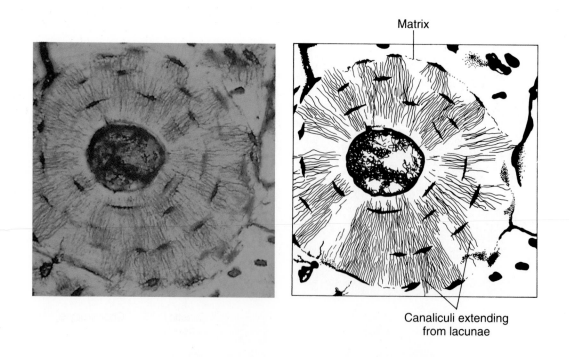

Canaliculi extending
from lacunae

■ **TABLE 4.7 Fluid Connective Tissue**

TISSUE	DESCRIPTION
Blood	Fluid matrix with three major cell types (formed elements): erythrocytes, leukocytes, and cell fragments, thrombocytes; fibers are in a soluble form
Lymph	Fluid matrix formed from intercellular or tissue fluid consisting mostly of water with some leukocytes and a lesser protein content than plasma

■ **FIGURE 4.18 Blood**

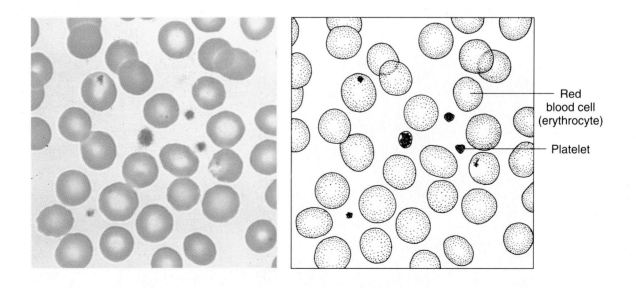

■ **TABLE 4.8 Muscle Tissue**

TISSUE	DESCRIPTION	LOCATION
Skeletal	Elongate, multinucleate cell with cross-striations and blunt ends; nuclei are peripheral, located just under cell membrane; exhibits voluntary control	Attached to skeleton
Smooth	Spindle-shaped cell with tapering end; a single, centrally placed nucleus and *no* cross-striations; exhibits involuntary control	Hollow body viscera such as stomach, bladder, uterus, intestines, and blood vessel wall
Cardiac	Single nucleus, elongated cell; forms a branched network of cells; cross-striations are present and the cell has a single nucleus; dark bands called **intercalated discs** are a prominent feature; control is involuntary	Heart

■ **FIGURE 4.19 Skeletal muscle**

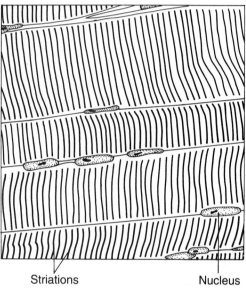

Striations   Nucleus

■ **FIGURE 4.20 Smooth muscle**

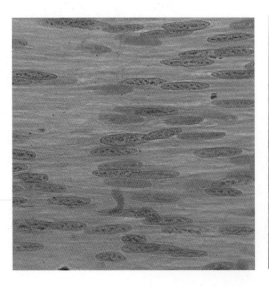
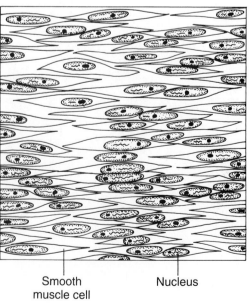

Smooth   Nucleus
muscle cell

FIGURE 4.21 Cardiac muscle

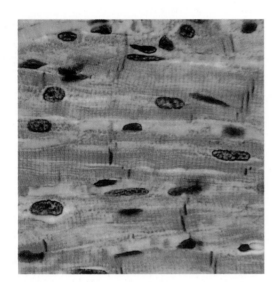
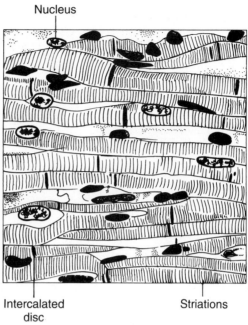

Nucleus

Intercalated disc

Striations

FIGURE 4.22 Neuron

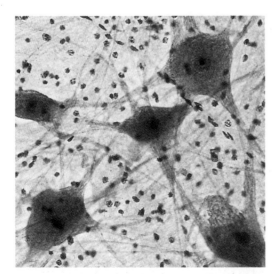
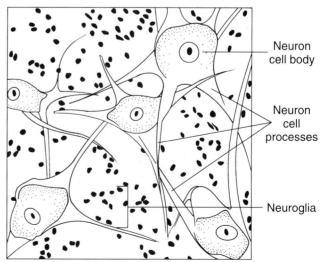

Neuron cell body

Neuron cell processes

Neuroglia

■ **FIGURE 4.23   Glial cells**

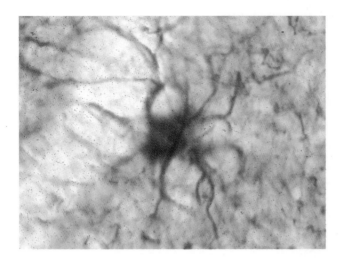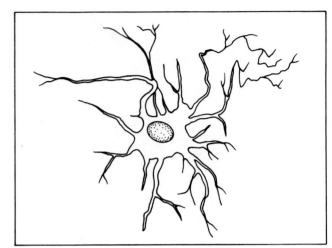

(a) Astrocyte (×527)

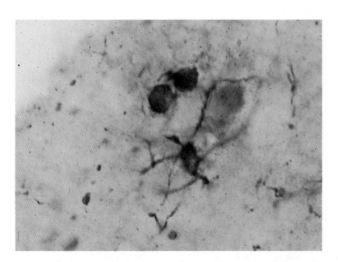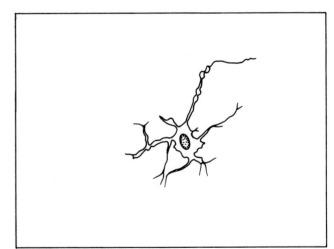

(b) Oligodendrocyte (×969)

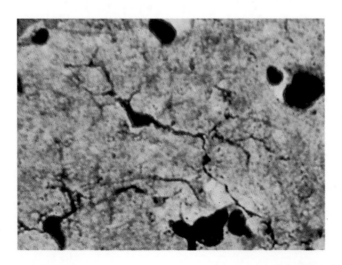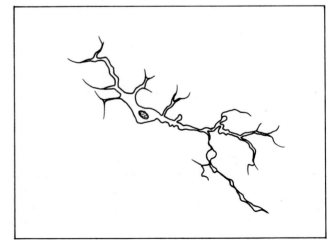

(c) Microglia (×969)

Skin

OVERVIEW

A special combination of epithelium and connective tissue comprises the skin. You will examine skin under the microscope and note the functions of the various layers.

OUTLINE

The Structure of Skin

Background

Although it is often thought of as only an outer protective covering, the skin is actually an organ that carries out several functions. These include temperature regulation, vitamin D synthesis, immune resistance (as a result of specialized cells of the immune system in the skin), and sensation (due to the presence of specialized nerve endings in the skin).

EXERCISE 1
Microscopic Examination of Skin

Discussion

The skin is composed of two major layers: an outer layer of epithelium called the **epidermis** and an underlying layer of connective tissue called the **dermis**. The epidermis is actually composed of five zones, the structure and characteristics of which are presented in Table 5.1. The dermis is also divided into two areas, and the characteristics are outlined in Table 5.1. See Figure 5.1.

Accessory structures associated with the skin include hair, nails and glands which produce sweat and oil. All these structures are derived from the epidermis but several permeate into the dermal layer. The dermis sits on a loose connective tissue layer which has varying amounts of adipose cells in it. This is called the hypodermis.

Materials

A microscope and prepared slide of human skin.

Procedure

1. Examine the slide using both low and high power.
2. Identify the different layers of the skin and any accessory structures that may be present in the section.
3. Label the figure in the Observations section.

■ **TABLE 5.1 Skin Layers**

LAYER	DESCRIPTION	FUNCTION
EPIDERMIS	Stratified squamous; five layers	Protection
Stratum corneum	Flat keratinized cells; thickness varies	Dead cells being sloughed off
Stratum lucidum	Narrow layer that is clear and found in thickened areas of epidermis such as soles of feet	Cells begin to degenerate
Stratum granulosum	Two to five layers of flattened cells with keratohyalin granules	Begins production of keratin, a waxy waterproofing material
Stratum spinosum	Several rows of irregularly shaped cells; together the stratum basale and stratum spinosum are called the **stratum germinativum**; both contain pigment producing cells called melanoblasts, the processes of which extend to upper layers	Cells may undergo mitosis; melanin production results in skin pigmentation
Stratum basale	Irregular layer of columnar cells; deepest of epidermal layer	Mitotic activity, which generates cells of upper layers and accessory structures
DERMIS	Dense, irregular connective tissue; two regions	Provides nourishment for epidermis; contains sensory receptors and derivatives
Papillary	Upper layer of dermis; convoluted layer with capillary networks and Meissner's corpuscles (touch receptors)	Nourishment and anchoring of lower epidermis
Reticular region	Contains arteries, veins and Pacinian corpuscles (pressure receptors); base of sweat and sebaceous glands derived from epidermis located here	Arteriovenous (connections) provides for bypass of capillaries promoting retention of body heat

Observations

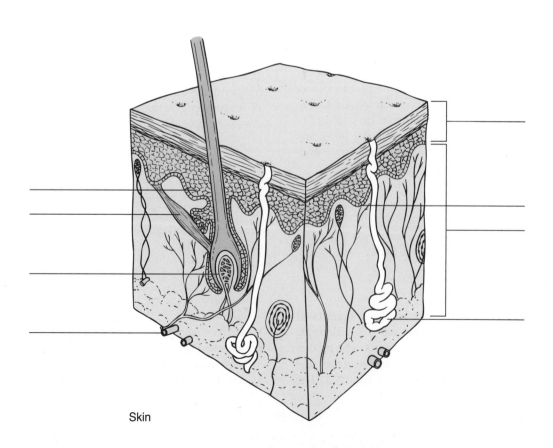

Skin

■ FIGURE 5.1 Structure of skin

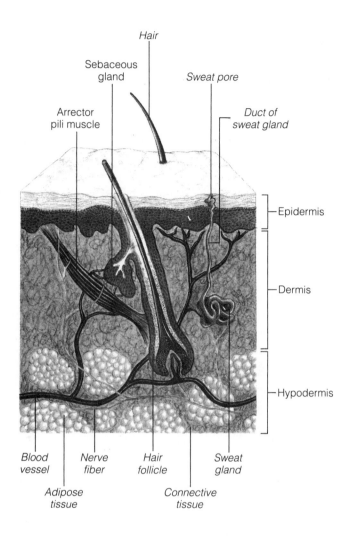

Conclusions

• Which layer of the skin or accessory structure gives rise to:

  a. Freckles _____

  b. Corns/calluses _____

  c. Sweat glands _____

The Skeletal System

O V E R V I E W

In this unit, you will investigate the macroscopic structure of a typical long bone and review the microscopic structure of compact bone tissue. Additionally, you will learn the names of the individual bones which comprise the axial and appendicular skeletons and the major surface markings which characterize these bones.

O U T L I N E

Bone Structure

Background

As indicated in Unit 4 there are two arrangements of bone tissue, **compact** and **cancellous**. Bones that form the skeleton are composed of both types, and the distribution of compact and cancellous bone varies. Generally, where a great deal of strength is required, such as in the shaft of long bones, the predominant type is compact bone. Where lighter weight bones are essential, such as in the skull, hollow areas in the bone called **sinuses** are present.

EXERCISE 1
Macroscopic Structure of Long Bone

Discussion

A typical long bone can be used to demonstrate the distribution of compact and cancellous bone. The shaft or **diaphysis** is a cylinder consisting primarily of compact bone. The innermost layer, around the medullary or marrow cavity, is composed of cancellous bone. The ends of the bone, the **epiphyses**, are composed of cancellous or spongy bone. They are covered with **articular cartilage**, which provides flexibility and a cushioning effect in joint motion. The outer covering on bone is a dense connective tissue called the **periosteum**, while the inner cavities are lined with a connective tissue called **endosteum**.

Materials

A sagittally cut long bone; models of longitudinally cut and cross-sectioned bone; charts or diagrams of long bone.

Procedure

1. Using your text as a guide, identify the parts of a typical long bone.
2. Label the diagram in the Observations section to the right.

Conclusions

* Are the spaces of cancellous bone and the space of medullary cavity filled with the same material in an adult? Explain.

_____

_____

_____

_____

_____

EXERCISE 2
Microscopic Structure of Bone

Discussion

Bone tissue consists of cells called **osteocytes**. The interstitial substance is composed of (1) fibers, (2) a complex of calcium carbonate and calcium phosphate salts, and (3) a protein-carbohydrate ground substance. In compact bone these components are organized into closely packed concentric rings or **Haversian systems**, which form a dense mass of tissue. Blood vessels enter compact bone from the **periosteum** through microscopic channels called **Volkmann's canals** and branch into these systems. The area between Haversian systems is filled with **interstitial lamellae**. In cancellous bone, the components are arranged in a "loose" pattern of interlacing macroscopic trabeculae or spicules, which resemble scaffolding. The large spaces are filled with red marrow.

A long bone

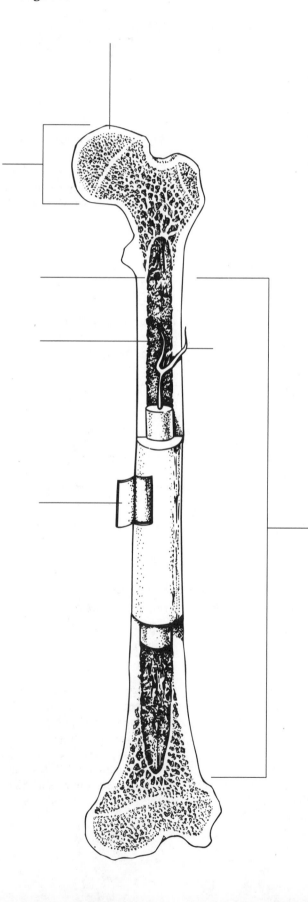

Materials

A microscope; prepared slides of ground bone.

Procedure

1. Review the material in Unit 4 on bone tissue.
2. Examine the slides and identify the following: Haversian canal, Volkmann's canal, lacunae, concentric lamellae, interstitial lamellae, and canaliculi.
3. Label the diagram in the Observations section below.

Observations

Compact bone

Conclusions

• Based upon your observation of compact bone tissue, where would you see osteocytes?

_____

_____

_____

_____

• Where are the blood vessels found in compact bone tissue?

_____

_____

_____

_____

EXERCISE 3
Bone Types and Markings

Discussion

As you examine the skeleton you will note that the bones can generally be classified into four principal types: long, short, flat and irregular. The construction of these different types is summarized in Table 6.1. Your examination of the bones will also reveal that there are numerous surface markings on the bones. These markings serve a variety of purposes such as sites for muscle attachment, articulation surfaces and passageways for blood vessels and nerves. Table 6.2 summarizes the variety of surface markings found on bones.

Materials

Assortment of bone types (whole and sectioned), charts and diagrams of bones and their markings.

Procedure

1. Using Table 6.1 and 6.2 as guides, examine the bones or bone groups listed in the Observation Table A.
2. Check which surface markings you observed and record the findings in the Observations section.

Conclusions

• What general description can you give for the overall structure of condyles, heads and tubercles?

_____

_____

_____

Observations

■ **TABLE A**

	TUBEROSITY	TUBERCLE	FORAMEN	HEAD	CONDYLE	SPINE	FOSSA
Skull							
Arm							
Hand							
Vertebrae							

■ **TABLE 6.1　Types of Bones**

TYPE	DESCRIPTION	EXAMPLES
Long	Bone length exceeds width; shaft is made of compact bone, ends made of cancellous bone; shaft has central marrow (or medullary) cavity	Humerus, radius, ulna, femur, tibia, fibula, phalanges
Short	Cube-shaped bone consisting of cancellous bone enclosed by a thin layer of compact bone	Carpals, tarsals
Flat	Flat and thin bone consisting of two plates of compact bone enclosing a core of cancellous bone	Cranial bones, sternum, ribs
Irregular	Shapes do not fit categories above; arrangement of compact and cancellous bone is similar to short bones	Vertebrae, ossicles, mandible
Sesamoid	Bone ensheathed in tendons	Patella

■ **TABLE 6.2　Markings of Bone**

MARKING	DESCRIPTION	EXAMPLE
PROCESSES FOR JOINT FORMATION		
Head	Rounded expanded end of bone beyond a constricted neck	Head of humerus
Condyle	Rounded projection found at point of articulation	Occipital condyle
Facet	Smooth flattened articular surface	Superior and inferior articulating facets of thoracic vertebrae
PROCESSES FOR ATTACHMENT OF MUSCLE		
Crest	Prominent, narrow ridge on bone	Iliac crest
Line	Bony ridge, less prominent than crest	Linea aspera
Spine	Sharp, slender projection	Vertebral spinous process
Trochanter	Large, blunt, rough process on femur	Greater trochanter
Tuberosity	Large, rounded process with rough surface	Deltoid tuberosity
Tubercle	Small, rounded process	Tubercle of humerus
DEPRESSIONS, CAVITIES, OPENINGS		
Fossa	Depression or opening that does not go through the bone	Olecranon fossa
Fovea	Shallow depression	Fovea capitas
Foramen	Hole for passage of blood vessels and nerves	Foramen ovale
Fissure	Narrow slit often between bones	Inferior orbital fissure
Meatus	Long tubelike passage	Auditory meatus
Sulcus/groove	Long, shallow depression	Intertubercular sulcus
Sinus	Air-filled cavity within bone	Sphenoid sinuses

Quick Quiz 1

Match the statement in column A with the term in column B.

A	B
_____ 1. smooth flattened articular surface	(a) diaphysis
	(b) interstitial lamellae
_____ 2. fills spaces between trabeculae	(c) endosteum
	(d) foramen
	(e) facet
_____ 3. shaft of long bone	(f) cancellous
_____ 4. hole for passage of blood vessel	(g) sinus
	(h) irregular bone
_____ 5. vertebrae or mandible for example	(i) interstitial ground substance
	(j) red marrow
_____ 6. fibers, calcium salts, protein-carbohydrate compound	
_____ 7. air-filled cavity	
_____ 8. found between Haversian units	
_____ 9. bone type made of trabeculae	
_____ 10. lines cavity in long bone	

❖ The Axial Skeleton

Background

The **axial skeleton** consists of the bones that are located through or around the midsagittal section of the body. They include the skull, vertebral column, rib cage, sternum, hyoid bone and ossicles. Figure 6.1 shows the organization of the skeleton and illustrates how the bones of the axial skeleton form the long axis of the body.

EXERCISE 1
The Skull

Discussion

The adult skull contains 28 bones, 6 of which are the auditory ossicles. Since the ossicles are in the middle ear and function in hearing, they are considered in the exercise on the ear. The remaining 22 bones can be conveniently divided into two main regions: the cranium, or brain case, and the face.

Surrounding and protecting the brain is a rigid bony case consisting of 8 bones: 1 frontal, 2 parietal, 2 temporal, 1 occipital, 1 sphenoid, and 1 ethmoid. There are 14 bones that comprise the face: 2 nasal, 2 maxillary, 2 zygomatic, 1 mandible, 2 lacrimal, 2 palatine, 2 inferior conchae, and 1 vomer. The nasal, maxillary, zygomatic, and mandible bones are superficial; the others are deep. Tables 6.3 and 6.4 summarize general information about the cranial and facial bones respectively. Table 6.5 provides more detailed information about the major bone markings which can be found on the bones of the skull.

Perhaps more than any other group of bones, those of the skull have a large number of **foramina** (singular, **foramen**) to allow for the entrance and exit of nerves and blood vessels to and from the brain. Table 6.6 summarizes the major foramina of the skull. A unique feature of some cranial and facial bones is the presence of air **sinuses**, which help to decrease the weight of the skull. They act as resonating chambers for the sound produced in speech. Additionally, they are lined with mucous membrane like that lining the nasal cavity, and the mucus they produce drains into this cavity. Some of the sinuses, such as those in the frontal bone, can be observed by holding the skull up to the light. The light areas seen in the bone are the sinuses. Table 6.7 summarizes the sinuses of the skull.

Except for the mandible, which forms a movable joint with the temporal bone, the bones of the skull form immovable joints called **sutures**. These are summarized in Table 6.8.

Many of the bones of the skull are formed by intramembranous ossification, which begins about the third month of fetal development and is not completed until over a year of postnatal development. Thus in fetus and newborn alike, there are nonossified membranous areas of the skull referred to as **fontanels**. These are described in Table 6.9 and illustrated in Figure 6.8.

Materials

A whole skull; "exploded" skull; sagitally sectioned skull; skull with removable cranium; fetal skull; skull charts and diagrams.

Procedure

1. Using available charts and Figures 6.2 through 6.7 as guides, locate each of the bones and sutures of the cranium and the face.
2. Using the text descriptions and Tables 6.3 through 6.5, identify the specific parts and markings of each bone.
3. Examine the fetal skull and locate the fontanels, using Table 6.9 and Figure 6.8 as guides. Note the degree of ossification and compare with the adult skull.

■ **FIGURE 6.1   The human skeleton**

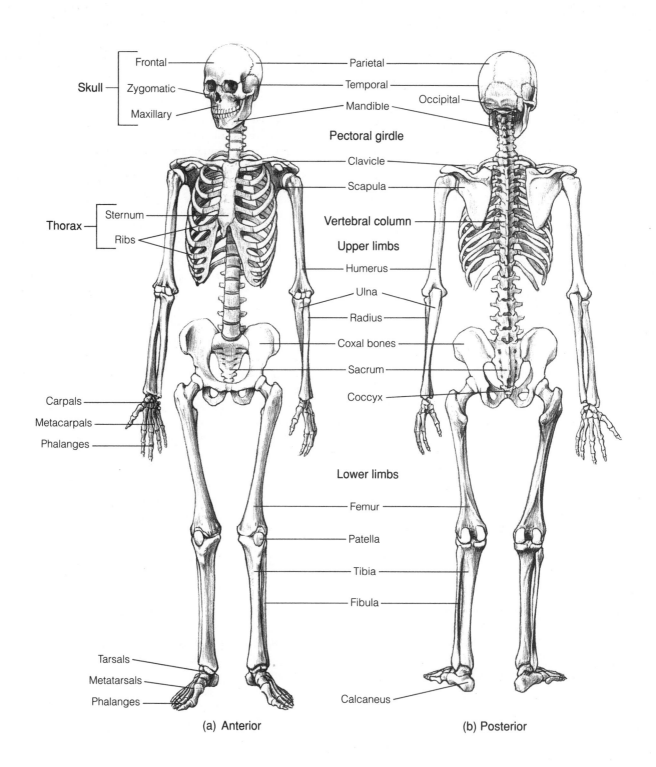

Skull {
Frontal
Zygomatic
Maxillary
}

Parietal
Temporal
Mandible
Occipital

Pectoral girdle
Clavicle
Scapula

Thorax {
Sternum
Ribs
}

Vertebral column

Upper limbs
Humerus
Ulna
Radius
Coxal bones
Sacrum
Coccyx

Carpals
Metacarpals
Phalanges

Lower limbs
Femur
Patella
Tibia
Fibula

Tarsals
Metatarsals
Phalanges

Calcaneus

(a) Anterior

(b) Posterior

■ **TABLE 6.3 Bones of the Cranium**

BONE	NUMBER	DESCRIPTION
Frontal	1	Forms anterior of cranium or forehead and superior portion of eye socket; contains two air sinuses; begins development as two pieces that completely fuse during early childhood (if a suture persists, it is called the **metopic** suture)
Parietal	2	Form the roof and sides of cranium; smooth outer surface with inner surface marked by grooves to accommodate brain's blood vessels; joined by **sagittal suture**
Occipital	1	Forms back and part of base of cranium; has **foramen magnum** through which spinal cord passes; articulates with axis of vertebral column by means of two condyles
Temporal	2	Form inferior sides of cranium; consists of four parts: **squamous** portion (side of cranium), **petrous** (wedge-shaped) portion, at base of cranium, housing inner ear, **mastoid process** (contains air sinuses communicating with middle ear), **tympanic** portion (below squamous portion forming part of auditory meatus), articulates with occipital, parietal, and sphenoid bones
Sphenoid	1	Butterfly-shaped bone forming floor of cranial cavity and articulating with all other cranial bones; body houses air sinus and sella turcica; greater and lesser wings project laterally from body; greater wings seen from both internal and external view of skull
Ethmoid	1	Located on the anterior floor of cranium; lateral labyrinths form walls of nasal cavity and have air cells; its **cribriform plate** forms roof of nasal cavities and is perforated, allowing passage of olfactory nerves; **crista galli** projects upward into cranial cavity serving as a point of attachment for dura mater; **perpendicular plate** projects downward forming part of nasal septum; **conchae** project inward from lateral part of ethmoid; posterior portion of eye orbit

■ **TABLE 6.4 Facial Bones**

BONE	NUMBER	DESCRIPTION
Nasal	2	Small, flat bones that form bridge of nose
Nasal conchae	2	Thin "scroll-shaped" bones on inferior portion of lateral wall of nasal cavity
Vomer	1	Thin bone that forms lower part of nasal septum; fuses with perpendicular plate of ethmoid
Maxillary	2	Large bones that form the upper jaw, floor of eye orbit and outer nasal cavity, and the hard palate; contain large sinus
Zygomatic	2	Form cheekbones, part of nose and mouth, and part of inferior portion of eye orbit
Palatine	2	Small, thin bones forming posterior portion of hard palate, floor of nasal cavity, and portion of eye orbit
Mandible	1	Largest and most mobile bone of face, forming lower jaw
Lacrimal	2	Posterior and lateral to nasal bones, form part of medial portion of eye orbit; has groove for lacrimal "tear" duct, which drains into the nasal cavity

■ **TABLE 6.5 Summary of Specific Features of Individual Skull Bones**

FRONTAL BONES	
Metopic suture	The line of junction between the two separate embryonic ossification centers. Generally not present in the adult skull.
Frontal sinuses	Mucous-membrane-lined air cavities located within the bone, close to the orbital cavities.
Supraorbital foramina or notches	Openings for blood vessels and nerves located just above the orbital cavities. They may appear as holes (foramina) or notches.
Glabella	The smooth area located between the two orbital cavities just above the nose.

■ **TABLE 6.5   Summary of Specific Features of Individual Skull Bones,** continued

OCCIPITAL BONE	
Foramen magnum	The opening through which the medulla oblongata of the brain stem leaves the skull to become continuous with the spinal cord.
Condyles	Smooth convex external projections on either side of the foramen magnum. They articulate with the first cervical vertebra.
Basioccipital	A narrow portion that extends anteriorly from the foramen magnum. It articulates with the sphenoid bone.
External occipital protuberance	A midline prominence on the outer surface, a short distance above the foramen magnum.
Nuchal lines*	Slight ridges on the external surface. *Medial nuchal line* Runs vertically between the external occipital protuberance and the foramen magnum. *Superior nuchal line* Extends laterally from the external occipital protuberance. *Inferior nuchal line* Extends laterally from the medial nuchal line at about its midpoint.
Internal occipital protuberance	A prominence on the inner surface of the bone. This marks the confluence of grooves for the sagittal, transverse, and occipital venous blood sinuses of the brain.
ETHMOID BONE	
Horizontal (cribriform) plate	The transverse portion that forms the roof of nasal cavity and floor of the anterior cranial cavity. The plate is perforated by the olfactory foramina to allow for the passage of the olfactory nerves (first cranial nerve).
Crista galli	A midline projection from the horizontal plate into the cranial cavity. It serves as the anterior point of attachment for the **falx cerebri**, a midline connective tissue septum that anchors the brain within the anterior cranial fossa.
Perpendicular plate	A downward projection from the midline of the undersurface of the horizontal plate. It forms the upper portion of the nasal septum. The remainder of the septum is formed by the vomer bone and hyaline cartilage.
Lateral masses	Thin-walled processes that extend downward from the lateral margins of the horizontal plate. They contain the **ethmoid sinuses**, which are mucous membrane-lined air cavities. The smooth lateral surfaces (*lamina orbitalis*) of the lateral masses form the medial walls of the orbital cavities.
Superior and middle conchae (turbinates)	Thin plates of bone that form the medial surfaces of the lateral masses. They also form part of the lateral walls of the nasal cavity. Recesses called **superior**, **middle** and **inferior meatuses** are located beneath the shelves of the conchae.
SPHENOID BONE	
Body	The central portion of the bone. It contains a large mucous-membrane-lined air sinus.
Sella turcica	A saddle-shaped depression on the superior surface of the body, bounded posteriorly by the **dorsum sellae**. It serves as the protective cavity for the pituitary gland.
Small wings	Sharp lateral projections from the superior portion of the body of the sphenoid. They form part of the posterior walls of the orbital cavities.
Optic foramina	Openings through the bases of each small wing for the passage of the optic nerves (second cranial nerves) into the orbital cavities.
Great wings	Large lateral projections from the body of the sphenoid. They form most of the posterior wall of the orbital cavity.
Superior orbital fissures	Slitlike openings between the great and small wings. They allow for the passage of the third, fourth, part of the fifth (ophthalmic division), and the sixth cranial nerves from the brain into the orbital cavity.
Foramen rotundum	The opening through the base of each of the great wings for the passage of the maxillary division of the fifth cranial nerves.
Foramen ovale	The opening through the base of each of the great wings for the passage of the mandibular division of the fifth cranial nerves.
Pterygoid processes	Two downward projections from the region where the great wings unite with the body. Each process consists of **medial** and **lateral plates**. The processes articulate anteriorly with the palatine bones.

■ **TABLE 6.5 Summary of Specific Features of Individual Skull Bones, continued**

TEMPORAL BONE

Squamous portion	The thin vertical projection that forms the anterior and superior portion of the bone. It meets with a parietal bone to form the squamous suture.
Zygomatic process	The anterior projection from the squamous portion. It articulates with the zygomatic (malar) bone to form the cheek (zygomatic arch).
Mandibular fossa	An oval depression on the inferior surface of the base of the zygomatic process. It articulates with the condyle of the mandible to form the temporomandibular joint.
Tympanic portion	Forms and surrounds the external acoustic meatus.
External acoustic meatus	The opening that leads into the middle-ear cavity from the exterior of the skull.
Petrous portion	A medial wedge of bone that forms the floor of the middle cranial fossa between the sphenoid and the occipital bones. It houses the middle- and inner-ear structures.
Internal acoustic meatus	The opening on the posterior surface of the petrous portion. It transmits the seventh cranial nerve as it travels to the facial structures, and the eighth cranial nerve as it travels to the inner ear.
Styloid process	A sharp spine that projects from the inferior lateral surface of the petrous portion. It serves as a point of attachment for the hyoid bone and for several ligaments and muscles of the pharynx and tongue.
Carotid canal	The passageway for the internal carotid artery as it travels through the petrous portion.
Jugular fossa	The depression for the internal jugular vein on the inferior surface of the petrous portion.
Stylomastoid foramen	The opening between the styloid process and mastoid process through which the seventh cranial nerve leaves the skull. (The nerve enters through the internal acoustic meatus.)
Jugular foramen	The large opening that allows for passage of several blood vessels, and the ninth, tenth, and eleventh cranial nerves. It is located at the junction of the petrous portion with the occipital bone.
Mastoid process	A prominent downward projection from the mastoid portion, just posterior to the external acoustic meatus.
Mastoid sinuses*	Mucous-membrane-lined air spaces within the mastoid process. These sinuses, which communicate with the middle-ear cavity, are the only cranial sinuses that do not drain into the nasal cavity.

MAXILLARY BONE (MAXILLA)

Maxillary sinus	A large mucous-membrane-lined cavity within the bone.
Frontal process	The vertical process that forms part of the bridge of the nose. It articulates above with the frontal bone, anteriorly with the nasal, and posteriorly with the lacrimal.
Zygomatic process	A rough triangular eminence that articulates with the zygomatic (malar) bone.
Alveolar process	The inferior border that holds the teeth. When the two maxillae are articulated with each other, their alveolar processes together form the alveolar arch.
Palatine process	The medial horizontal shelf that runs from the inner surface of the alveolar process. It joins with the palatine process of the other maxillary bone to form most of the hard palate.
Anterior nasal spine	A pointed process just below the nasal cavity. It joins with the nasal spine of the other maxillary bone to form a point of attachment for the cartilage portion of the nasal septum.
Infraorbital foramen	The opening just below the margin of the orbit. It transmits blood vessels and nerves.
Orbital surface	The smooth, flat surface that forms the floor of the orbit.

MANDIBLE

Body	The curved, horizontal portion that forms the chin.
Rami	Two perpendicular projections that join the posterior lateral margins of the body at approximately right angles.
Mandibular symphysis	The vertical midline fusion between the two embryonic ossification centers that form the body.
Alveolar border	The superior edge of the body that contains the sockets for the teeth.
Mental foramina	Two openings on the external surface of the body that allow for the passage of blood vessels and nerves.
Angle	A sharp curve on the posterior inferior portion of the ramus.
Mandibular foramina	Openings on the inner surfaces of each ramus for the passage of blood vessels and nerves.
Coronoid processes	Thin upward projection on the anterior surface of each ramus. They provide attachment for the temporalis muscle.
Mandibular condyles	Smooth convex surface on the superior borders of each ramus. They articulate with the mandibular fossae of the temporal bones.
Mandibular notches	Deep depression between the coronoid process and the condyle of each ramus.

■ **FIGURE 6.2 Skull: Front view**

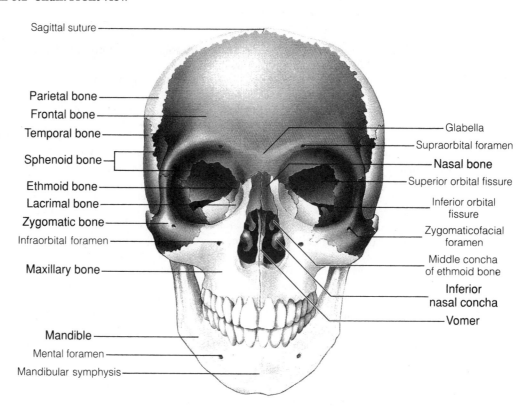

Sagittal suture

Parietal bone

Frontal bone

Temporal bone

Sphenoid bone

Ethmoid bone

Lacrimal bone

Zygomatic bone

Infraorbital foramen

Maxillary bone

Mandible

Mental foramen

Mandibular symphysis

Glabella

Supraorbital foramen

Nasal bone

Superior orbital fissure

Inferior orbital fissure

Zygomaticofacial foramen

Middle concha of ethmoid bone

Inferior nasal concha

Vomer

■ **FIGURE 6.3 Skull: Lateral view**

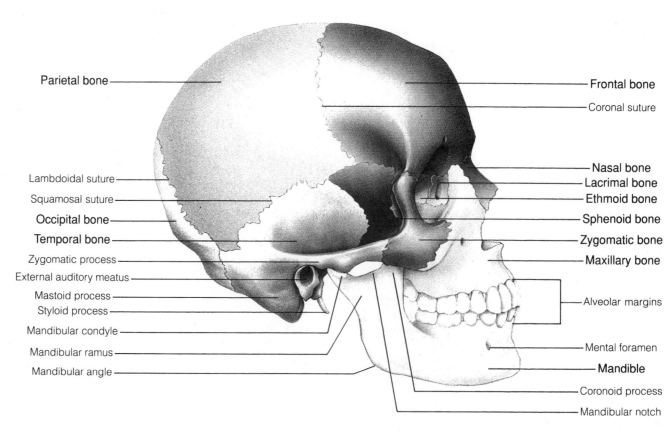

Parietal bone

Lambdoidal suture

Squamosal suture

Occipital bone

Temporal bone

Zygomatic process

External auditory meatus

Mastoid process

Styloid process

Mandibular condyle

Mandibular ramus

Mandibular angle

Frontal bone

Coronal suture

Nasal bone

Lacrimal bone

Ethmoid bone

Sphenoid bone

Zygomatic bone

Maxillary bone

Alveolar margins

Mental foramen

Mandible

Coronoid process

Mandibular notch

■ **FIGURE 6.4 Skull: Sagittal view**

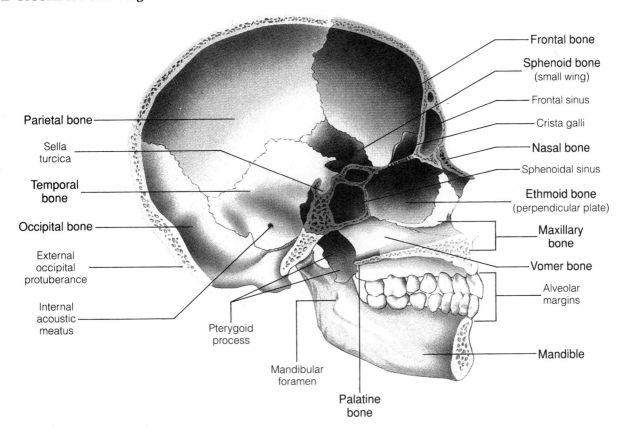

Parietal bone

Sella turcica

Temporal bone

Occipital bone

External occipital protuberance

Internal acoustic meatus

Pterygoid process

Mandibular foramen

Palatine bone

Frontal bone

Sphenoid bone (small wing)

Frontal sinus

Crista galli

Nasal bone

Sphenoidal sinus

Ethmoid bone (perpendicular plate)

Maxillary bone

Vomer bone

Alveolar margins

Mandible

■ **FIGURE 6.5 Skull: Inferior view**

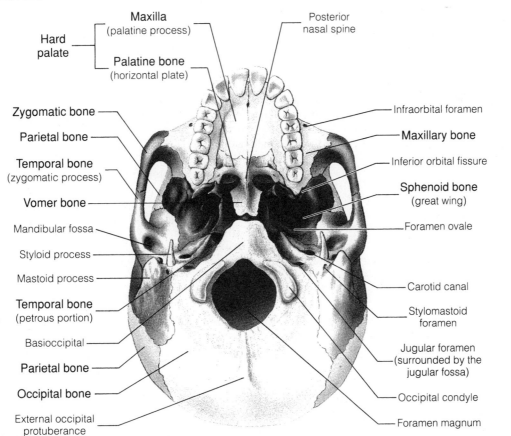

Hard palate

Maxilla (palatine process)

Palatine bone (horizontal plate)

Posterior nasal spine

Zygomatic bone

Parietal bone

Temporal bone (zygomatic process)

Vomer bone

Mandibular fossa

Styloid process

Mastoid process

Temporal bone (petrous portion)

Basioccipital

Parietal bone

Occipital bone

External occipital protuberance

Infraorbital foramen

Maxillary bone

Inferior orbital fissure

Sphenoid bone (great wing)

Foramen ovale

Carotid canal

Stylomastoid foramen

Jugular foramen (surrounded by the jugular fossa)

Occipital condyle

Foramen magnum

■ **FIGURE 6.6 Floor of the cranial cavity**

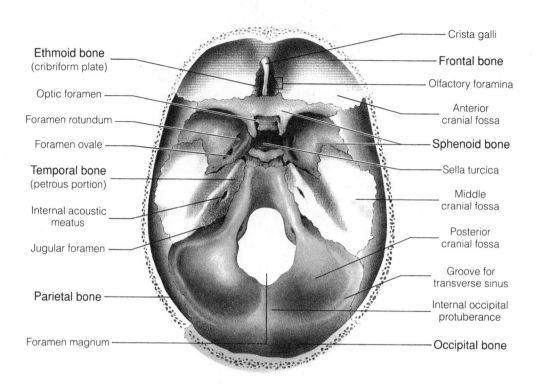

Crista galli

Frontal bone

Ethmoid bone
(cribriform plate)

Olfactory foramina

Optic foramen

Anterior
cranial fossa

Foramen rotundum

Sphenoid bone

Foramen ovale

Sella turcica

Temporal bone
(petrous portion)

Middle
cranial fossa

Internal acoustic
meatus

Posterior
cranial fossa

Jugular foramen

Groove for
transverse sinus

Parietal bone

Internal occipital
protuberance

Foramen magnum

Occipital bone

■ **TABLE 6.6 Major Skull Foramina**

FORAMEN	BONE LOCATION	STRUCTURE PASSING THROUGH FORAMEN
Carotid canal	Temporal	Internal carotid artery
Infraorbital	Maxilla	Maxillary branch of trigeminal nerve
Jugular	Between temporal and occipital	Internal jugular vein; glossopharyngeal, vagus, and accessory cranial nerves
Lacerum	Between temporal and occipital	Communicates with carotid canal
Magnum	Occipital	Spinal cord from medulla; accessory nerve, vertebral, and spinal arteries
Mandibular	Mandible	Nerves to lower teeth
Mental	Mandible	Mental nerve and blood vessels
Optic	Sphenoid	Optic nerve, ophthalmic artery
Ovale	Sphenoid	Mandibular branch of trigeminal nerve
Rotundum	Sphenoid	Maxillary branch of trigeminal nerve
Spinosum	Sphenoid	Meningeal vessels
Stylomastoid	Temporal	Facial nerve
Supraorbital	Frontal	Supraorbital nerve and artery

FIGURE 6.7 Ethmoid bone (top), sphenoid bones (bottom)

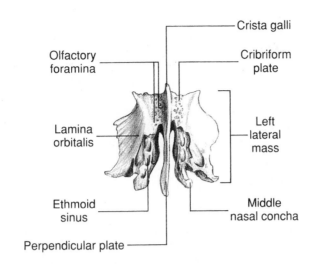

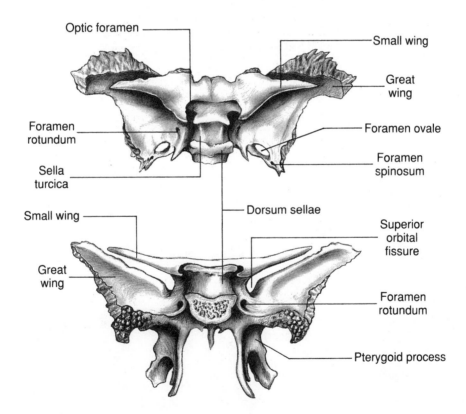

■ **TABLE 6.7 Sinuses of Skull Bones**

SINUS[a]	LOCATION
Frontal	Spaces on each side of midsagittal plane; drains into nasal cavity
Sphenoidal	Large space in body of sphenoid; drains into nasal cavity
Ethmoidal	Cluster of "air cells" in labyrinths of ethmoid; drains into nasal cavity
Maxillary	Large spaces in middle of maxilla; drains into nasal cavity
Mastoid	Spaces within mastoid process of temporal bone; drains into middle ear

[a]The first four sinuses are called **paranasal** sinuses.

■ **TABLE 6.8 Major Cranial Sutures**

SUTURE	LOCATION
Sagittal	Union of parietal bones
Coronal	Union of frontal and parietal bones
Lambdoidal	Union of occipital and parietal bones
Squamosal	Union of temporal and parietal bones
Occipitomastoidal	Union of temporal and occipital bones
Sphenosquamosal	Union of temporal and sphenoid bones
Metopic	If present, junction of frontal bones (usually not found after 2 years of age)

■ **TABLE 6.9 Fontanels of Fetal Skull**

FONTANEL	LOCATION AND DESCRIPTION
Frontal (anterior)	At juncture of parietal bones with frontal bone; closes by 2 years of age
Occipital (posterior)	At juncture of parietal bones with occipital bone; closes by 3 months of age
Sphenoidal (anterolateral)	Juncture of frontal, parietal, temporal, and sphenoid bones; closes by 3 months of age
Mastoid (posterolateral)	Juncture of parietal, occipital, and temporal bones; closes by 1 year of age

■ **FIGURE 6.8 Fetal skull**

(a) Superior view (b) Lateral view

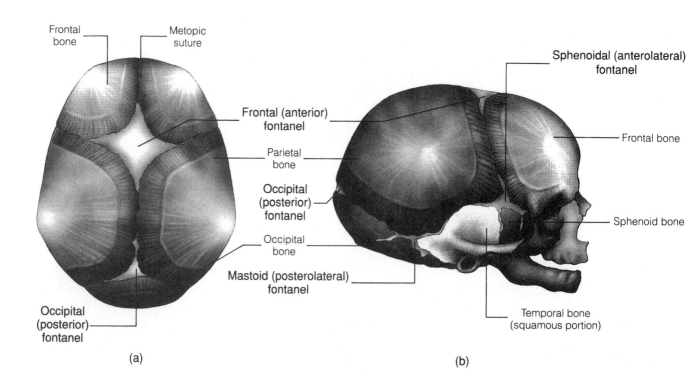

Conclusions

- What bones form the framework of the external nose and nasal cavity? Indicate what part they form.

_____

_____

_____

- List all bones forming the orbit of the eye.

_____

_____

_____

- How does the fetal skull compare with the adult skull in general?

_____

_____

_____

EXERCISE 2
The Vertebral Column

Discussion

Although the vertebral column starts its development in the fetus as 33 separate bones, ultimate fusion of the lower vertebrae results in an adult vertebral column numbering 26 bones. These bones are classified into groups according to location: 7 cervical, 12 thoracic, 5 lumbar, 1 sacral, and 1 coccygeal. Each of the major groups of bones has its own distinct characteristics, which are summarized in Table 6.10. However, as a whole, all vertebrae have certain shared characteristics. The anterior rounded central portion is called the **centrum** or body. Projecting posteriorly from the centrum are bony projections which enclose the **vertebral foramen** and form what is known as the **neural arch**. Extending laterally from the neural arch are the **transverse processes**. The portion of the neural arch between the centrum and the transverse process is called the **pedicle** and the portion that is found between the transverse process and **spinous process** at the midline is called the **lamina**. Projecting from the vertebrae are 7 processes: 2 **transverse processes**, which are lateral extensions from the junction of the pedicle and lamina; 2 **superior** and 2 **inferior articulating processes**, which project vertically near the junction of the pedicle and lamina; and 1 **spinous process**, which extends posteriorly from the midline of the vertebrae. Refer to Figure 6.10.

When the vertebral column is articulated it is neither stiff nor straight. Between the vertebrae is a fibrocartilage disc, known as the **intervertebral disc**, which cushions the joint between adjacent vertebrae. The vertebral column also exhibits several curves (Figure 6.9), which develop during early postnatal development as a baby begins to hold its head up, sit up, and ultimately walk.

Materials

Articulated and disarticulated vertebral columns.

Procedure

1. Examine the various **vertebrae** noting the features they have in common.
2. Examine the individual bones of the vertebral column and note their unique characteristics (Table 6.10) so you can identify the section of origin (cervical, thoracic, lumbar, sacrum, coccyx). Refer to Figures 6.10–6.12.
3. Examine the articulated vertebral column, noting the way the vertebrae articulate with each other and also the different curvatures shown in Figure 6.9.

■ **TABLE 6.10 Bones of the Vertebral Column**

TYPE OR NAME	NUMBER	DESCRIPTION
Cervical	7	Greatly reduced body; C_1 is the **atlas**, which lacks a body and spinous process; C_2 is the **axis**, which has an upward projection from the body called the **odontoid process (dens)**; C_3 through C_5 have forked spinous process; all have transverse foramina
Thoracic	12	Larger than cervical vertebrae; long, pointed, posteriorly directed spinous processes; bodies and transverse processes have facets for rib articulation
Lumbar	5	Largest and strongest vertebrae; processes are short and thick
Sacrum	1	Wedged-shaped bone formed from fusion of 5 sacral vertebrae; **medial sacral crest** is remnant of spinous processes
Coccyx	1	Formed by fusion of 2–4 rudimentary vertebrae, degree of fusion may vary; triangular shape

■ **FIGURE 6.9 Vertebral column and curves**

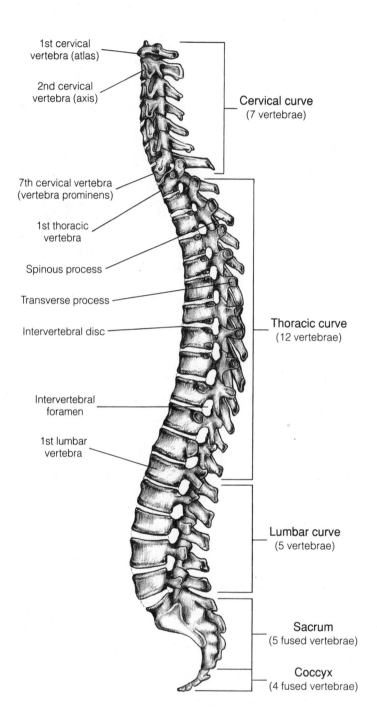

1st cervical
vertebra (atlas)

2nd cervical
vertebra (axis)

Cervical curve
(7 vertebrae)

7th cervical vertebra
(vertebra prominens)

1st thoracic
vertebra

Spinous process

Transverse process

Intervertebral disc

Thoracic curve
(12 vertebrae)

Intervertebral
foramen

1st lumbar
vertebra

Lumbar curve
(5 vertebrae)

Sacrum
(5 fused vertebrae)

Coccyx
(4 fused vertebrae)

■ **FIGURE 6.10 Typical vertebrae**

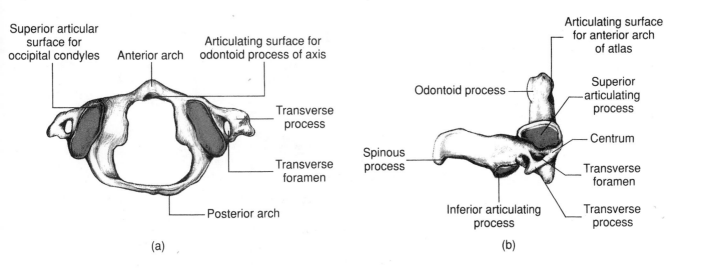

Superior articular surface for occipital condyles

Anterior arch

Articulating surface for odontoid process of axis

Transverse process

Transverse foramen

Posterior arch

(a)

Articulating surface for anterior arch of atlas

Odontoid process

Superior articulating process

Centrum

Spinous process

Transverse foramen

Inferior articulating process

Transverse process

(b)

■ **FIGURE 6.11 Two typical lumbar vertebrae**

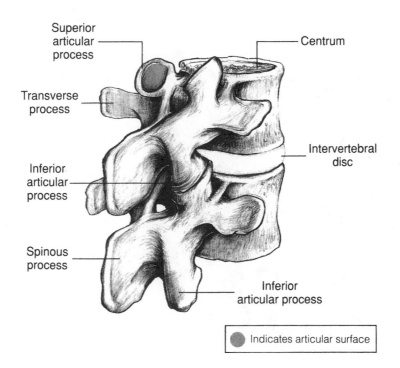

Superior articular process

Centrum

Transverse process

Inferior articular process

Intervertebral disc

Spinous process

Inferior articular process

● Indicates articular surface

■ **FIGURE 6.12 Sacrum and coccyx**

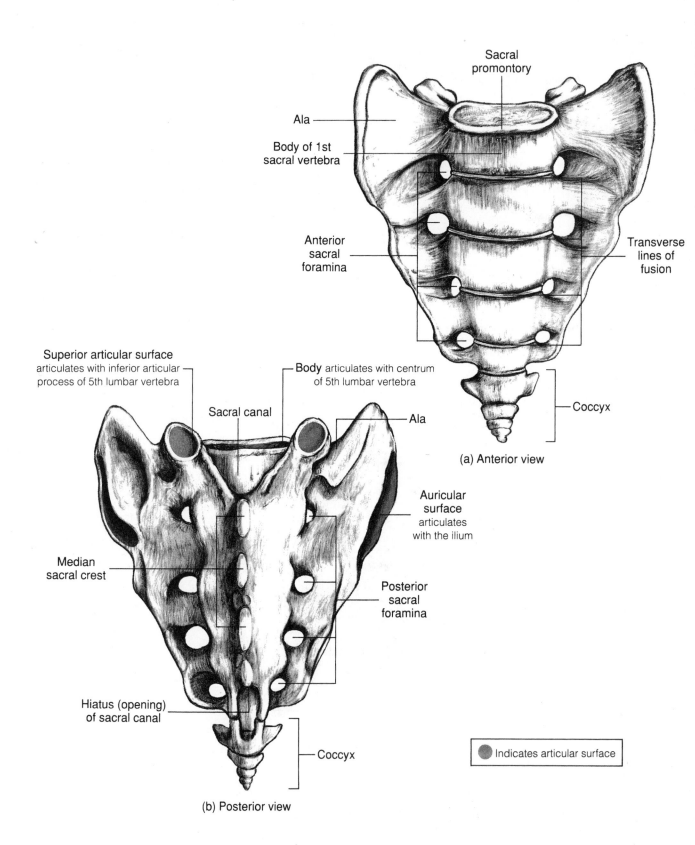

Sacral promontory

Ala

Body of 1st sacral vertebra

Anterior sacral foramina

Transverse lines of fusion

Coccyx

(a) Anterior view

Superior articular surface
articulates with inferior articular process of 5th lumbar vertebra

Body articulates with centrum of 5th lumbar vertebra

Sacral canal

Ala

Auricular surface articulates with the ilium

Median sacral crest

Posterior sacral foramina

Hiatus (opening) of sacral canal

Coccyx

(b) Posterior view

● Indicates articular surface

Conclusions

- What is the function of the following:

  a. Odontoid process (dens) _____

  b. Facets of thoracic vertebrae _____

  c. Intervertebral foramina _____

- If the laminae fail to fuse, what do you think would happen?

  _____

  _____

  _____

- Look up the medical terms for the following exaggerated curvatures and indicate what vertebrae are involved.

  a. Hunchback _____

  b. Swayback _____

- With what specific structures do the superior articulating processes of the atlas articulate?

  _____

  _____

EXERCISE 3
The Rib Cage

Discussion

The thorax is formed by 12 pairs of ribs, the sternum, and 12 thoracic vertebrae, and they protect the vital organs of the chest and some participate in ventilatory movements. The ribs are elongated, flat, curved bones that articulate posteriorly with the thoracic vertebrae. Typically, a **rib** has (1) a posterior rounded **head** that articulates with a vertebra, (2) a **neck**, a narrowed protion lateral to the head, which runs into (3) the long, thin, curved **shaft**. Just below the neck is a small rounded bulge called the **tubercle**, which also serves in articulation with the vertebrae. The **costal cartilage** is a band of hyaline cartilage that is involved in the anterior attachment of the rib to the sternum.

The 12 pairs of ribs are classified according to the type of anterior attachment they exhibit. Rib pairs 1 through 7 attach, via the costal cartilage, directly to the sternum and are called **true ribs**. Rib pairs 8 through 10 are at-tached, via their costal cartilage, to the costal cartilage of the rib immediately above and thus are referred to as **false ribs**. The last two pairs, 11 and 12, are called **floating ribs**, since they have no anterior attachment but rather are simply embedded in the muscle of the body wall.

In the anterior midline of the thorax is the **sternum**, which serves as the point of attachment for the true ribs. It is a flat bone that is somewhat dagger shaped, consisting of three portions: manubrium, body, and xiphoid process. The point at which the manubrium joins the body is an anatomical landmark, the sternal angle. The second rib articulates with the sternum here and deep to it the aorta arches, the trachea bifurcates, and the lungs pass laterally to their apices. The lower portion of the sternum lies over the soft liver, and is to be avoided in cardiac compressions during cardiopulmonary resuscitation, to prevent damage to the liver.

Materials

An articulated skeleton; disarticulated ribs and sternum.

Procedure

1. Examine the articulated skeleton noting the organization of the rib cage and how the three components, ribs, sternum, and vertebrae, articulate with each other.
2. Examine a typical rib and the sternum and identify the features illustrated in Figures 6.13 and 6.14. Be able to identify left and right ribs. A rib, if properly oriented, passes anteriorly and curves inferiorly at its distal end.
3. Articulate a rib with the vertebral column.

Conclusions

- How do most of the ribs articulate with the vertebrae? Explain the articulation of a typical rib in detail based upon your observations.

  _____

  _____

  _____

  _____

  _____

■ **FIGURE 6.13 Thorax: anterior view**

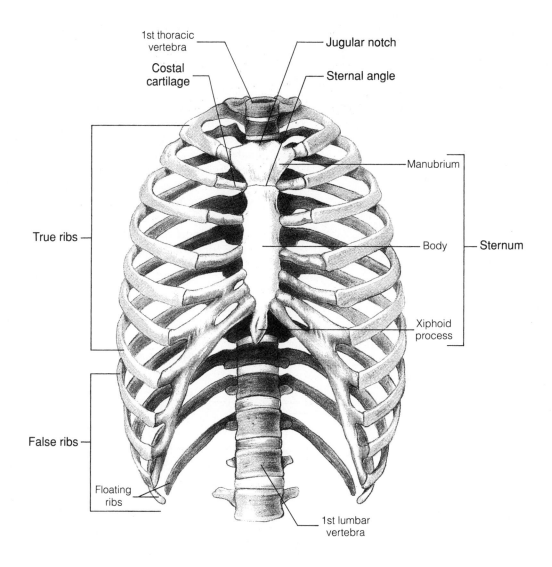

■ **FIGURE 6.14 A typical rib**

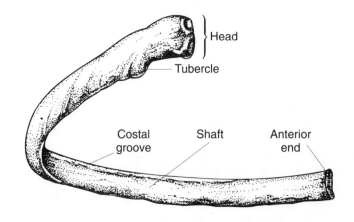

EXERCISE 4
Miscellaneous Bones

Discussion

Perhaps the most unusual bone of the skeleton is the **hyoid** bone, since it does not form direct articulations with any other bones. It is suspended in the anterior neck below the mandible by ligaments attached to the styloid process of the temporal bone. The hyoid bone serves as a point of attachment for the musculature of the tongue.

In addition, the auditory **ossicles** are six tiny bones located within the temporal bone. There are three in each ear and they are called the malleus, incus and stapes. Their role in hearing and examination of ear models will be carried out later in the unit on Special Senses.

Materials

An articulated skeleton; hyoid bone; model of ear showing ossicles.

Procedure

1. Note the location of the hyoid bone on the articulated skeleton.
2. Examine the hyoid and identify the features illustrated in Figure 6.15.
3. Identify the ossicles on ear model. Refer to Figure 6.16.

Conclusions

• What are the ligaments that suspend the hyoid bone called?

_____

_____

• To what structure of the respiratory system would the hyoid bone be anterior?

_____

_____

■ **FIGURE 6.15 Hyoid bone**

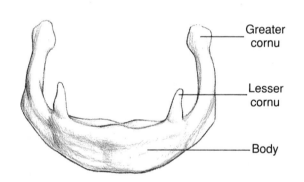

Greater cornu

Lesser cornu

Body

■ **FIGURE 6.16 Auditory ossicles of the middle ear**

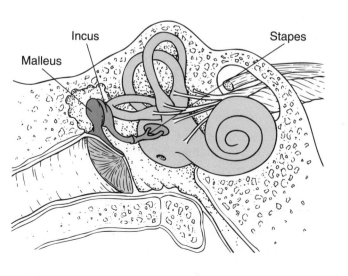

Incus

Malleus

Stapes

❖ Quick Quiz 2

Match the terms in column A with the appropriate bone in column B.

	A	B
_____	1. odontoid process	(a) sphenoid
_____	2. petrous portion	(b) rib
_____	3. sella turcica	(c) vertebrae
_____	4. mental foramen	(d) axis
_____	5. cribriform plate	(e) temporal
_____	6. laminae	(f) sternum
_____	7. xiphoid process	(g) ethmoid
_____	8. spinous process	(h) mandible
_____	9. costal groove	
_____	10. pterygoid process	

❖ The Appendicular Skeleton

The **appendicular skeleton** consists of the bones that are the framework of the upper and lower limbs and the bones which form their attachments to the axial skeleton. These include the bones of the arm and the shoulder or **pectoral girdle** and the bones of the leg and the hip or **pelvic girdle**.

EXERCISE 1
The Pectoral Girdle

Discussion

More commonly known as the shoulder blades and collar bones, respectively, the **scapulae** and **clavicles** form the

---

■ FIGURE 6.17 Human shoulder articulation (left)

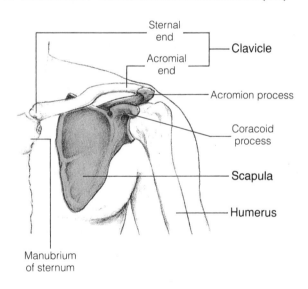

■ FIGURE 6.18 Clavicle

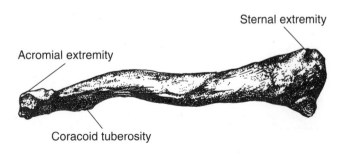

Anterior view

---

pectoral girdle and serve as points of articulation for the upper extremity. These bones are illustrated in Figures 6.18 and 6.19. (If you can properly orient a bone in/on the skeleton, you should be able to identify whether it is a left or right bone. This should be part of your work with the appendicular skeleton.)

Materials

An articulated skeleton; disarticulated scapula and clavicle.

Procedure

1. Examine the articulated skeleton and observe how the upper extremity articulates with the pectoral girdle. Use Figure 6.17 as a guide.
2. On the articulated skeleton, note where the pectoral girdle is joined with the axial skeleton.
3. Using Figures 6.17 through 6.19 examine the scapula and clavicle and identify the features illustrated in the figures.

Conclusions

• How many articulations does the pectoral girdle form with the axial skeleton. Where are they?

_____

_____

_____

• With what part of the pectoral girdle does the upper extremity form a direct articulation?

_____

_____

EXERCISE 2
The Upper Extremity

Discussion

Each arm consists of three bones: the **humerus** in the upper arm and the **radius** and the **ulna** in the forearm. The humerus is a typical long bone and the radius and ulna are also long bones, with the ulna being slightly longer than the radius. The hand consists of the wrist, palm, and fingers, which together have a total of 27 bones. Comprising the wrist are 8 **carpal** bones, which are small, irregularly shaped bones arranged in two rows.

■ **FIGURE 6.19 Scapula (left)**
(a) Anterior view (b) Posterior view

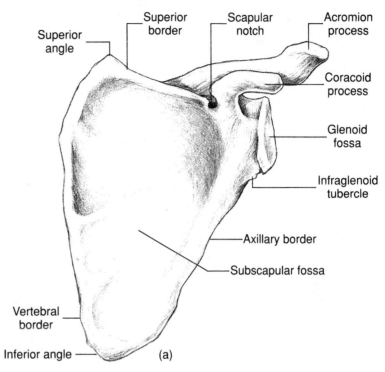

Superior angle
Superior border
Scapular notch
Acromion process
Coracoid process
Glenoid fossa
Infraglenoid tubercle
Axillary border
Subscapular fossa
Vertebral border
Inferior angle

(a)

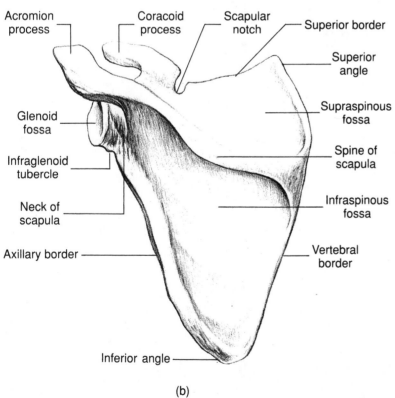

Acromion process
Coracoid process
Scapular notch
Superior border
Superior angle
Supraspinous fossa
Glenoid fossa
Spine of scapula
Infraglenoid tubercle
Neck of scapula
Infraspinous fossa
Axillary border
Vertebral border
Inferior angle

(b)

The **5** bones of the palm are called **metacarpals**, while the fingers have a total of 14 bones, referred to as **phalanges**.

Materials

An articulated skeleton; disarticulated humerus, radius, and ulna; articulated hand.

Procedure

1. Examine the articulated skeleton and observe how the bones of the upper extremity articulate with each other and with the pectoral girdle.
2. Using Figures 6.20 through 6.21 as guides, examine the humerus, radius, and ulna and identify the features illustrated in these figures.
3. Examine the articulated hand and identify the bones illustrated in Figure 6.22.

■ **FIGURE 6.20 Humerus (right)**
(a) Anterior view (b) Posterior view

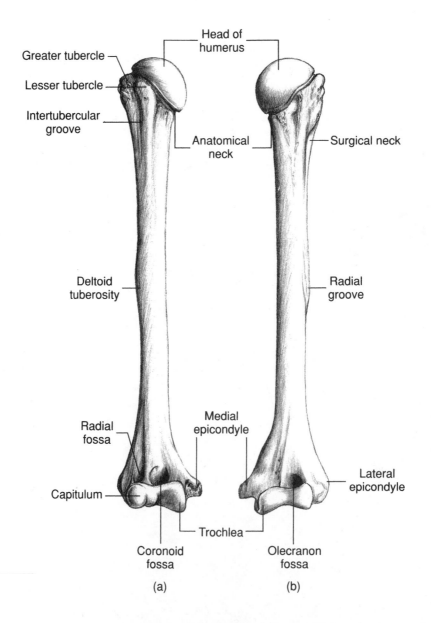

Greater tubercle
Lesser tubercle
Intertubercular groove
Head of humerus
Anatomical neck
Surgical neck
Deltoid tuberosity
Radial groove
Radial fossa
Medial epicondyle
Lateral epicondyle
Capitulum
Trochlea
Coronoid fossa
Olecranon fossa

(a)   (b)

■ FIGURE 6.21 Radius, ulna (right)

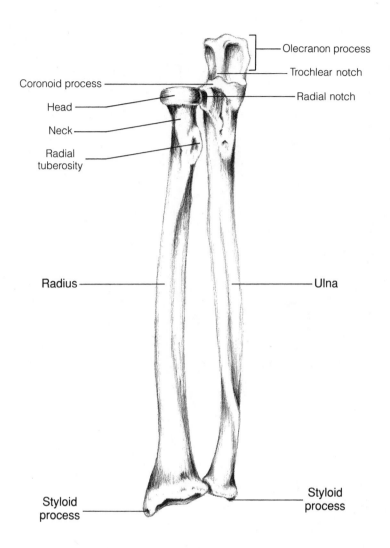

Olecranon process

Trochlear notch

Coronoid process

Radial notch

Head

Neck

Radial
tuberosity

Radius

Ulna

Styloid
process

Styloid
process

■ **FIGURE 6.22   The hand (right)**

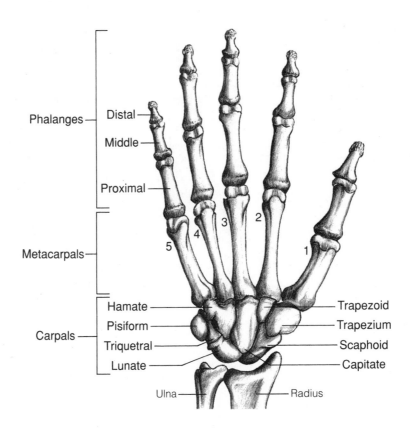

Phalanges — Distal

Middle

Proximal

Metacarpals

5   4   3   2   1

Carpals — Hamate — Trapezoid
Pisiform — Trapezium
Triquetral — Scaphoid
Lunate — Capitate

Ulna — Radius

Conclusions

- Which of the bones of the forearm is lateral and which is medial in the anatomic position?

  _____

  _____

- What is the purpose of the coronoid fossa?

  _____

  _____

  _____

- With what structure does the capitulum articulate?

  _____

- Since the metacarpals are numbered from 1 to 5 starting on the lateral side, what is the number of the metacarpal associated with the index finger?

  _____

- What bone do you see projecting from the medial side of the "heel" of your hand?

  _____

  _____

EXERCISE 3
The Pelvic Girdle

Discussion

Although male and female skeletons are quite similar, distinct differences are observable in the pelvis. The female pelvis is broader but more shallow than the male pelvis and has a wider pubic arch. These features are evident in Figure 6.23. The pelvic girdle (pelvis) is formed by the two hip bones (os coxae or innominate bones), which articulate anteriorly at the pubic symphysis and posteriorly with the sacrum. Each hip bone begins its development as three separate bones: the **ilium**, the **ischium**, and the **pubis**. By adolescence, however, these bones unite at the **acetabulum**. Figure 6.24 shows articulation with the lower extremity and Figure 6.25 provides two views of the pelvic bones.

Materials

An articulated skeleton; disarticulated os coxae; male and female pelvises.

■ **FIGURE 6.23 Human pelvis (male and female)**

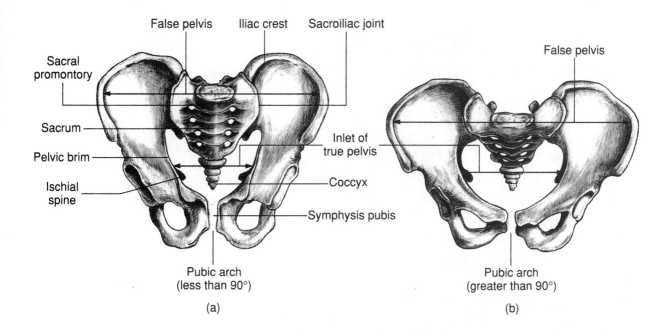

False pelvis   Iliac crest   Sacroiliac joint

Sacral
promontory

Sacrum

Pelvic brim

Ischial
spine

Inlet of
true pelvis

Coccyx

Symphysis pubis

Pubic arch
(less than 90°)

(a)

False pelvis

Pubic arch
(greater than 90°)

(b)

Procedure

1. Examine the male and female pelvises and observe the differences between them.
2. Examine the articulated skeleton and note how the pelvis and lower extremity articulate with the axial skeleton.
3. Examine the disarticulated os coxae and determine which area is the ilium, which the ischium, which the pubis, and which the acetabulum.
4. Identify the features of the os coxae illustrated in Figure 6.25.

Conclusions

• Are the lines of fusion in the acetabulum still visible?

_____

• Look up the structure(s) that passes through the obturator foramen.

_____

_____

■ **FIGURE 6.24 Hip articulation**

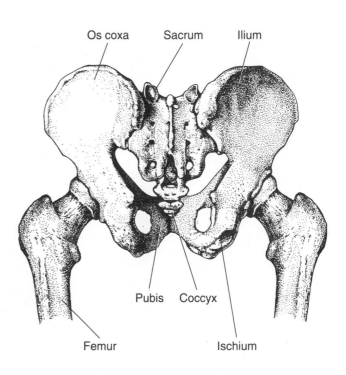

Os coxa   Sacrum   Ilium

Pubis   Coccyx

Femur   Ischium

■ **FIGURE 6.25 Os coxae (right)**
(a) Medial view (b) Lateral view

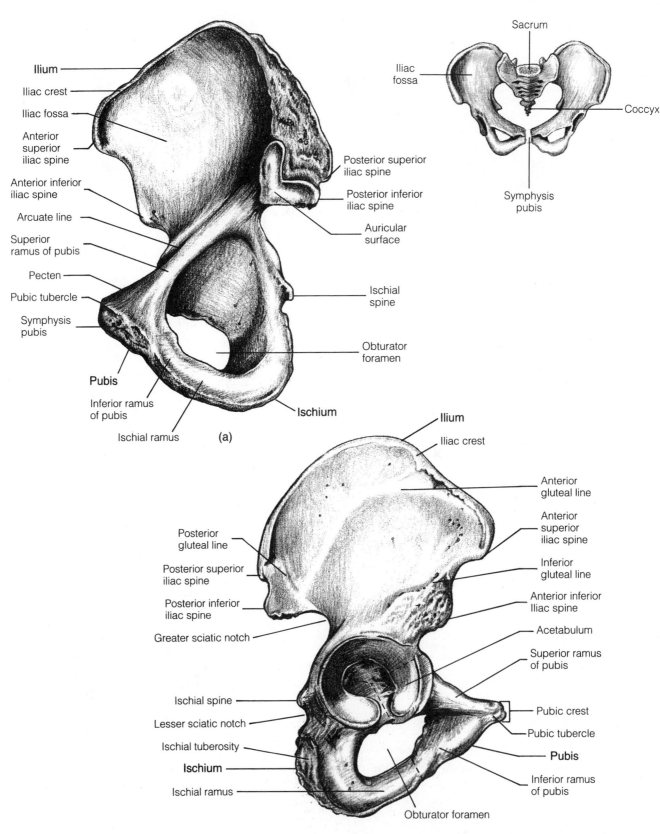

Ilium

Iliac crest

Iliac fossa

Anterior superior iliac spine

Anterior inferior iliac spine

Arcuate line

Superior ramus of pubis

Pecten

Pubic tubercle

Symphysis pubis

Pubis

Inferior ramus of pubis

Ischial ramus

Posterior superior iliac spine

Posterior inferior iliac spine

Auricular surface

Ischial spine

Obturator foramen

Ischium

(a)

Sacrum

Iliac fossa

Coccyx

Symphysis pubis

Ilium

Iliac crest

Anterior gluteal line

Anterior superior iliac spine

Inferior gluteal line

Anterior inferior Iliac spine

Acetabulum

Superior ramus of pubis

Pubic crest

Pubic tubercle

Pubis

Inferior ramus of pubis

Posterior gluteal line

Posterior superior iliac spine

Posterior inferior iliac spine

Greater sciatic notch

Ischial spine

Lesser sciatic notch

Ischial tuberosity

Ischium

Ischial ramus

Obturator foramen

(b)

EXERCISE 4
The Lower Extremity

Discussion

There are 30 bones in each of the lower extremities. The lower extremity has 4 bones and the ankle and foot have a total of 26. The bones in the leg are the **femur** in the upper leg, the **tibia** and **fibula** in the lower leg, and the **patella** (a sesamoid bone) lying at the front of the knee joint. The 7 bones of the ankle are called **tarsals**. They are comparable to the carpals of the wrist but are heavier and arranged to withstand the pressure of walking. The remainder of the foot is quite similar to the hand. The 5 **metatarsals** are in the sole of the foot and form the arches; the proximal, medial, and distal **phalanges** form the toes.

Materials

An articulated skeleton; disarticulated femur, tibia, fibula, and patella; articulated foot.

Procedure

1. Examine the articulated skeleton and observe how the bones of the lower extremity articulate with each other.

■ **FIGURE 6.26 Femur (right)**
(a) Anterior view (b) Posterior view

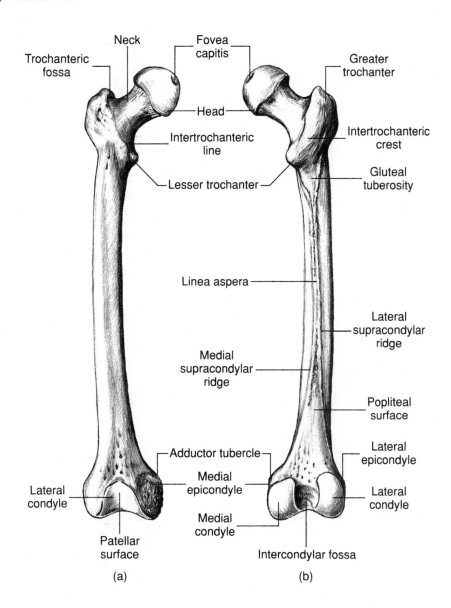

(a)   (b)

2. Using Figures 6.26 through 6.28 as guides, examine the femur, tibia, fibula, and patella and identify the features illustrated in these figures.

3. Examine the articulated foot and identify the bones illustrated in Figure 6.28.

Conclusions

● With what bones does the fibula articulate at its proximal and distal end?

_____

_____

● With what do the articular facets on the posterior surface of the patella articulate?

_____

_____

● Based upon your observation of the articulated foot, how many arches are there? Explain.

_____

_____

_____

■ **FIGURE 6.27 Tibia and fibula (right)**
(a) Anterior view (b) Posterior view

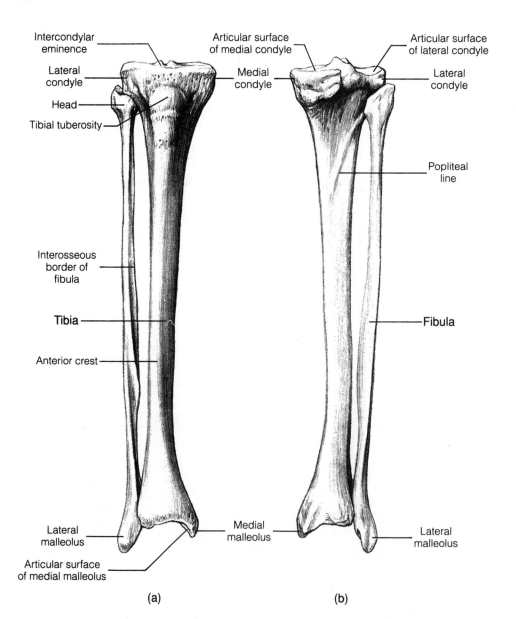

Intercondylar eminence
Lateral condyle
Head
Tibial tuberosity
Interosseous border of fibula
Tibia
Anterior crest
Lateral malleolus
Articular surface of medial malleolus
Articular surface of medial condyle
Medial condyle
Medial malleolus
Articular surface of lateral condyle
Lateral condyle
Popliteal line
Fibula
Lateral malleolus

(a)

(b)

■ FIGURE 6.28 The foot (right)

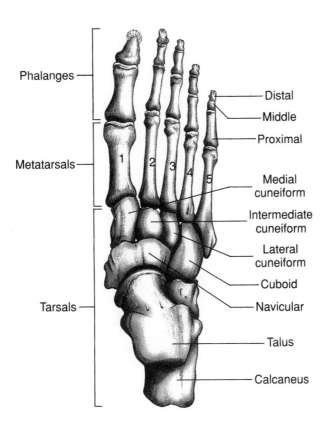

UNIT 7

Articulations

OVERVIEW

After studying the bones of the skeleton, it is now appropriate to study how the individual bones are put together to form the functional framework of the body. You will now learn how the bones of the skeleton join to one another or articulate. In this unit, you will compare the major types of articulations, learn the characteristics of a typical synovial joint and identify the various types of motion possible at synovial joints.

OUTLINE

Types of Articulations

Exercise 1   Types of Joints
Exercise 2   A Typical Synovial Joint

Movements

Exercise 1   Movable Joint Types and Possible Movements

❖ Types of Articulations

Background

As you examined the articulated skeleton in the preceding exercises, you undoubtedly became aware of the wide variation that is possible in the types of articulations or joints. Typically, joints are classified on either a structural basis or on the degree of movement they allow. On a structural basis, three types of joints are apparent depending on the type of tissue and its organization in the formation of the joint: **fibrous, cartilagenous** and **synovial**. Using the degree of mobility as a criterion, it is also possible to identify three types of joints: **synarthroses** (immovable), **amphiarthroses** (slight mobility) and **diarthroses** (freely movable). Table 7.1 summarizes information about joints with regard to their overall structure and mobility.

EXERCISE 1
Types of Joints

Discussion

Joints provide us with the capability to move from place to place, but also to move various bones in relationship to one another. As a result we are able to carry out a wide variety of activities in our daily lives. Although you have, in studying the skeleton, observed numerous types of articulations, you should now review these articulations in light of additional information.

Materials

Articulated skeleton, skull, articulation models.

85

■ **TABLE 7.1 Types of Joints**

JOINT	DESCRIPTION	EXAMPLE
FIBROUS		
Suture	Bone are held together by a thin layer of fibrous tissue and also by interlocking projections of the bones	Cranial sutures
Syndesmosis	Bones are held together by dense fibrous connective tissue called a ligament	Tibia/Fibula joint
Gomphosis	Teeth held to alveolar sockets by a thin fibrous membrane called the periodontal ligament	Teeth/Mandible
CARTILAGENOUS		
Synchondrosis	Bone surfaces are connected by hyaline cartilage (may be permanent or temporary)	Ribs/Sternum Epiphyseal Plate
Symphysis	Bone ends covered with hyaline cartilage and connected with a disc of fibrocartilage	Pubic Symphysis
SYNOVIAL	Articulating ends of bone are covered with hyaline cartilage and are held together with a double layered joint capsule; the inner membrane secretes a lubricating fluid into the joint cavity; many variations possible	Humerus/ Scapula

Procedures

1. For the joints listed in the Observation Table A, examine the skeleton and/or models and observe the structural organization of the articulation. (Use Figure 6.1 in the previous unit as a guide.)
2. Record your observations in Table A.
3. Fill in the last column.

Conclusions

● Based upon your observations, what is the relationship of the structure of a joint to its function; e.g., skull–sutures/protection.

_____

_____

_____

Observations

■ **TABLE A Types of Joints**

JOINT	OBSERVATIONS	TYPES OF JOINTS
Parietal/parietal		
Mandible/temporal		
Tibia/fibula		
Radius/ulna		
Occipital/atlas		
Ribs/sternum		
Elbow		

EXERCISE 2
A Typical Synovial Joint

Discussion

As indicated in Figure 7.1 a typical synovial joint has a joint cavity between the articulating ends of the bones, each of which has a covering of hyaline cartilage. A joint capsule encloses these structures and the inner **synovial membrane** secretes a lubricating, synovial fluid, which also provides nourishment for the articular cartilage and keeps the articulating surfaces slightly apart. Figure 7.2 illustrates the various types of synovial joints. It can be readily seen that the articulating surfaces of synovial joints have a variety of shapes.

Materials

Fresh animal knee joints (from butcher), whole and sagittally sectioned; model of human synovial joints.

Procedure

1. Examine the models and animal joints as illustrated in Figure 7.1. Identify the features of a synovial joint, on the models.
2. Compare the fresh specimen with the model of a similar human joint.

Conclusions

- Based on your observations, why are knee injuries so common among athletes?

  _____

  _____

  _____

■ **FIGURE 7.1 Typical synovial joint**

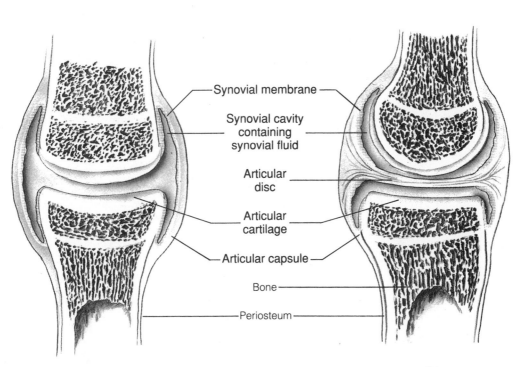

Synovial membrane

Synovial cavity containing synovial fluid

Articular disc

Articular cartilage

Articular capsule

Bone

Periosteum

(a)   (b)

■ **FIGURE 7.2 Various human synovial joints**

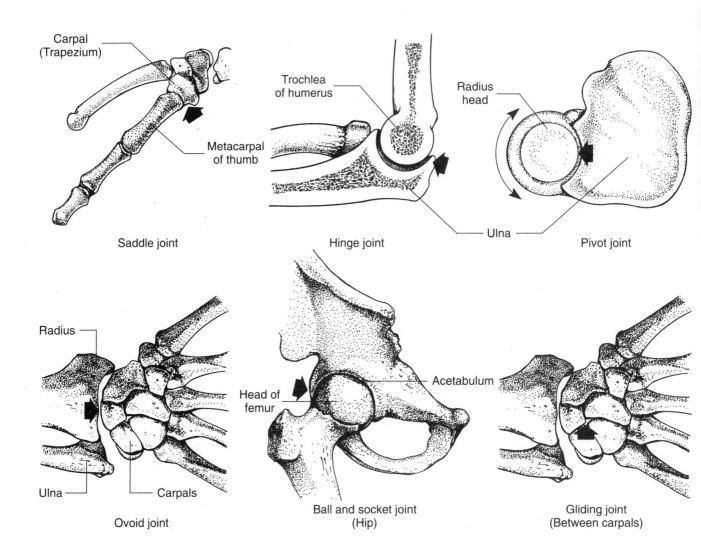

Carpal
(Trapezium)

Metacarpal
of thumb

Saddle joint

Trochlea
of humerus

Ulna

Hinge joint

Radius
head

Ulna

Pivot joint

Radius

Ulna

Carpals

Ovoid joint

Head of
femur

Acetabulum

Ball and socket joint
(Hip)

Gliding joint
(Between carpals)

❖ Movements

Background

As shown in Figure 7.3 there are a variety of movable or diarthritic joints all of which are synovial in construction. There are also a variety of movements possible at these joints. Usually, these movements are considered in relation to the specific axes of the body as a whole or to that of a particular extremity. For example, lifting the arm straight out to the side moves the arm away from the midsagittal plane of the body, while bending the elbow decreases the angle between the bones of the forearm and upper arm. Table 7.2 summarizes the types of diarthritic joints and Table 7.3 provides information about the types of movement possible at diarthritic joints. Figure 7.4 illustrates human diarthritic joints as in the body with ligaments attached.

■ **TABLE 7.2 Types of Diarthroses**

JOINT	DESCRIPTION	NUMBER OF AXES OF MOVEMENT	EXAMPLE
Ball and Socket	Ball-shaped head fits into a cup-shaped depression	Triaxial	Humerus/Scapula
Condyloid	Oval-shaped surface fits into an elliptical cavity	Biaxial	Metacarpal/proximal phalange
Saddle	First bone's articular surface is concave in one direction and convex in the other while the second bone is just the opposite	Biaxial	Thumb carpometatarsal joint
Hinge	Convex surface of one bone fits smoothly into concave surface of the second bone	Uniaxial	Humerus/Ulna
Pivot	One bone turns around another	Uniaxial	Atlas/Axis
Gliding	Opposite bone surfaces are flat or slightly curved for sliding motion in all directions	Nonaxial	Between Tarsals

■ **TABLE 7.3 Movements of Diarthroses**

MOVEMENT	DESCRIPTION	EXAMPLE
ANGULAR		
Extension	Increases angle between two bones	Straightening arm at elbow
Flexion	Decreases angle between two bones	Bending arm at elbow
Abduction	Movement of extremity away from the midline	Lifting arm out to side
Adduction	Movement of extremity toward the midline	Dropping arm
CIRCULAR		
Circumduction	A sequential combination of flexion, abduction, extension and adduction in which a cone is delineated	Drawing a circle with arm extended
Rotation	Movement of a bone around a central axis	Shake head to say "No"
SPECIAL		
Supination	Outward rotation of forearm	Turn palms up
Pronation	Inward rotation of forearm	Turn palms down
Inversion	Turn sole of foot inward	As described
Eversion	Turn sole outward	As described
Depression	Lower a part	Drop lower jaw
Protraction	Move a part forward	Push mandible forward
Retraction	Return from protraction	Move mandible back

■ **FIGURE 7.3 Angular and circular movements of synovial joints**

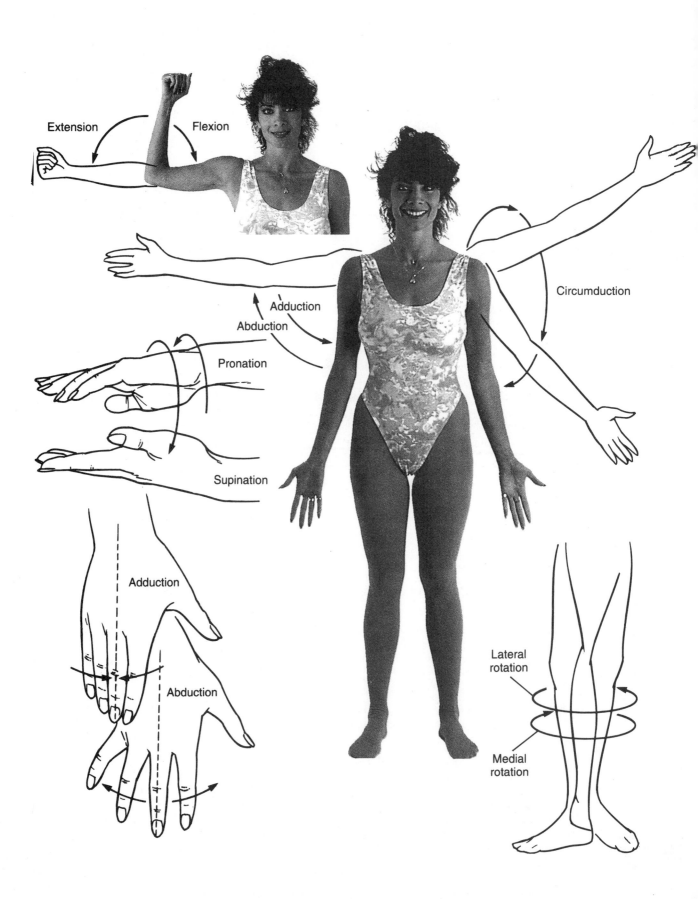

■ **FIGURE 7.3 Angular and circular movements of synovial joints (Continued)**

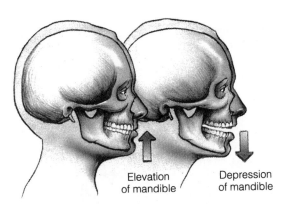

Elevation
of mandible

Depression
of mandible

Elevation and depression

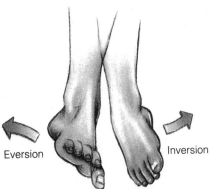

Eversion   Inversion

Eversion and inversion

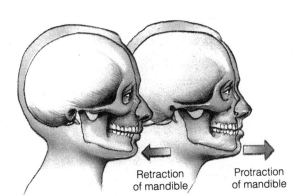

Retraction
of mandible

Protraction
of mandible

Protraction and retraction

■ **FIGURE 7.4 Human joints with ligaments**

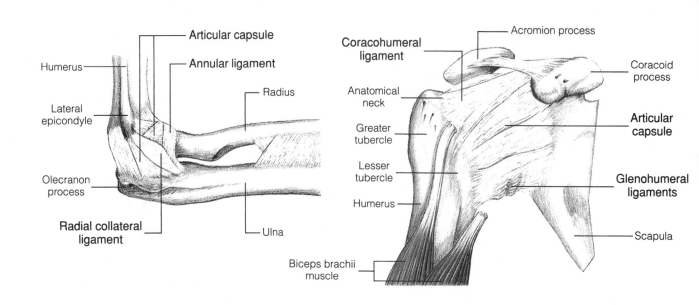

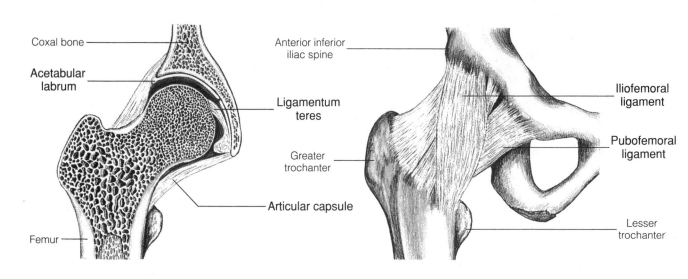

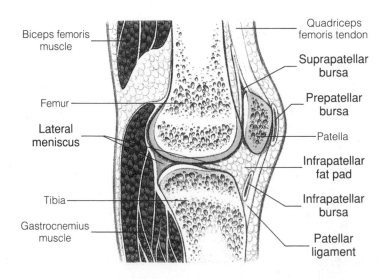

EXERCISE 1
Movable Joint Types and Possible Movements

Discussion

Although diathroidal joints have certain common characteristics, there are many variations, and not all such joints exhibit the same type of mobility. Movements at joints can be affected by the shape of the joint, as well as by the arrangement of muscles and ligaments associated with these joints. Thus, diathroidal joints fall into six main types.

Materials

Models of knee, elbow, and ankle joints; articulated skeleton.

Procedure

1. For each of the joints listed in the Observations section, examine available models as well as the articulated skeleton and determine the type of diarthroidal joint it is. Record the results in Table B.

2. Using the articulated skeleton, determine the types of movement possible at these joints. Record the results in Table B.
3. Since the articulated skeleton does not have muscles and ligaments to limit motion, try the movements on your lab partner and compare with the skeleton. Record the results in Table B.
4. For each of these joints determine the number of axes of movement allowed. Record the results in Table B.

Conclusions

- Indicate the type of movement or combination of movements in each of the following:

  a. Bending the head back to look at the sky

  _____

  b. Spreading the fingers part

  _____

  c. Hitting an overhead smash in tennis

  _____

  d. Tapping your toes

  _____

Observations

TABLE B Diarthritic Joints

JOINT	TYPE OF DIARTHROSIS	MOVEMENTS POSSIBLE ON SKELETON	MOVEMENTS POSSIBLE WITH LAB PARTNER	NUMBER OF AXES
Shoulder				
Elbow				
Wrist				
Thumb				
Hip				
Knee				
Ankle				

Skinning the Fetal Pig

OVERVIEW

In this unit you will skin the fetal pig to prepare it for the dissection of the major muscle groups.

OUTLINE

Skinning the Fetal Pig

Exercise 1 Examining External Features of the Fetal Pig

Exercise 2 Skinning the Fetal Pig

Skinning the Fetal Pig

Background

The fetal pigs used for anatomical studies have been preserved with chemicals that may have irritating properties and an odor. Before you start working on your specimen you may want to do the following:

1. The area in which you are working should have good air movement.
2. Plastic gloves may be used to protect your hands. Wash your hands well at the end of the lab session.

The purpose of the dissections performed throughout the course is to observe the various systems of the fetal pig as a close parallel to those of the human. Therefore, there are certain measures you can take to ensure good care of the specimen.

1. Identify your pig with a clearly marked tag; use *only* this pig for dissection.
2. If possible, save the skin you remove and wrap the pig in this at the end of each lab session to keep it moist.

3. At the end of the lab session, place the pig in a plastic bag, remove the excess air, and tie up the bag.

EXERCISE 1
Examining External Features of the Fetal Pig

Discussion

The pigs you will dissect are fetal pigs, close to the end of their 112-115 day gestation period. Therefore, you will observe fetal structures not present in an adult specimen. The specimens have been obtained from slaughter houses and then prepared for your use. You will notice a slit in the neck region where the blood has been removed and a latex substance injected into the blood vessels to facilitate their identification.

The pigs that other students in the class dissect will vary in sex, size and color. However, you will note that certain features are common to all the specimens.

Materials

A preserved and injected fetal pig; dissecting tray and instruments.

■ FIGURE 8.1 External ventral views of posterior region of male and female fetal pig
Refer to color photo gallery.

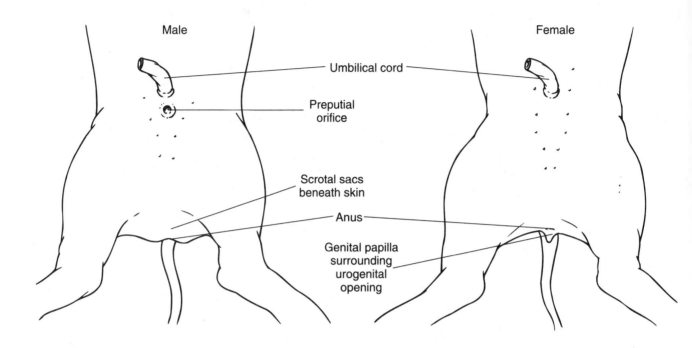

Be sure to review safety precautions before handling fresh or preserved specimens.

Procedure (Figure 8.1)

1. Obtain a fetal pig for your group and lay it on a dissecting tray on its dorsal surface with the ventral surface facing you.
2. The body of the fetal pig is divided into the **head**, **neck**, **trunk**, and **tail**.
3. Depending on the age of the fetal pig you may observe a layer of peeling embryonic skin, the **epitrichium**. This layer is lost as the hair develops.
4. On the head, observe a pair of **nostrils** or the **external nares** at the tip of the snout, and the flattened flap of skin on the ear, the **pinna**.
5. The trunk can be divided into the anterior **thorax** and the posterior **abdomen**.
6. Nipples are present in both sexes. Note the small **teats** or mammary papillae.
7. Observe the **umbilical cord** on the ventral surface of the abdomen.
8. In both sexes the **anus** is found ventral to the tail.
9. Distinguish the sex of your specimen and if appropriate, observe both sexes. See Figure 8.1.

   a. In the male, you will observe a small opening just posterior to the umbilical cord called the **preputial orifice** or urogenital opening, marking the termination of the penis on the body surface. If you gently pull the hindlimbs apart you can observe the **penis** lying beneath the skin. Observe the **scrotal sacs** ventral to the anus.
   b. In the female, ventral to the anus, observe a small projection of skin, the **genital papilla**, which surrounds the urogenital opening.

Conclusions

* What is the sex of your pig? _____
* What is the most obvious difference between the male and female pig?

  _____

  _____

EXERCISE 2
Skinning the Fetal Pig

Discussion

There are several ways of removing the skin. These instructions will begin with an incision on the dorsal surface. The skin on the fetal pig is rather thin so be careful not to cut into the underlying muscles.

Materials

A preserved and injected fetal pig; dissecting tray and instruments.

Be sure to review safety precautions before handling fresh or preserved specimens.

Procedure (Figure 8.2)

1. Place the pig in the dissecting tray on its ventral surface with the dorsal surface facing you.
2. With a scalpel make a small shallow longitudinal incision through the skin in the midorsal neck region. Note the thickness of the skin and be careful not to cut into underlying structures. Cautiously continue the incision down the midorsal line to the tail.
3. Make an incision around the neck just under the ear. It is not necessary to remove the skin from the head.
4. Make additional incisions down the lateral surface of each leg and around the wrist and ankle. The skin can be left on the feet, tail, and genital area.
5. Gently separate the skin from the underlying fascia and muscles by using a blunt probe. Pull back the skin to the lateral surface and then turn the pig over.
6. Complete the skinning of the ventral surface of the pig. Cut around the umbilical cord.
7. When you have completed the skinning of the pig, you may wish to wrap your specimen in the skin and place it in a plastic bag with an identification tag attached to it.
8. Make sure the dissecting tray, dissecting instruments and lab table tops are clean. Discard any remnants of the pig according to the directions given by your instructor.

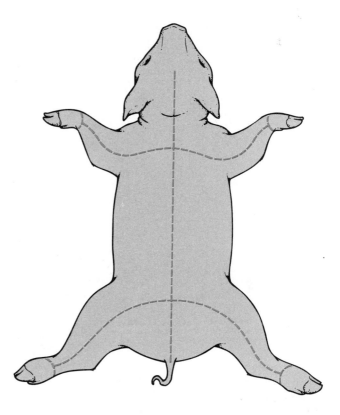

■ **FIGURE 8.2 Dorsal surface of a fetal pig showing directional cutting lines**

The Anatomy of the Muscular System

OVERVIEW

Once you have skinned the fetal pig you are ready to locate the major muscles; then you will become familiar with selected human muscles and their actions.

OUTLINE

Muscle Anatomy

EXERCISE 1
Microscopic Examination of Muscle Tissue

Discussion (Figure 9.1)

The belly of a muscle is wrapped in a connective tissue sheath called the **epimysium**. This is a deep fascia, and it appears silvery white and glistens in a fresh preparation. (You may have observed this in chicken or beef before it is cooked.) The muscle belly is subdivided into sections known as **fascicles** by less dense connective tissue called **perimysium**. Fascicles are bundles of individual muscle fibers or cells, the **myofibers**. Each myofiber is wrapped in a delicate connective tissue sheath, the **endomysium**.

Skeletal muscle is considered striated muscle. The regular arrangement of the proteins actin and myosin is reflected in the **dark A bands** (where these proteins overlap) and the **lighter I bands** (where only actin appears.) It is a **multinucleate** tissue, and it is under voluntary control.

■ **FIGURE 9.1   Skeletal muscle**
(a) Cross section of skeletal muscle. (b) Microscopic view of skeletal muscle tissue.

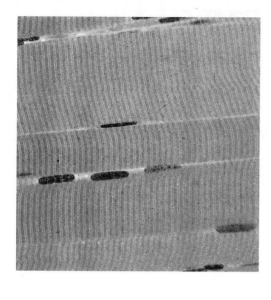

(a)

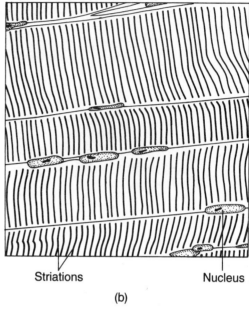

Striations   Nucleus

(b)

Materials

A microscope; c.s. and l.s. slides of skeletal muscle tissue.

Procedure

1. Review the material regarding skeletal muscle in Table 4.8.
2. Identify the structures in Figure 9.1b on the slide of skeletal muscle under the microscope. Also identify the A bands and the I bands.

Note: In Exercises 2–11, the muscles that are described as being located "ventrally" are ones that are generally visible when the fetal pig is lying on its dorsal surface. The reverse is true for those muscles described as being "dorsal."

Each exercise includes a "Discussion–Procedure" section, a detailed table, and diagram. Corresponding color photos are in a section near the back of the book. Read through the "Discussion–Procedure" section *and* the chart thoroughly *before* you do any dissection.

EXERCISE 2
Superficial Muscles of the Ventral Thigh: The Fetal Pig

Discussion–Procedure (Table 9.1; Figure 9.2)

Be sure to review safety precautions before handling fresh or preserved specimens.

As you identify muscles on the fetal pig, you should become familiar with the terms referring to the attachments of the muscle to the bone or to another muscle. Muscles are attached either by means of a **tendon** (a cordlike attachment) or directly into the bone, or to a sheet of connective tissue which runs either to bone or to another muscle (an **aponeurosis**.)

The attachments are at the two ends of a muscle with the **belly** of the muscle in between. The more movable attachment of the muscle is called the **insertion** and the less movable attachment is called the **origin**.

Two broad, thin superficial muscles are visible on the ventral thigh.

1. Carefully trim the fat and fascia from the surface of the thigh to expose the two superficial muscles.
2. Running diagonally and covering the anterior part of the thigh is a narrow thin muscle, the **sartorius**.
3. Covering the posterior aspect of the medial thigh is the **gracilis**, a broad thin muscle.

TABLE 9.1 Superficial Muscles of the Ventral Thigh of the Fetal Pig

MUSCLE	DESCRIPTION	ACTION
Sartorius	A flattened band of muscle on the medial surface of the thigh; it has its origin on the ilium and extends to the proximal femur	Adducts thigh; extends leg
Gracilis	Broad thin superficial muscle of the medial thigh; runs from its origin near the pubic symphysis to the proximal tibia	Adducts thigh

FIGURE 9.2 Superficial muscles of the right ventral thigh of the fetal pig

Refer to the color photo gallery.

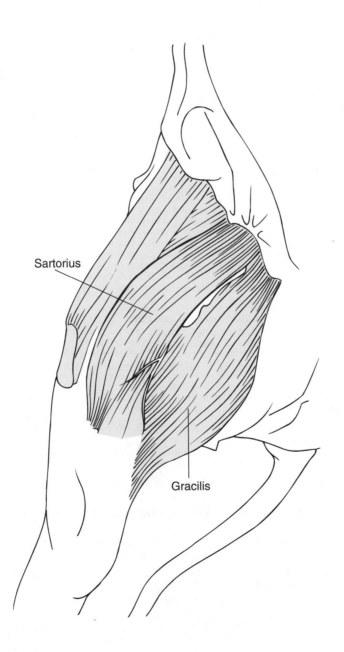

EXERCISE 3
Deep Muscles of the Ventral Thigh: The Fetal Pig

Discussion–Procedure **(Table 9.2; Figure 9.3)**

Be sure to review safety precautions before handling fresh or preserved specimens.

Frequently muscles of the lower extremity span two joints, in which case they can move both joints. Thus the concept of a more fixed and a more movable attachment is not strict here.

1. Free the edges of the superficial muscles, the sartorius and gracilis. Bisect and reflect them to expose the deeper muscles.
2. Lying beneath the gracilis is the large **semimembranosus**.
3. Next to the semimembranosus is a tendon shaped muscle, the **semitendinosus**.
4. Lying anterior to the semimembranosus is the **adductor**. In the pig the adductor is not subdivided as it is in the human and some other mammals.
5. A smaller muscle, the **pectineus**, lies anterior to the adductor.

6. Observe the femoral artery extending across the medial thigh. Anterior to the artery lies the small medial **psoas major** and lateral **iliacus**. In the human these two muscles fuse to form the iliopsoas.
7. Identify the muscles that are collectively referred to as the **quadriceps femoris**. The **vastus lateralis** is the large muscle on the anterolateral surface on the thigh. Lying medial to it, is the thick **rectus femoris** followed by the **vastus medialis**. The **vastus intermedius** is beneath the rectus femoris and can be seen by transecting the rectus femoris.

Conclusions

• Which muscles comprise the "hamstrings"?

_____

_____

• Which muscles comprise the quadriceps femoris?

_____

_____

■ **TABLE 9.2   Deep Muscles of the Ventral Thigh of the Fetal Pig**

MUSCLE	DESCRIPTION	ACTION
Semimembranosus	Large muscle; originates on the ischium and inserts on the distal end of femur and proximal tibia	Extends hip and adducts the leg
Semitendinosus	Thick band of muscle under the posterior edge of the biceps; extends from first caudal vertebra and ilium to the tibia	Extends thigh and flexes leg
Adductor	Large muscle anterior to the semimembranosus; has its origin on the ischium and inserts along the femur	Adducts thigh
Pectineus	Small muscle next to the adductor; originates on the ventral pubis and inserts on the femur	Adducts thigh and flexes hip
Psoas major	Small muscle originating on lumbar vertebrae with the iliacus and inserts on the femur	Flexes hip and rotates thigh
Iliacus	Small muscle originating on the ilium and inserts on the femur with the psoas major	Flexes hip and rotates thigh
Vastus lateralis	Large muscle on the lateral surface partially covered by the biceps femoris; originates on the femur and inserts on the patella	Extends leg
Rectus femoris	Thick muscle which partly lies on the vastus lateralis; extends from ilium to the patella where it inserts with the vastus lateralis	Extends leg
Vastus medialis	Lies medial to the rectus femoris; originates from the femur and inserts on the patella	Extends leg
Vastus intermedius	Lies deep between the vastus lateralis and rectus femoris; originates on the femur and inserts on the patella	Extends leg

■ **FIGURE 9.3 Deep muscle of the right ventral thigh of the fetal pig**
Refer to the color photo gallery.

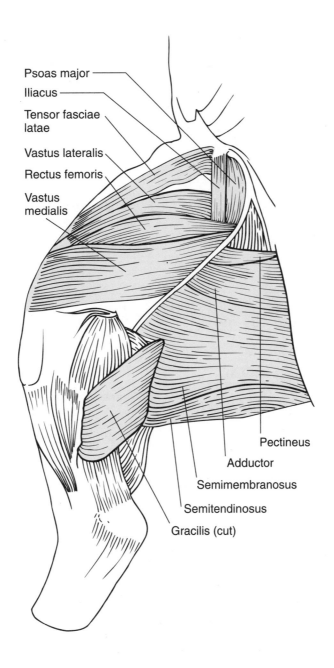

Psoas major

Iliacus

Tensor fasciae latae

Vastus lateralis

Rectus femoris

Vastus medialis

Pectineus

Adductor

Semimembranosus

Semitendinosus

Gracilis (cut)

EXERCISE 4
Selected Muscles of the Lower Hind Limb: The Fetal Pig

Discussion–Procedure (Table 9.3; Figure 9.4)

Be sure to review safety precautions before handling fresh or preserved specimens.

These muscles are primarily flexors and extensors of the foot.

1. Locate the **gastrocnemius**, the large calf muscle seen on the posterior surface, extending from the femur to the calcaneus.
2. Beneath the gastrocnemius is the **soleus**.
3. Locate the most ventral muscle of the lower leg, the **tibialis anterior** and the deeper, **peroneus longus**.

EXERCISE 5
Superficial Muscles of the Buttocks and Dorsal Thigh: The Fetal Pig

Discussion–Procedure (Table 9.4; Figure 9.5)

Be sure to review safety precautions before handling fresh or preserved specimens.

Many of the pelvic muscles extend across both the hip and knee joints.

1. Carefully remove the fat and fascia from the buttocks and thigh.
2. On the anterolateral surface of the thigh lies a small muscle, the **tensor fasciae latae** which has its origin on the ilium. Extending from the muscle is a sheet of fascia which inserts near the knee. This muscle covers a large part of the vastus lateralis identified earlier.
3. A large muscle, the **biceps femoris**, lies posterior to the tensor fascia latae and covers much of the lateral thigh.
4. Between the biceps femoris and tensor fasciae latae, find the **gluteus maximus** which may be partly fused to the biceps femoris. This is a rather small muscle in the pig compared to its larger size in the human.
5. The more obvious muscle and lying deeper is the **gluteus medius** which is located anterior to the gluteus maximus.

Conclusions

• Contrast the gluteus maximus in the pig and human.

_____

• Which muscle is the "main actor" in flexing the foot?

_____

■ **TABLE 9.3   Selected Muscles of the Lower Hind Limb of the Fetal Pig**

MUSCLE	DESCRIPTION	ACTION
Gastrocnemius	Large calf muscle; originates as two heads upon the distal end of the femur; inserts on calcaneous via Achilles tendon	Extends the foot
Soleus	Lies beneath the gastrocnemius; has its origin on the fibula and inserts on the calcaneus with the gastrocnemius	Extends the foot
Tibialis anterior	Ventral muscle that lies upon the tibia; has its origin on the proximal end of the tibia and fibula; inserts on the metatarsal	Flexes the foot
Peroneus longus	Deeper muscle that originates from the tibia and distal end of the femur; inserts on the metatarsals	Flexes the foot

■ **FIGURE 9.4 Selected muscles of the right lower hind limb of the fetal pig**
Refer to the color photo gallery.

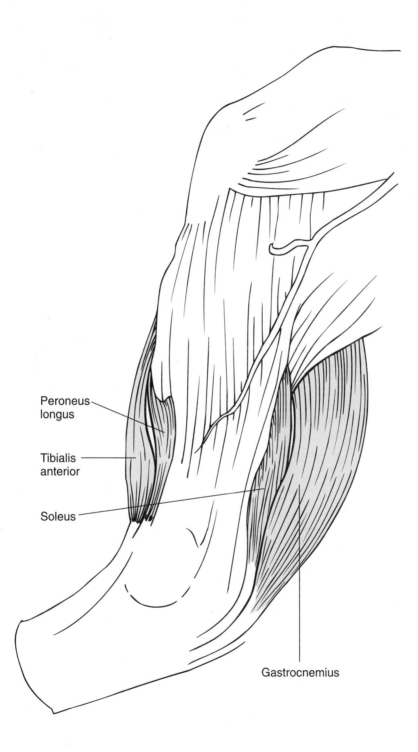

Peroneus longus

Tibialis anterior

Soleus

Gastrocnemius

■ **TABLE 9.4 Superficial Muscles of the Buttocks and Dorsal Thigh of the Fetal Pig**

MUSCLE	DESCRIPTION	ACTION
Tensor fasciae latae	Triangular muscle which continues as a sheet of connective tissue; most inferior muscle on the lateral thigh; extends from the ilium to the fascia which attaches to the patella and tibia	Extends leg
Biceps femoris	Large muscle covering most of lateral surface of the thigh; originates on the sacrum and ischium and inserts on the lower femur and upper tibia	Abducts thigh, flexes leg
Gluteus maximus	Relatively small thin muscle posterior to the tensor fasciae latae; has its origin on the last sacral and first caudal vertebrae and inserts into the tensor fasciae latae	Abducts thigh
Gluteus medius	Thicker than gluteus maximus and lying beneath it; originates on ilium and inserts on the femur	Abducts and extends thigh

■ **FIGURE 9.5 Superficial muscles of the right buttocks and dorsal thigh of the fetal pig**
Refer to the color photo gallery.

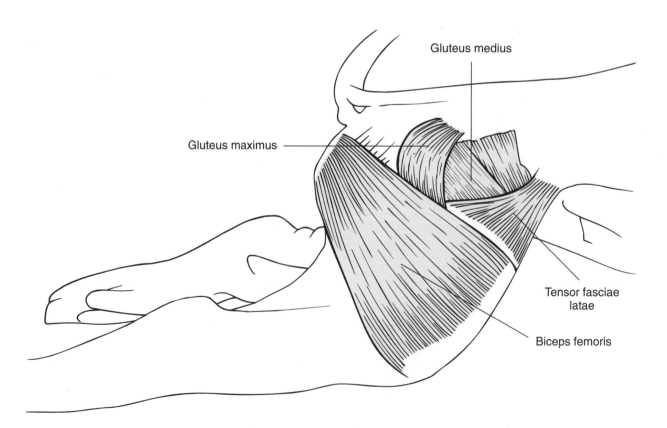

EXERCISE 6
Superficial Muscles of the Back and Shoulder: The Fetal Pig

Discussion–Procedure (Table 9.5; Figure 9.6)

Be sure to review safety precautions before handling fresh or preserved specimens.

Before you can identify the muscles of the back and shoulder, it may be necessary to "pick" away at the connective tissue which may obscure the separations of these muscles.

■ **TABLE 9.5 Superficial Muscles of the Back and Shoulder of the Fetal Pig**

MUSCLE	DESCRIPTION	ACTION
Trapezius	Extensive superficial muscle on the upper shoulder and back; it originates on the occipital bone and thoracic vertebrae and runs to its insertion on the spine of the scapula	Elevates scapula; draws scapula laterally
Latissimus dorsi	Broad muscle on the dorsal and lateral body surface; fibers run downward and anteriorly from the posterior thoracic and lumbar vertebrae and extend to the humerus	Moves forearm dorsally and posteriorly; flexes shoulder
Brachiocephalic	Band of muscle that covers dorsal surface of the neck extending from its origin on the mastoid process to the humerus	Moves forearm anteriorly
Deltoid	Broad muscle on the anterior part of the shoulder; originates from the scapula and inserts on the posterior humerus	Flexes shoulder; abducts arm

■ **FIGURE 9.6 Superficial muscles of the right back and shoulder of the fetal pig**
Refer to the color photo gallery.

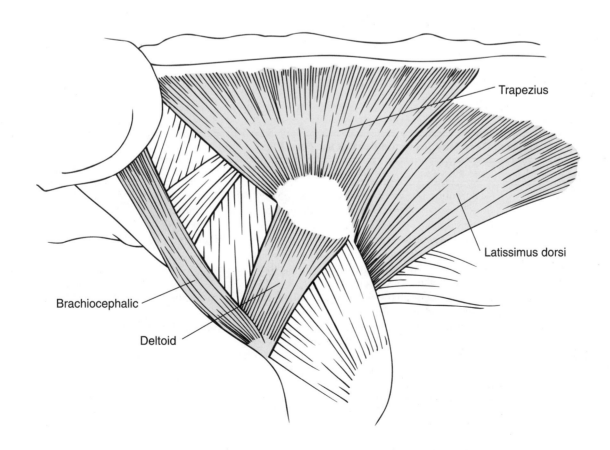

1. In the upper part of the body in the area of the face, neck and shoulder, you will see a broad muscle attached to the skin. This is the **cutaneous muscle**, found in the pig but not in the human. You may have to cut this to expose other muscles in the region.
2. On the dorsal area of the upper back, shoulder and neck, locate the thin triangular muscle, the **trapezius**.
3. Posterior to the trapezius is the **latissimus dorsi**.
4. Lying anterior to the trapezius and extending diagonally from the skull to the humerus, is a band of muscle called the **brachiocephalic**.
5. The **deltoid** is found covering the anterior part of the shoulder.

EXERCISE 7
Selected Deep Muscles of the Back: The Fetal Pig

Discussion–Procedure **(Table 9.6; Figure 9.7)**

Be sure to review safety precautions before handling fresh or preserved specimens.

1. To identify the deep muscles of the back, transect and reflect the trapezius. Beneath the trapezius lies the **rhomboideus** muscle which extends from the scapula to the cervical and thoracic vertebrae. The rhomboideus is divided into three distinct muscles. The most anterior muscle of this group in the band-like **rhomboideus capitis**, a muscle not present in the human. Lying next to this is the **rhomboideus cervicus**, followed by the **rhomboideus thoracis**.
2. Below and ventral to the rhomboideus capitis lies the **splenius** which covers a large portion of the dorsal and lateral aspects of the neck.
3. The surface of the scapula is covered by two muscles. The spine of the scapula separates the upper **supraspinatus** from the lower **infraspinatus**.
4. Cut and reflect the latissimus dorsi. Along the posterior border of the scapula is the **teres major**.

TABLE 9.6 Selected Deep Muscles of the Back of the Fetal Pig

MUSCLE	DESCRIPTION	ACTION
Rhombiodeus capitis	Most anterior of the group of muscles, the rhomboideus; band of muscle extending from the occipital bone to the verterbral border of the scapula	Draws scapula anteriorly and holds it in place
Rhomboideus cervicus	Deep muscle lying adjacent to the rhomboideus capitis; originates on the cervical vertebrae and inserts on the dorsal scapula	Draws scapula anteriorly and holds it in place
Rhomboideus thoracis	Most posterior of the group; originates on the thoracic vertebrae and inserts on the dorsal scapula	Draws scapula anteriorly and holds it in place
Splenius	Thick muscle beneath the rhomboideus; originates on the thoracic vertebrae and inserts on the occipital bone	Extends head and neck
Supraspinatus	Below the trapezius and above the spine of the scapula occupying the supraspinatus fossa; inserts on the humerus	Extends shoulder
Infraspinatus	Found below the spine of the scapula and occupies the infraspinatus fossa; inserts on the humerus	Abducts arm; rotates arm laterally
Teres major	Thick muscle lying posterior to infraspinatus; originates on the scapula and extends to the humerus	Flexes shoulder; adducts arm

■ **FIGURE 9.7 Selected right deep muscles of the back of the fetal pig**
Refer to the color photo gallery.

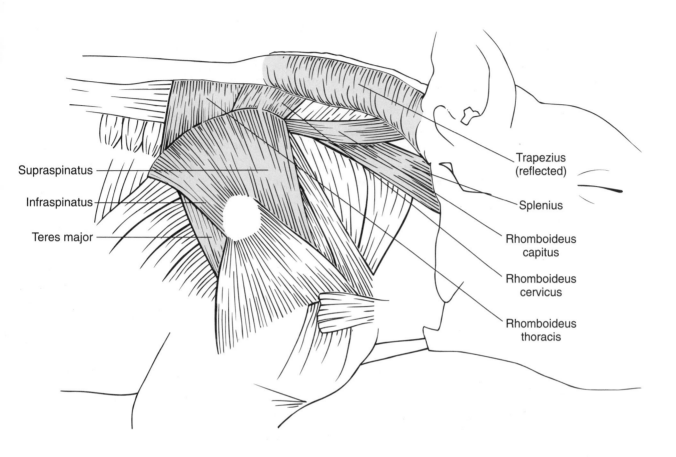

EXERCISE 8
Selected Muscles of the Forelimb: The Fetal Pig

Discussion–Procedure **(Table 9.7; and Figure 9.8)**

Be sure to review safety precautions before handling fresh or preserved specimens.

1. Carefully clean away the fascia to expose the muscles of the forelimb. On the posterior surface lies the large **triceps brachii** consisting of three heads. Only two of these can be seen on the surface, the long and lateral heads.
2. Anterior to the lateral head of the biceps is the **brachialis** which covers a portion of the ventrolateral surface of the humerus and extends to the ulna.
3. Lift the pectoralis and locate the **biceps brachii**.

■ **TABLE 9.7 Selected Muscles of the Forelimb of the Fetal Pig**

MUSCLE	DESCRIPTION	ACTION
Triceps brachii	Large muscle covering most of upper forelimb and is divided into three heads, the long and lateral heads seen on the surface; extends from the upper end of the humerus and scapula to the ulna	Extends the forearm
Brachialis	Located on the ventrolateral surface of the humerus; arises from the humerus and extends to the radius and ulna	Flexes the forearm
Biceps brachii	Smaller muscle consisting of two heads located on the ventromedial surface of the humerus and is partly covered by the pectoralis; it has its origin on the scapula and inserts on the radius and ulna	Flexes the forearm

4. Several muscles can be identified in the lower fore-limb which act as extensors and flexors. The extensors are generally found on the lateral side and the flexors on the medial side.

Conclusions

• Identify a muscle that is in the neck and shoulder area not present in the human, but present in the pig.

_____

• Identify a deep muscle that is on the back of the pig but not found in the human.

_____

• What is the largest muscle in the forelimb of the pig?

_____

■ **FIGURE 9.8 Selected muscles of the right forelimb of the fetal pig**
Refer to the color photo gallery.

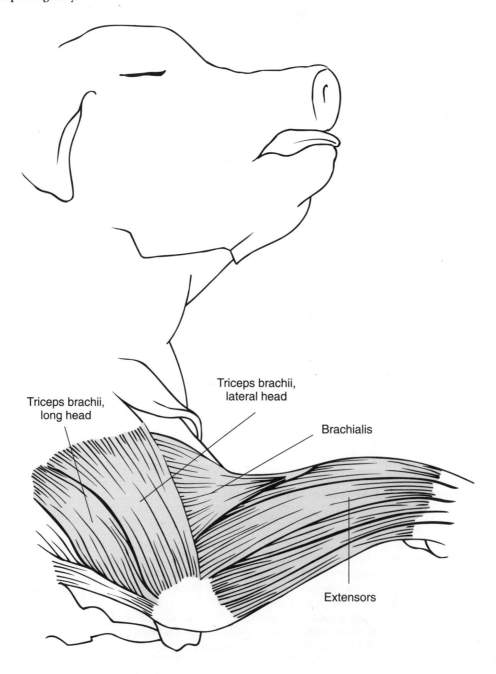

EXERCISE 9
Selected Muscles of the Superficial Chest and Ventral Neck: The Fetal Pig

Discussion–Procedure **(Table 9.8; Figures 9.9 and 9.10)**

Be sure to review safety precautions before handling fresh or preserved specimens.

It is more difficult to separate the superficial ventral chest muscles than the other muscles studied. They exist as thin sheets and are basically the pectoral muscles.

1. With the pig lying on its dorsal surface, carefully remove any fascia covering the muscles. Gently extend the forelimbs to help you locate the chest muscles.
2. The **superficial pectoralis** or **pectoralis major** is a fan-shaped muscle originating on the upper sternum and extending to the humerus.
3. Below and deeper lies the **deep pectoralis** or **pectoralis minor**. It may seem to be composed of two parts since its anterior portion inserts on the scapula and the posterior portion inserts on the humerus.
4. Locate the large **latissimus dorsi** originating on the dorsal surface and continuing to the lateral surface of the thorax.
5. If you have not done so remove the skin and expose the muscles of the ventral surface of the neck. Be careful not to damage any of the blood vessels.
6. Locate the V-shaped **sternomastoid** extending diagonally from the sternum to the skull.
7. Running along the mid-ventral portion of the neck is a narrow pair of muscles, the **sternohyoids**. Just beneath and lateral to these are the **sternothyroids**.
8. Another V-shaped muscle in the neck is the **digastric** found extending from the jaw to the base of the skull.
9. Locate the muscle whose fibers run transversely across the midline and under the digastric. This is the **mylohyoid**.
10. The large round muscle found on the side of the face is the **masseter**.

■ **TABLE 9.8 Selected Muscles of the Superficial Chest and Ventral Neck of the Fetal Pig**

MUSCLE	DESCRIPTION	ACTION
Superficial pectoralis (Pectoralis major)	Fan-shaped muscle that extends from the sternum to the humerus	Adducts forearm
Deep pectoralis (Pectoralis minor)	Lies deep to the pectoralis superficialis with its fibers running anteriorly and laterally; extends from the sternum to the humerus	Adducts and retracts forearm
Latissimus dorsi	A muscle covering a large portion of the back and extending to the lateral and ventral surface; originates from the thoracic and lumbar vertebrae and inserts on the humerus	Retracts forearm and flexes shoulder
Sternomastoid	Long band of muscle extending from the sternum to the mastoid process	Flexes head
Sternohyoids	Narrow pair of muscles with its fibers running longitudinally along the midventral neck from the sternum to the hyoid bone	Depresses hyoid
Sternothyroids	Pair of muscles lying deep and lateral to the sternohyoids and covering the trachea; extends from the sternum to the thyroid cartilage	Moves the larynx posteriorly
Digastric	V-shaped muscle running under the chin from the mastoid process to the mandible	Depresses mandible
Mylohyoid	Runs under the chin transversely, deep to the digastric; extends from the base of the skull to the mandible	Raises floor of the mouth
Masseter	Round mass of muscle posterior to the angle of the jaw; originates on the zygomatic arch and inserts on the mandible	Elevates mandible

■ **FIGURE 9.9 Selected muscles of the chest of the fetal pig**
Refer to the color photo gallery.

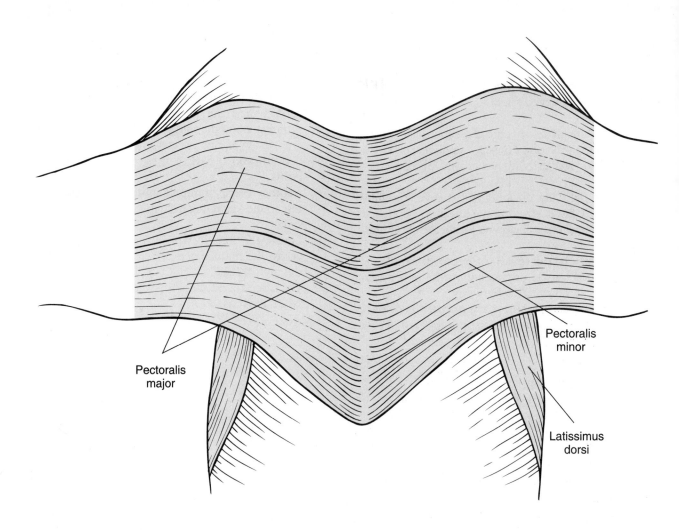

Pectoralis
major

Pectoralis
minor

Latissimus
dorsi

■ **FIGURE 9.10 Selected muscles of the neck of the fetal pig**
Refer to the color photo gallery.

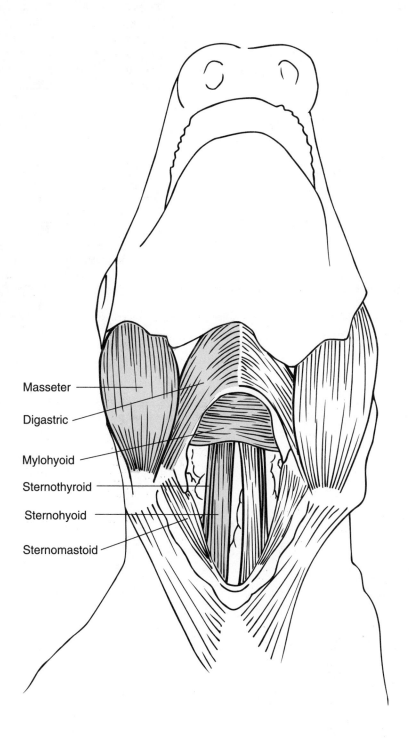

Masseter

Digastric

Mylohyoid

Sternothyroid

Sternohyoid

Sternomastoid

EXERCISE 10

Selected Deep Muscles of the Ventral Chest: The Fetal Pig

Discussion–Procedure **(Table 9.9; Figure 9.11)**

Be sure to review safety precautions before handling fresh or preserved specimens.

The muscles identified in this exercise are used as muscles of inspiration.

1. To locate the deep muscles of the ventral chest, transect and reflect the latissimus dorsi. Seen on both the lateral and ventral surfaces is a large fan-shaped muscle, the **serratus ventralis**.
2. The **external intercostals** can be located between the ribs. To expose the **internal intercostals** bisect the external intercostals. The fibers of the internal intercostals run perpendicular to those of the externals.

■ **TABLE 9.9** **Selected Deep Muscles of the Ventral Chest of the Fetal Pig**

MUSCLE	DESCRIPTION	ACTION
Serratus ventralis	A fan-shaped muscle originating as strips from the ribs and passing to the scapula	Pulls scapula posteriorly and downward
External intercostals	Located between the ribs and run ventrally and posteriorly	Raise rib cage during inhalation
Internal intercostals	Deeper rib muscles whose fibers run at right angles to the external intercostals	Depresses rib cage

■ **FIGURE 9.11** **Selected deep muscles of the ventral chest of the fetal pig**
Refer to the color photo gallery.

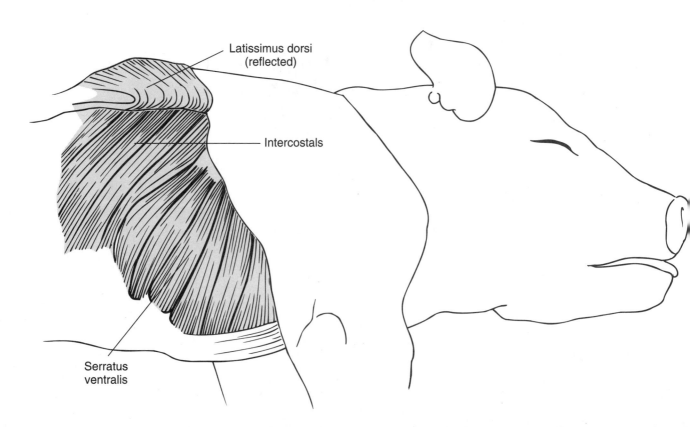

Latissimus dorsi (reflected)

Intercostals

Serratus ventralis

EXERCISE 11
Muscles of the Abdomen: The Fetal Pig

Discussion–Procedure **(Table 9.10; Figure 9.12)**

Be sure to review safety precautions before handling fresh or preserved specimens.

The abdominal wall is composed of several layers of muscles whose fibers run in different directions.

1. The lateral wall of the abdomen in the pig is comprised of three layers of muscles as in the human. Locate the outermost layer, the **external oblique** whose fibers run across the abdominal wall in a posterior and ventral direction, eventually inserting along a midventral line of connective tissue, the linea alba.
2. Make a thin incision through the external oblique and separate it from a thin sheet of underlying muscle, the **internal oblique**. Note its fibers run in the opposite direction.

■ **TABLE 9.10 Muscles of the Abdomen of the Fetal Pig**

MUSCLE	DESCRIPTION	ACTION
External oblique	The outermost thin sheet of muscle covering the ventral and lateral abdominal wall; its fibers run ventrally and posteriorly and insert on the linea alba	Compresses abdomen and flexes trunk
Internal oblique	Lies beneath the external oblique and its fibers run ventrally and anteriorly, opposite to those of the external oblique	Compresses abdomen and flexes trunk
Transverse abdominis	Third abdominal muscle layer lying beneath the internal oblique with its fibers running transversely from the lower rib and lumbar vertebrae to the linea alba	Compresses abdomen and flexes trunk
Rectus abdominis	Two parallel bands of muscle with its fibers running longitudinally along the mid-ventral abdomen from the pubis to the sternum and costal cartilages	Compresses abdomen and flexes trunk

■ **FIGURE 9.12 Muscles of the abdomen of the fetal pig**
Refer to the color photo gallery.

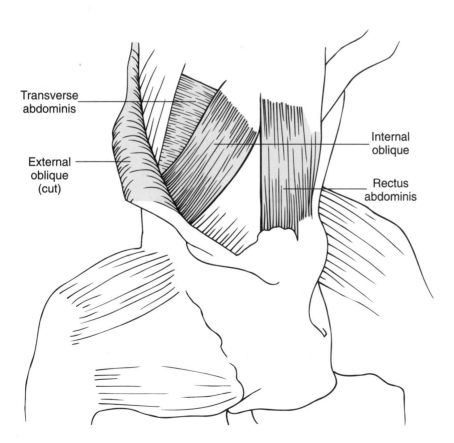

3. Below the internal oblique lies the third layer, the **transverse abdominis**. Its fibers run almost in the same direction as the external oblique.
4. Running on each side of the linea alba lies a band of muscle, the **rectus abdominis**.

Conclusions

- Which muscle is used to open the jaw?

  _____

- Name the muscles that raise and lower the rib cage.

  _____

  _____

- Which muscles would compress the abdomen?

  _____

  _____

Quick Quiz 1

1. Which of these best describes the pectoralis major in the pig?
   (a) It is a fan-shaped muscle.
   (b) It lies superior to the pectoralis minor.
   (c) It is a superficial muscle of the chest.
   (d) all of these
2. Which of these is a deep muscle?
   (a) gracilis
   (b) masseter
   (c) transverse abdominis
   (d) biceps femoris
3. Which describes skeletal muscle?
   (a) voluntary and smooth
   (b) involuntary and smooth
   (c) involuntary and striated
   (d) voluntary and striated
4. Which is the normal insertion of the sternomastoid?
   (a) mastoid process
   (b) clavicle
   (c) sternum
   (d) none of these
5. Which is the normal origin of the pectoralis major?
   (a) humerus
   (b) sternum
   (c) humeralis
   (d) oleocranon process

Muscle Location: Human Muscles

Background

A muscle or muscle group can often be identified by how it is used. For example, the biceps brachii bulges on the upper arm as the forearm is pulled toward the upper arm (flexion). This effect becomes more pronounced if there is a weight being lifted or if there is resistance applied against the action. The firm contracting muscle can then be palpated easily. Muscle action that involves movement is said to be **isotonic**. If the resistance is so great that no shortening occurs, the contraction is said to be **isometric**.

EXERCISE 1
Location of Selected Muscles: The Human

Discussion

In general, the muscle groups of the pig are fairly similar to those of the human. Human muscle groups commonly referred to are illustrated in Figures 9.13 and 9.14.

Materials

A model of the human torso; models of limb muscle; charts of human muscles.

Procedure

1. Using the materials for this exercise, identify the muscles indicated in Figures 9.13 and 9.14.

■ FIGURE 9.13 Selected human muscles — anterior view

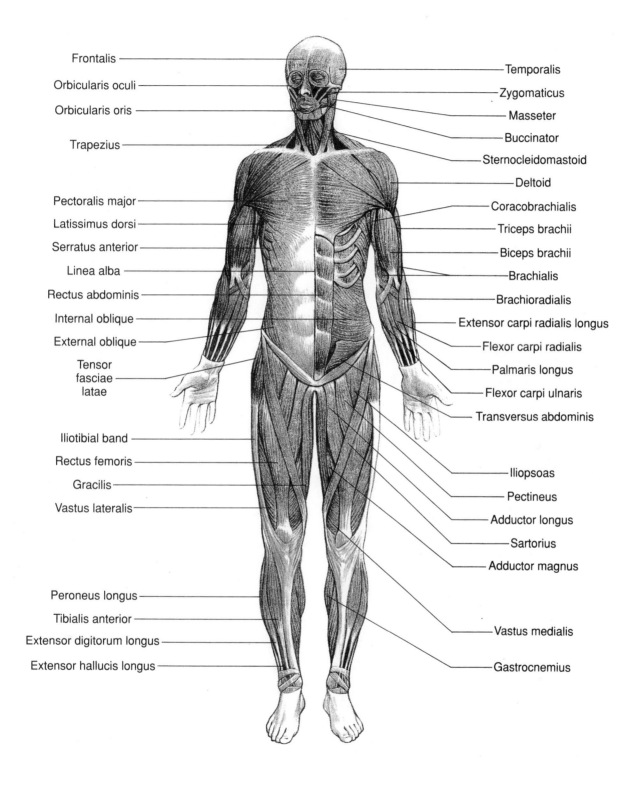

Frontalis

Orbicularis oculi

Orbicularis oris

Trapezius

Pectoralis major

Latissimus dorsi

Serratus anterior

Linea alba

Rectus abdominis

Internal oblique

External oblique

Tensor
fasciae
latae

Iliotibial band

Rectus femoris

Gracilis

Vastus lateralis

Peroneus longus

Tibialis anterior

Extensor digitorum longus

Extensor hallucis longus

Temporalis

Zygomaticus

Masseter

Buccinator

Sternocleidomastoid

Deltoid

Coracobrachialis

Triceps brachii

Biceps brachii

Brachialis

Brachioradialis

Extensor carpi radialis longus

Flexor carpi radialis

Palmaris longus

Flexor carpi ulnaris

Transversus abdominis

Iliopsoas

Pectineus

Adductor longus

Sartorius

Adductor magnus

Vastus medialis

Gastrocnemius

■ **FIGURE 9.14 Selected human muscles — posterior view**

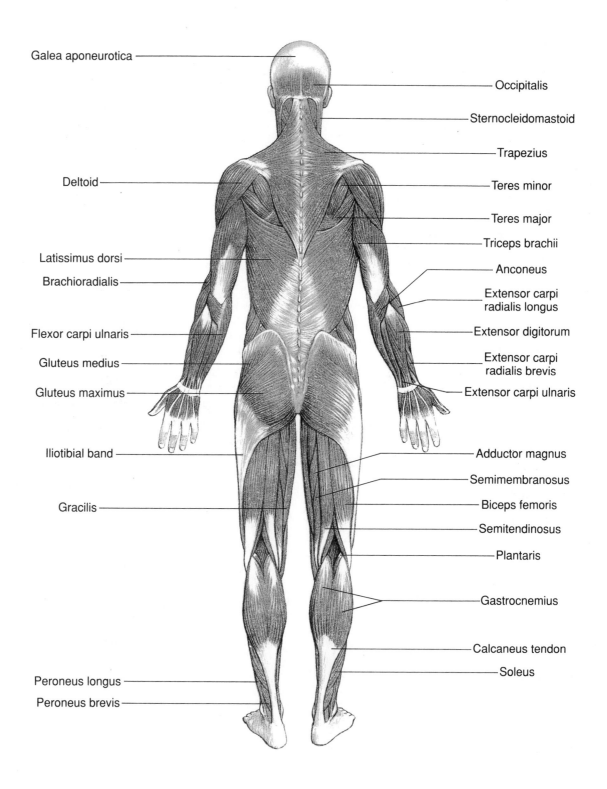

Galea aponeurotica

Occipitalis

Sternocleidomastoid

Trapezius

Deltoid

Teres minor

Teres major

Triceps brachii

Latissimus dorsi

Anconeus

Brachioradialis

Extensor carpi radialis longus

Flexor carpi ulnaris

Extensor digitorum

Gluteus medius

Extensor carpi radialis brevis

Gluteus maximus

Extensor carpi ulnaris

Iliotibial band

Adductor magnus

Semimembranosus

Gracilis

Biceps femoris

Semitendinosus

Plantaris

Gastrocnemius

Calcaneus tendon

Soleus

Peroneus longus

Peroneus brevis

Muscle Physiology

OVERVIEW

In this unit you will study the process of muscle contraction by observing the single muscle twitch, as well as the various responses of muscle under changing conditions.

OUTLINE

The following exercises involve experimentation on living material. A commercial audio-visual tape of this subject may be substituted if appropriate.

Muscle Physiology

Background

A good way to understand the many factors that affect muscle response is to isolate a muscle and observe it during its simplest contraction: the **single muscle twitch**. When an electrical stimulus is applied to a muscle (or the nerve supplying the muscle), it causes the cell membrane (sarcolemma) to depolarize. Following this electrical event is a mechanical event, the contraction, which can be observed with the naked eye. If a mechanical event is converted to an electrical impulse by means of a transducer, the contraction can be recorded on a physiograph or biograph or similar instrument.

Several factors can alter the mechanical response of a whole muscle. Among them are temperature, load on the muscle, length of the muscle, physiological status, and frequency and strength of the stimuli. For skeletal muscle

the entire event is finished anywhere from 15 to 100 milliseconds (msec), depending on the area of the body from which the muscle is prepared.

EXERCISE 1
The Single Muscle Twitch

Discussion

When the calf muscle (gastrocnemius) of the frog is held between two clamps, one fixed and one movable, and stimulated electrically for a short time (measured in milliseconds), it will contract once, then relax. The same type of response will occur if the nerve serving the muscle (in this case the sciatic nerve) is stimulated electrically. The resulting graph of the contraction (a **myograph**) is recorded and illustrated in Figure 10.1. Notice the labels on the axes. The graph exhibits a latent period, followed by a period of contraction and a period of relaxation. The

■ **FIGURE 10.1 A muscle twitch**

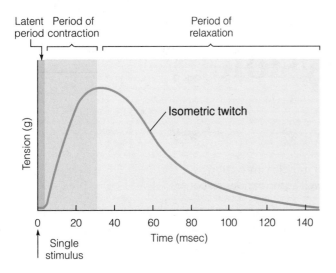

time period is rapid for the single muscle twitch for skeletal muscle (longer for smooth and cardiac muscle).

Materials

A pithing probe; amphibian Ringer's solution (room temperature); scissors; dissecting instruments; biograph (or similar instrument), transducer, and stimulator; thread; ringstand with fixed clamp and movable arm.

Be sure to review safety precautions before handling fresh or preserved specimens.

Procedure

A. Preparing the Frog (Figure 10.2)

For this experiment a fresh preparation of a muscle is required. A frog will be pithed in both the spinal and cerebral areas, and then the gastrocnemius muscle hooked up to clamps and stimulated. The freeing of the gastrocnemius muscle is shown in Figure 10.2.

1. Pith the frog in the manner described by your instructor. Follow along with Figure 10.2 for steps 2–4.
2. Peel back the skin on the leg by carefully cutting it free where the femur joins the trunk, and pulling it completely off the leg, like a stocking.
3. Locate and free the sciatic nerve from along the dorsal side of the thigh. Tie a string around the nerve so that it may be lifted from the thigh muscle.
4. Gently free the belly of the gastrocnemius muscle with a dissecting needle and follow its insertion to the heel. Carefully cut the Achilles tendon free from the underside of the foot. This will become the "movable end" of your muscle setup.

5. Separate the thigh from the body of the frog by means of a transverse cut along the leg up near the groin area. Immerse this preparation in a petri dish with Ringer's solution.
6. Suspend the preparation from the fixed clamp of your setup, as in Figure 10.3. The fixed clamp should tighten around the femur that forms the upper portion of the joint. Being sure not to damage the sciatic nerve, tie a thin string onto the tendon at the free end of the muscle and hook this to the movable clamp. This will move as the muscle responds to a stimulus.
7. Moisten the preparation by dripping Ringer's solution onto it frequently.

B. Stimulation of Frog Muscle (Figure 10.3)

1. Turn the power switch "on" and make sure the pens are recording movements of the movable arm.
2. Turn the stimulator "on" and the mode switch to "single."
3. Set the stimulator at the "15 milliseconds" duration. This will allow the electrical impulse to last for 15 msec.
4. Set the voltage to "0.1 to 0.9 V."
5. Turn the paper speed to "25 mm/sec."
6. Stimulate the frog muscle at the lowest voltage reading (0.1 V). If you do not observe a muscle twitch, increase the voltage of the single stimulus by tenths, until a single twitch curve is recorded by the pen. The curve should be about 1.5 cm high at the peak of contraction. If it is not, repeat the stimulus procedure, and alter the "sensitivity" knob until a curve of this height is recorded.
7. Apply Ringer's solution frequently to moisten the preparation.
8. Note the shape, timing, and characteristics of the curve.
9. Record the curve in the space provided in the Observations section.
10. Continue stimulating the muscle with increasing voltages up to 0.9 mV.
11. Record your observations.

C. Stimulation of Frog Nerve (Figure 10.3)

1. Gently remove the electrodes from the belly of the muscle and place them under the sciatic nerve as you carefully lift the nerve with the thread tied to it.
2. With all settings as in Part B, repeat the procedure you followed for stimulating the muscle as above. Note the voltage at which the single muscle twitch occurs.
3. Record your observations.

■ **FIGURE 10.2   Preparing the gastrocnemius muscle**

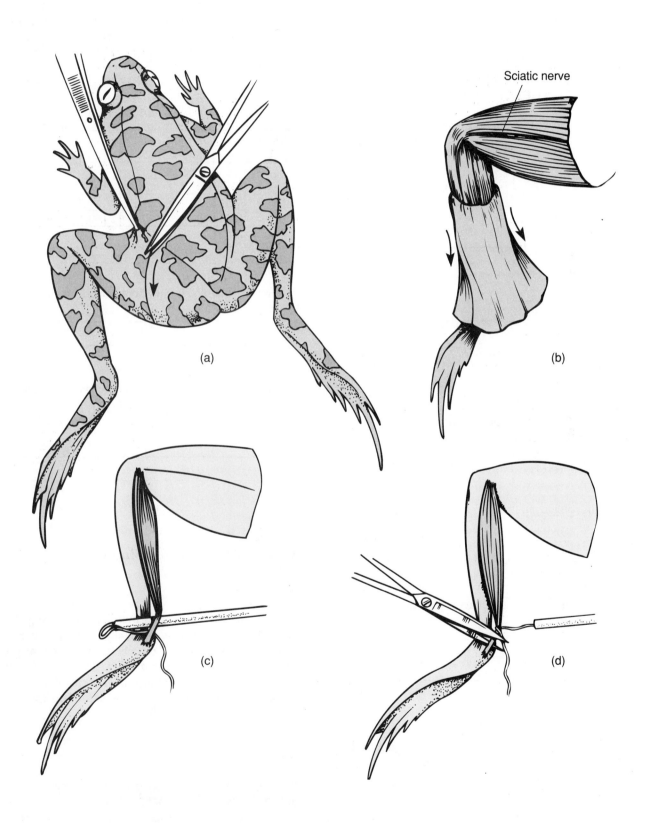

■ **FIGURE 10.3 Setup for the gastrocnemius muscle suspension**

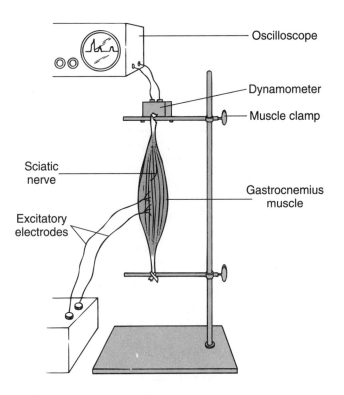

Oscilloscope

Dynamometer

Muscle clamp

Sciatic nerve

Gastrocnemius muscle

Excitatory electrodes

Observations

Draw the graph of a muscle twitch as you observed it on graph 1. Label (1) latent period, (2) period of contraction, (3) period of relaxation and (4) peak of contraction. Include the appropriate time and voltage scales and muscle name.

Minimum voltage required:

a. To stimulate the muscle directly _____

b. To stimulate the muscle via the sciatic nerve

_____

Response of the muscle as voltage is increased

_____

_____

_____

Conclusions

● Look up "recruitment" and explain the change in muscle response as the voltage is increased.

_____

_____

_____

Strength of contraction

Time   **Graph 1**

● How do you explain the latent period?

_____

_____

_____

● How do you explain the difference between the slope of contraction and the slope of relaxation?

_____

_____

_____

EXERCISE 2
Factors Affecting the Single Muscle Twitch

Discussion

Athletes are aware that their muscles must be kept warm when they are sitting on the sidelines waiting to enter a game. The response they can elicit from a "warmed up" muscle is greater than that from a "cool" muscle. Temperatures, as well as the nutritional status of the muscle, the load on the muscle, and the length of the muscle, all affect muscle performance.

Materials

Same as for Exercise 1; plus amphibian Ringer's solution at 5–7°C; 20–22°C; 35–37°C.
   Be sure to review safety precautions before handling fresh or preserved specimens.

Procedure

You will be noting the strength of contraction and the length of the twitch.

1. Using the same setup for direct muscle stimulation as in Exercise 1, stimulate the calf muscle with the *minimal* voltage previously found to give a *maximal response*.
2. Drizzle room-temperature Ringer's solution (about 20–22°C) on the muscle and obtain a single muscle twitch. Record on graph 1 and in the Observations section.

3. Now drizzle Ringer's solution that has been heated to 35–37°C over the muscle several times. Stimulate the muscle again and elicit a single twitch. Record on graph 2 and in the Observations section.
4. Drizzle Ringer's solution that is about 5–7°C over the muscle several times and elicit a single twitch as before. Record on graph 3 and in the Observations section.
5. Stretch the muscle slightly by moving the "fixed end" clamp a small distance. Using the same voltage as in step 1, elicit a single twitch. Record on graph 4 and in the Observations section.

Observations

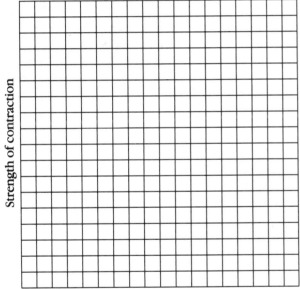

Time   **Graph 1**

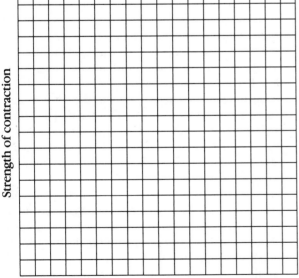

Time   **Graph 2**

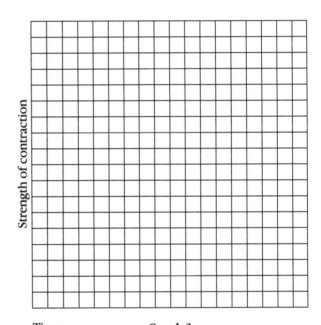

Time   **Graph 3**

Time   **Graph 4**

Count the number of boxes on your graph to indicate:

Height of muscle twitch obtained at 37°C _____ boxes
Time span of twitch _____

Height of muscle twitch obtained at 22°C _____ boxes
Time span of twitch _____

Height of muscle twitch obtained at 5°C _____ boxes
Time span of twitch _____

Height of muscle twitch obtained for unstretched muscle _____ boxes
Time span of twitch _____

Height of muscle twitch obtained for stretched muscle _____ boxes
Time span of twitch _____

Conclusions

- At which temperature does the muscle give a maximal strength of contraction? Explain.

  _____

  _____

  _____

  _____

- If the muscle were warmed to 65°C, what would the resulting contraction be like? Explain.

  _____

  _____

  _____

  _____

- Whereas a "load" has a limited effect of stretching a skeletal muscle because of its fixed attachments, cardiac muscle has no such fixed attachments. How is cardiac muscle stretched?

  _____

  _____

  _____

  _____

- What does this mean for the heart?

  _____

  _____

  _____

  _____

EXERCISE 3
Various Responses to Repetitive Stimuli

Discussion

When the calf muscle of the frog is set up as in Figure 10.3 and stimulated repeatedly with the same voltage, yet allowed complete relaxation between contractions, a phenomenon known as **treppe** occurs. (The absolute refractory period, ARP, must be passed in order to elicit any subsequent response.) Treppe is an increasing strength of contraction of a rested muscle, to a maximum level, given the conditions just noted. However, when a second stimulus follows so rapidly on the first stimulus that complete relaxation of the muscle is impossible, a **summation** of two contractions can be seen. Finally, if the stimuli are repeated extremely rapidly, a sustained contraction called **tetany** is produced. Normal body movements are tetanic contractions. In **incomplete tetanus** the contraction shows small relaxation curves; in **complete tetanus** the graph of contraction is apparently smooth. See Figure 10.4.

Materials

As for Exercise 1.
   Be sure to review safety precautions before handling fresh or preserved specimens.

Procedure

1. Use the same setup for direct muscle stimulation as in Exercise 1. Stimulate the calf muscle with the *minimal* voltage previously found to give a *maximal response*. Obtain a single muscle twitch, allow relaxation, then stimulate again. Repeat this with the same voltage, about five times. Moisten with Ringer's solution. Record results in graph 1.
2. Using the same voltage, stimulate the muscle, again eliciting a single muscle twitch response. Then stimulate the muscle with two stimuli, one rapidly following the other so as to disallow *complete relaxation*. Moisten with Ringer's solution. Record results in graph 2.
3. Turn the stimulator mode switch to constant. Once again using the same voltage, stimulate the muscle with increasing frequency until tetany is obtained (about 5–7 stimuli per second). Note the frequency at which incomplete and complete tetany are elicited. Continue the stimulus until the muscle fatigues. Record your results in the Observations section in graph 3.

FIGURE 10.4 Record of response of an isolated skeletal muscle stimulated with increasing frequencies of stimuli of sufficient intensity to produce a maximal response from the muscle

(a) Single muscle twitches. (b) Partial or incomplete tetanus (summation of twitches). (c) Complete tetanus.

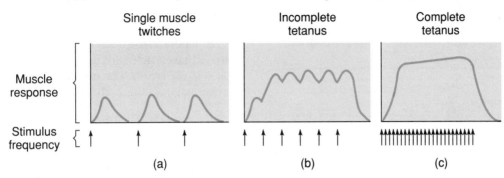

Observations

1. Record the results from step 1.
2. Record the results from step 2.
3. Record the results from step 3. (Label incomplete and complete tetany on the graph and note the frequency at which these responses were elicited.)

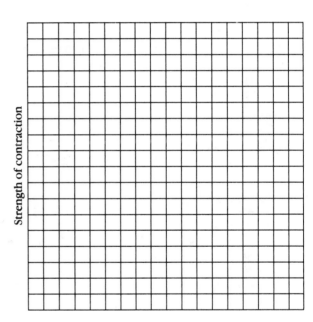

Time   **Graph 2**

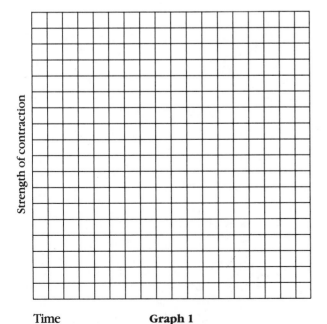

Time   **Graph 1**

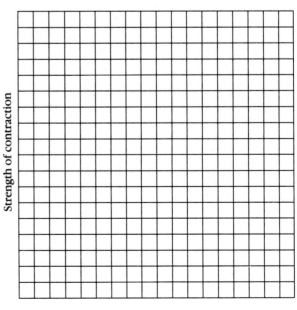

Time   **Graph 3**

Conclusions

● How do you define incomplete and complete tetanus?

_____

_____

_____

_____

_____

_____

● What are some possible explanations for fatigue in a muscle?

_____

_____

_____

_____

_____

_____

EXERCISE 4
Action of Selected Human Muscles

Discussion

Certain actions cause contraction of muscles that can be easily **palpated** (felt with the fingers) when resistance is applied to the action. The stickmen in the accompanying sketches show actions with their opposing forces.

Materials

Three lab partners.

Procedure

1. One lab partner should perform the actions indicated by label "A" (action) in the stick figures.
2. The second lab partner should resist the movement by exerting pressure in the direction marked "R" (resistance).
3. While this is occurring, the third partner should identify and palpate the major muscle involved. In the space provided in the Observations section, add the appropriate arrows to the stick figures and record the muscle name below each stick figure.

Observations

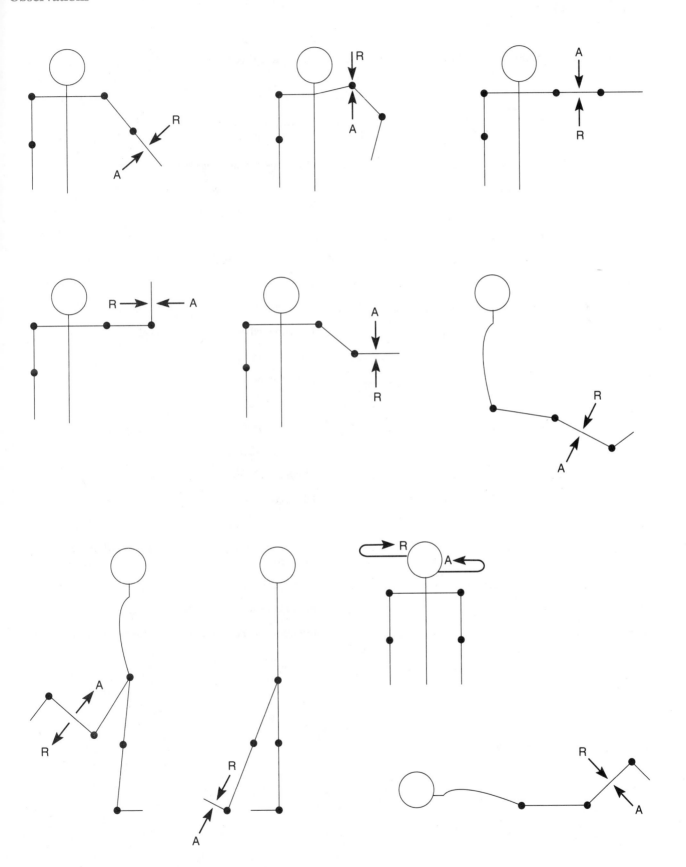

Quick Quiz 1

Complete the Table A. You may need to use your lecture text to complete this.

■ **TABLE A   Human Muscles**

	MUSCLE NAME	ORIGIN	INSERTION	ACTION CHARACTERISTIC
1	_____	_____	_____	Back muscle that brings arm down in swim stroke
2	_____	_____	_____	Raises your eyebrows
3	_____	_____	_____	Produces "pucker" in kissing
4	_____	_____	_____	Lifts your rib cage
5	_____	_____	_____	Turns your head left
6	_____	_____	_____	Used in hugging
7	_____	_____	_____	Upper arm muscle used to turn doorknob clockwise
8	_____	_____	_____	Is segmented on a weight lifter's belly
9	_____	_____	_____	Clenches your jaw
10	_____	_____	_____	Powerful muscle of the rump
11	_____	_____	_____	Shrugs your shoulders
12	_____	_____	_____	"Tailor's muscle"
13	_____	_____	_____	Antagonist to muscle 7
14	_____	_____	_____	Thigh muscle used to "kick" a football
15	_____	_____	_____	Lifts a chair "stiff-armed"
16	_____	_____	_____	Points your toes down
17	_____	_____	_____	Upper arm muscle that lifts humerus in swimming the backstroke
18	_____	_____	_____	Points your toes up
19	_____	_____	_____	Buttocks muscle where shots are given
20	_____	_____	_____	Inferior border of this muscle involved in inguinal hernia

EXERCISE 5
Human Muscle Activity

Materials

A rubber ball; sphygomanometer; clock.

Procedure

1. Squeeze a rubber ball in your hand until fatigue causes you to stop. Count the number of contractions (squeezes) and the time it takes to fatigue. Record the results in the Observations section.
2. After resting for several minutes, repeat this procedure with a sphygmomanometer on your arm with the pressure pumped up to about 90 mm Hg. Record the results.
3. Once again, after sufficient rest, repeat step 1 with the forearm submerged in cold water. Record the results.

Observations

Contractions per minute, room temperature,
  No sphygmomanometer   _____

Contractions per minute, room temperature,
  With sphygmomanometer   _____

Contractions per minute, cold temperature,
  No sphygmomanometer   _____

Conclusions

- Why is a difference observed with and without the sphygmomanometer? Explain.

  _____

  _____

  _____

  _____

  _____

- What happens to heart muscle when its blood supply is blocked? Explain.

  _____

  _____

  _____

  _____

  _____

  _____

- Why does a pitcher wear his jacket between innings in a ball game? Explain. (See results in parts A and B.)

  _____

  _____

  _____

  _____

  _____

Anatomy of the Nervous System

O V E R V I E W

In this unit, you will examine several of the structural aspects of the nervous system. Microscopic and macroscopic observations of the major components of the nervous system will be carried out.

O U T L I N E

Nervous System: Microscopic Anatomy

Background

The nervous system is the control and coordinating center for the body's activities. It interprets incoming information, processes that information and, in turn, sends out control signals, allowing the body to maintain a homeostatic state. This basic functioning is carried out through the activity of cells called neurons, which conduct "informational" impulses.

As indicated in Unit 4, the tissue of the nervous system has two types of components, **neurons**, which conduct impulses and **neuroglial** cells, which are supportive and protective elements within the nervous system.

EXERCISE 1
Microscopic Anatomy of Nervous Tissue

Discussion

In Unit 4, a typical neuron was described as having a cell body or soma from which processes extend. These proc-

esses are the axon, which transmits impulses *away from* the cell body, and the dendrites, which transmit impulses *toward* the cell body. There are a variety of neuron types within nervous tissue. Using a structural classification based on the number of processes off the cell body, three main types of neurons can be identified: **unipolar**, which have one main process that branches into an axon and a dendrite; **bipolar**, having two processes; and **multipolar**, the most common type, having many branches. See Figure 11.1. The cell body of the neuron contains organelles like other cells, but also has neurofibrils that extend into the processes, and chromatophilic substance (**Nissl bodies**), which are dark-staining arrangements of rough ER and free ribosomes. While most cell bodies are located in the Central Nervous System (CNS), clusters of them can be found in the periphery. These are termed **ganglia**.

The processes of many neurons are enveloped by a sheath called the **myelin sheath** and **neurilemma**. These coverings result from certain cells wrapping around the outside of the process. The space between these cells is called a **node of Ranvier**. Cells which have this outer wrapping are called myelinated neurons.

Four major types of neuroglial cells are found within the central nervous system: astrocytes, oligodendrocytes, microglia and ependymal cells. Table 11.1 summarizes information about the structure and function of these cells and Figure 11.2 shows three of them.

The nerves that make up the peripheral nervous system (PNS) are a collection of neuron processes (axons and dendrites) arranged in related bundles. Several bundles are packed together to form a nerve that is macroscopic. Within the nerve, the individual myelinated and unmyelinated fibers are wrapped by a delicate connective tissue called the **endoneurium**. Single bundles of fibers are wrapped together by a stronger connective tissue called the **perineurium**, and all the bundles of a nerve are wrapped and bound together by the **epineurium**.

Materials

A microscope; prepared slides of myelinated multipolar neuron; c.s. spinal cord; c.s. nerve.

Procedure

1. Examine the slide of multipolar neurons and locate the axon, dendrites, soma, nucleus, and Nissl bodies. Draw a typical multipolar neuron in the space provided on page 138.
2. Examine the spinal cord cross section and try to identify several types of neuroglial cells. Draw several examples on page 138.
3. Examine the nerve cross section and identify the different connective tissues associated with the nerve. Also identify the axon, myelin sheath, and neurilemma. Label these on the diagram on page 138.

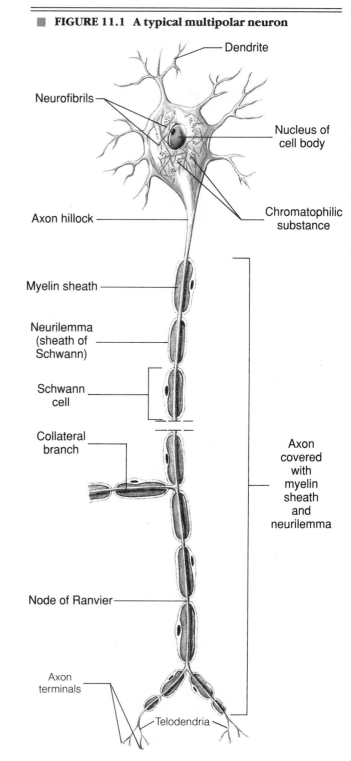

■ **FIGURE 11.1   A typical multipolar neuron**

Dendrite

Neurofibrils

Nucleus of cell body

Chromatophilic substance

Axon hillock

Myelin sheath

Neurilemma (sheath of Schwann)

Schwann cell

Collateral branch

Axon covered with myelin sheath and neurilemma

Node of Ranvier

Axon terminals

Telodendria

■ **TABLE 11.1   Neuroglial Cells of the Central Nervous System**

CELL TYPE	STRUCTURE	FUNCTION
Astrocyte	Stellate shape; many processes entwining neurons and blood vessels	Support
Oligodendrocyte	Small cell; few, short processes	Forms myelin sheath
Microglia	Tiny cell with few processes; may migrate	Phagocytosis
Ependymal	Small, cuboidal shaped; line brain ventricles	Assist in formation of cerebrospinal fluid (CSF)

■ FIGURE 11.2 Neuroglial cells

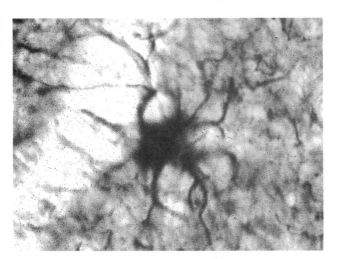 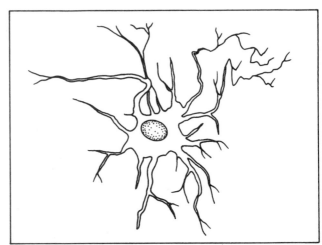

(a) Astrocyte (×527)

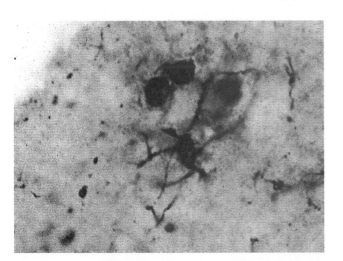 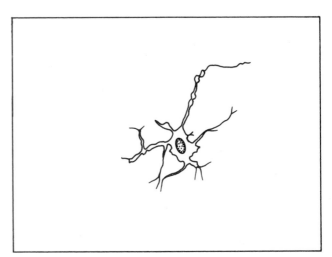

(b) Oligodendrocyte (×969)

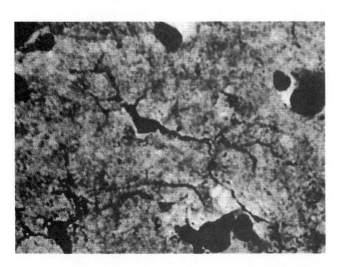 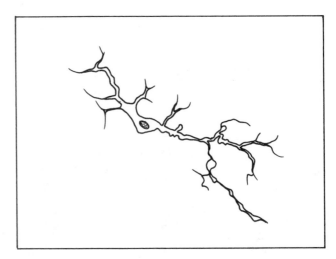

(c) Microglia (×969)

Observations

Typical Neuron

Neuroglial Types

Diagram of Nerve: Cross Section

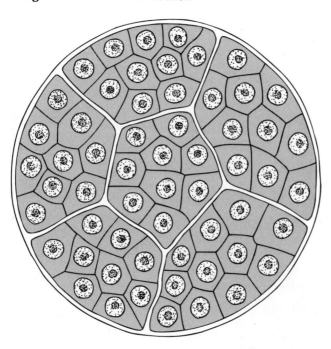

Conclusions

• What types of cells did you find in the spinal cord cross section?

_____

_____

_____

_____

_____

• Were you able to distinguish cell type in the nerve cross section? Explain.

_____

_____

_____

_____

_____

_____

Quick Quiz 1

Match the numbered structure in column A with the lettered phrase in column B.

A	B
_____ **1.** axon	(a) chromatophilic substance
_____ **2.** neuron	(b) connective tissue around bundle of neurons
_____ **3.** Nissl bodies	(c) conducting unit of nerve tissue
_____ **4.** perineurium	(d) transmits away from soma
_____ **5.** myelin	(e) fatty sheath material

Nervous System: Macroscopic Anatomy

Background

Anatomically, the nervous system is organized into two main divisions: the **central nervous system (CNS)**, consisting of the brain and spinal cord, and the **peripheral nervous system (PNS)**, comprised of the nerves which enter and exit from the central nervous system. The nerves connect the CNS with the rest of the body. In humans, the PNS has 12 pairs of cranial and 31 pairs of spinal nerves.

EXERCISE 1
Anatomy of the Brain

Discussion

The brain, weighing about 3 lb. in the adult human, has three principal divisions: the forebrain or prosencephalon, the midbrain or mesencephalon, and the hindbrain or rhombencephalon. Since nervous tissue is so delicate, the brain is protected by the bony cranial case and also covered by the connective tissue **meninges** (singular, meninx). The meninges consist of three layers: (from outside inward) the tough dura mater, the arachnoid membrane, and the delicate, vascular pia mater. Cavities within the brain, called ventricles, are filled with cerebrospinal fluid, which also circulates between the meningeal layers and provides a cushioning effect that helps protect the brain.

Largest and most prominent of the forebrain parts is the **cerebrum** or **telencephalon**, which functions in interpretation of sensory impulses, and controls voluntary activities. It consists of a core of white matter surrounded by a convoluted layer of gray matter called the **cerebral cortex**. The elevations are called **gyri** and the shallow depressions are called **sulci**. The entire cerebrum is divided midsagittally into two hemispheres that are connected by a band of transverse white fibers called the **corpus callosum**. Each hemisphere is divided into four lobes by depressions called **fissures**. The four lobes, which take their names from the bones lying above them, are named the frontal, parietal, occipital, and temporal lobes. Within each hemisphere is a cavity that is part of the ventricular system of the brain, discussed later.

Deep within the cerebral hemisphere are paired masses of gray matter called the **basal ganglia**: the caudate nuclei, the lentiform nuclei, and amygdala. The lentiform nuclei are subdivided into the putamen and the globius pallidum. Collectively, the basal ganglia provide a steadying effect on voluntary movement.

Beneath the cerebral hemispheres are the structures that form the diencephalon, namely the **thalamus**, the **hypothalamus**, and the **pineal gland**. The thalamus is a pair of oval bodies of gray matter located on each side of the third ventricle and connected by the **intermediate mass**, a band of fibers running through the third ventricle. This area serves as a relay station for sensory information traveling from the spinal cord to the cerebral cortex and has some involvement in motor tracts leaving the cerebral cortex. In the dorsomedial portion of the thalamus is a small oval mass known as the **pineal gland**. Although its function is the subject of controversy, it is thought to be involved in cyclic activities. The mass of gray matter below the thalamus is the hypothalamus, which forms the floor of the third ventricle. Although

---

▓ TABLE 11.2 Human Brain: Major Divisions

Prosencephalon: Forebrain		
Telencephalon	Cerebral hemispheres	
Diencephalon	Thalamus,[a] hypothalamus, pineal gland	
Mesencephalon: Midbrain	Corpora quadrigemia, Cerebral peduncles	
Rhombencephalon: Hindbrain		Brain stem
Myelencephalon	Medulla	
Metencephalon	Pons, Cerebellum	

[a]Some consider the thalamus to be the superior aspect of the brain stem.

▓ TABLE 11.3 Major Cerebral Fissures

FISSURE	LOCATION
Longitudinal	Deep, vertical fissure; separates the cerebral hemispheres
Transverse	Deep fissure; separates cerebrum and cerebellum
Fissure of Rolando (central sulcus)	Begins at middle of longitudinal fissure and extends inferiorly toward the fissure of Sylvius (see below); separates frontal and parietal lobes
Fissure of Sylvius	Horizontal fissure along the superior border of the temporal lobe; separates temporal and frontal lobes

physically it is a small part of the brain, the functions it carries out are largely responsible for "normal living." These include body temperature regulation, control of food intake, hormone production, water regulation, and regulation of endrocrine function. It has a direct connection, the **infundibulum**, to the **pituitary gland** below.

Extending from the lower border of the diencephalon is the midbrain (mesencephalon), consisting of the corpora quadrigemina, cerebral peduncles, cerebral aqueduct, and red nucleus. The **corpora quadrigemina**, on the dorsal portion, are four rounded prominences: the larger **superior colliculi** and smaller **inferior colliculi** which serve as reflex centers. The superior colliculi are involved in movement of the eyes in response to stimuli, while the inferior colliculi serve as reflex centers for movement of the head in response to auditory stimuli. The cerebral peduncles are located in the ventral portion of the midbrain and serve as a connection station for tracts between the forebrain, hindbrain, and spinal cord. Near the center of the midbrain is a mass of gray matter called the red nucleus because of deposits of reddish pigment. It is part of the reticular formation and is a key component in the maintenance of posture.

The hindbrain (rhombencephalon) consists principally of the pons varolli, medulla oblongata, and cerebellum. The **pons** is a rounded bulge that separates the midbrain from the rest of the hindbrain. It is composed mainly of white matter and houses two main respiratory centers as well as serving as a bridge relaying impulses from the medulla to higher brain centers. Inferior to the pons is the **medulla**, which is a continuation of the spinal cord above the foramen magnum. The medulla consists of a core of gray matter that is broken up into nuclei, surrounded by white matter organized into tracts. The nuclei serve as centers from which cranial nerves arise as relay stations for sensory tracts; and as main control centers for cardiac, vasomotor and respiratory functions. Like other areas of the brain the tracts of the medulla are both ascending and descending, forming communication paths between the spinal cord and various superior parts of the brain.

The second largest portion of the brain is the **cerebellum**, which lies under the occipital lobe of the cerebrum, and is somewhat like an appendage connected to the brain by the **cerebellar peduncles**. The cerebellum is a mixture of gray and white matter, the latter being organized in treelike fashion. The gyri and sulci of the cerebellum are straighter and more narrow than those of the cerebrum. In humans the hemispheres of the cerebellum are connected by a central constricted area called the **vermis**. Functionally, the cerebellum is mainly a reflex center for the coordination of skeletal muscle activity.

As indicated earlier, there is within the brain a series of spaces called ventricles that are interconnected and serve in the formation and circulation of most of the CSF. The right and left or **lateral ventricles** are located within the cerebrum and communicate with the third ventricle via the foramen of Monro. The **third ventricle**, located between the two halves of the thalamus, is connected to the fourth ventricle via the **cerebral aqueduct**, which runs through the midbrain. The wedge-shaped **fourth ventricle** is bordered by the pons, the medulla, and the cerebellum and communicates directly with the central canal of the spinal cord and with the meningeal subarachnoid space via the **foramina of Luschka** and **foramen of Magendie**. In each ventricle, CSF is secreted by a network of capillaries called the **choroid plexus**. As a result of these open communications, the CSF is maintained at a constant pressure and serves as a "shock absorber" to protect the brain from mechanical injury.

Materials

A preserved sheep brain; human brain models; dissecting instruments; dissecting pan.

Be sure to review safety procedures before handling preserved or fresh specimens.

Procedure

A. Surface Anatomy

1. Place the preserved sheep brain in the dissecting pan, dorsal side up.
2. Note the convolutions of the cerebrum and cerebellum.
3. Locate the longitudinal fissure, the fissure of Rolando, the fissure of Sylvius, and the transverse fissure.
4. Identify the four lobes of each cerebral hemisphere.
5. Separate the cerebrum and cerebellum by pushing the cerebellum downward. This will expose the corpora quadrigemina and pineal body.
6. Pull the left and right hemispheres apart slightly and note the corpus callosum running between.
7. Using the human brain model, compare the dorsal surface anatomy with that of the sheep brain.
8. Place the preserved sheep brain dorsal side down in the pan.
9. Locate the ventral side of the forebrain, brain stem, and midbrain.
10. Note the location of the following cranial nerves or tracts: olfactory, optic, oculomotor, trigeminal, vagus, and hypoglossal nerve roots.
11. Locate the pituitary, infundibulum, and mammillary body.
12. Again, make comparisons with the human models.
13. Using Figures 11.3 through 11.7 as guides, be able to locate and identify on the sheep and human brain the parts identified.

■ **FIGURE 11.3 The human brain**
(a) Whole; lateral view. (b) Ventral view.

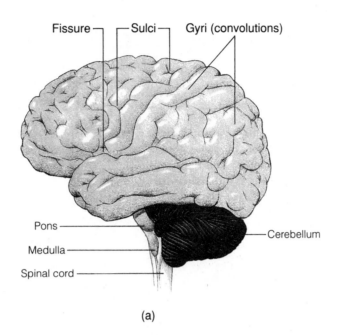

(a)

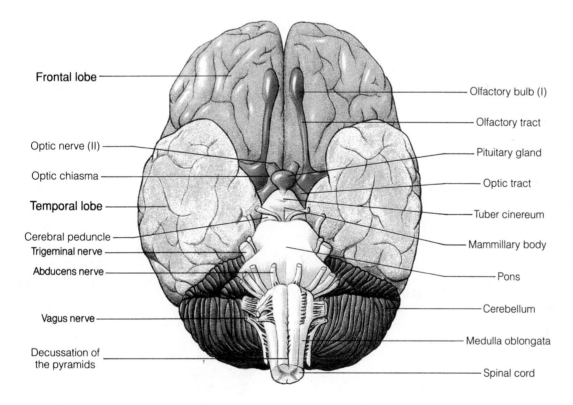

(b)

■ **FIGURE 11.4 The human brain**
(a) Coronal section. (b) Sagittal view.

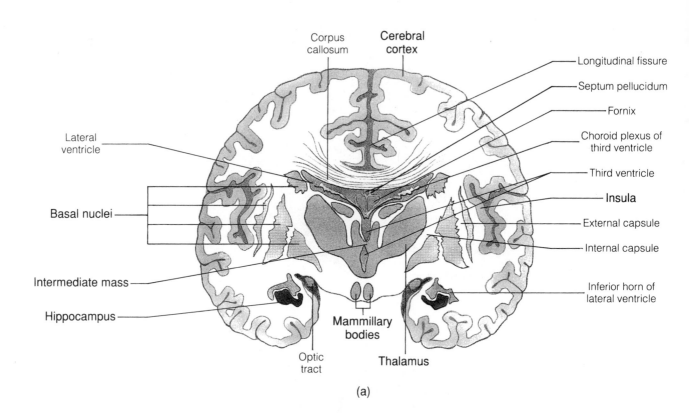

(a)

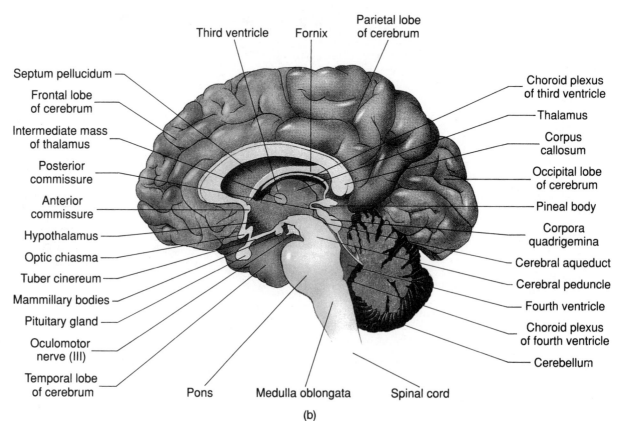

(b)

■ **FIGURE 11.5 The sheep brain: dorsal view**
Refer to the color photo gallery.

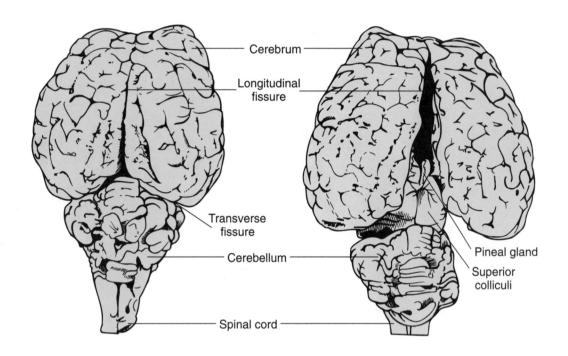

■ **FIGURE 11.6 The sheep brain: ventral view**
Refer to the color photo gallery.

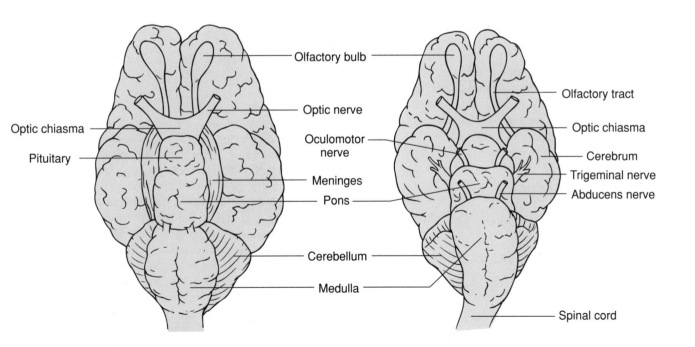

■ **FIGURE 11.7 The sheep brain: sagittal view**
Refer to the color photo gallery.

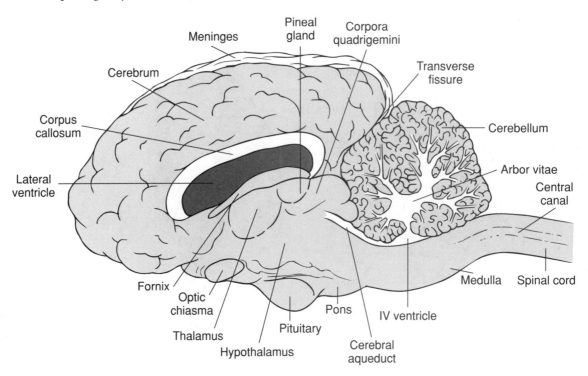

B. Midsagittal Section

1. If sagittally sectioned sheep brains are not available, make a midsagittal section following the longitudinal fissure.
2. Examine the medial surface of one of the halves and identify the structures illustrated in Figure 11.4.
3. Locate the structures on the human brain model as well.

C. Frontal Section

1. With the dorsal side down, cut across one of the halves of the brain about the middle of the thalamus.
2. Identify the lateral ventricle, third ventricle, caudate nucleus, putamen, globus pallidus, thalamus, internal capsule, corpus callosum, fornix.

Conclusions

• From what part of the brain do the following cranial nerves appear to emerge?

 a. Oculomotor _____

 b. Abducens _____

 c. Spinal accessory _____

 d. Trigeminal _____

• Were there any observable differences between the sheep and human brain? Explain.

 _____

 _____

 _____

 _____

EXERCISE 2
Anatomy of the Human Spinal Cord

Discussion

The human spinal cord is a cylindrical mass of nerve tissue extending from the medulla oblongata to the second lumbar vertebra. Like the brain, the spinal cord is encased in meningeal coverings. It has two enlargements, the **cervical enlargement** (extending from vertebrae C_4 to T_2) and the **lumbar enlargement** (extending from vertebrae T_{10} to T_{12}). Below the lumbar enlargement, the spinal cord tapers to the **conus medullaris**, which gives rise to the **filum terminale**, a fibrous extension of the pia mater that extends and attaches to the coccyx. Arising from the spinal cord are pairs of spinal nerves: 8 cervical, 12 thoracic, 5 lumbar, 5 sacral, and 1 coccygeal. The cat has 38 or 39 pairs of spinal nerves: 8, 13, 7, 3, and 7 or 8. Basically, these nerves exit through the intervertebral foramina in the following manner: spinal nerves C_1 through C_7 exit above the cervical vertebrae of the same number. Spinal nerve C_8 exits between vertebrae C_7 and T_1, and the remaining nerves exit caudal to the vertebrae of the same number as the nerve (e.g., spinal nerve T_2 exits between vertebrae T_2 and T_3). The roots of the lower spinal nerves pass inferiorly in the vertebral canal before reaching the appropriate intervertebral foramen and exiting to the periphery. These descending nerve roots give the lower end of the spinal cord the appearance of a horse's tail, which is referred to as the **cauda equina**.

On the ventral surface of the spinal cord there is a deep fissure called the **anterior medial fissure**. A more shallow groove on the dorsal surface is called the **posterial medial fissure**. These two grooves effectively divide the spinal cord into left and right halves.

Examination of a cross section of the spinal cord, illustrated in Figure 11.8, shows that it consists of a core of gray matter surrounded by white matter. The core of gray matter consists of nerve cell bodies, and unmyelinated fibers; it is shaped roughly like an "H." The crossbar of the H is called the **gray commissure**, in the center of which is the **central canal**, filled with CSF. The vertical bars of the H are referred to as **horns** or columns and designated as anterior or posterior. The white matter of the spinal cord is primarily composed of myelinated axons and is organized into ascending and descending tracts. These tracts, illustrated in Figure 11.9, conduct impulses to and from the brain, respectively.

Spinal nerves are "pathways" through which "information" enters and leaves the spinal cord. Each spinal nerve has an anterior or **ventral root** leading away from the spinal cord and a posterior or **dorsal root** leading toward the spinal cord. The dorsal root has an enlargement that contains the cell bodies of sensory neurons,

■ **FIGURE 11.8 Cross section of the spinal cord**

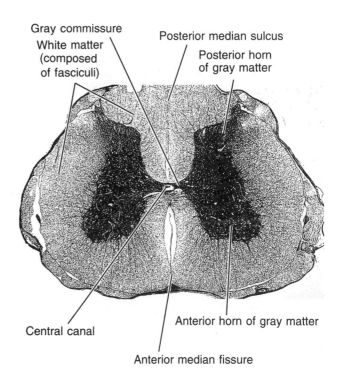

■ **FIGURE 11.9 Major tracts of the spinal cord**

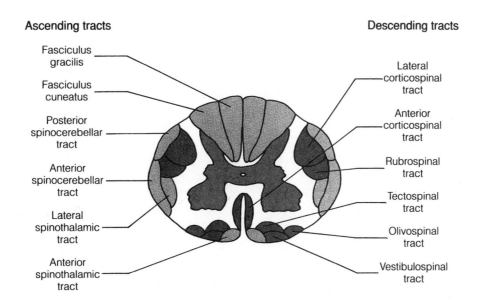

both visceral and somatic. This enlargement is called the **dorsal root ganglion**. Sensory information enters the spinal cord via the dorsal root and motor impulses leave via the ventral root. Upon leaving the intervertebral foramen, a spinal nerve branches into what are called **rami**. There are four rami associated with a spinal nerve: a **meningeal ramus**, which supplies the meninges, a **dorsal ramus**, which supplies the deep muscles and skin of the back; a **ventral ramus**, which supplies the skin and muscle of the anterior and lateral portions of the trunk and limbs, and a **visceral ramus**, which is part of the autonomic nervous system. The ventral rami, except for those in the thoracic region (T_2–T_{12}) intermix with each other and form networks of nerve fibers called **plexuses**. These plexuses allow a redistribution of nerve fibers and include the cervical, brachial, lumbar, sacral, pudendal, and coccygeal plexuses.

Materials

A model of spinal cord and spinal nerves; cross-sectional model of spinal cord; preserved spinal cord

Procedure

1. Examine the preserved specimen of the spinal cord, noting the enlargements, and the branching of the spinal nerves.
2. Examine the cross-sectional model of the spinal cord and identify those structures illustrated in Figure 11.8.
3. Label the accompanying diagram.

Spinal Cord

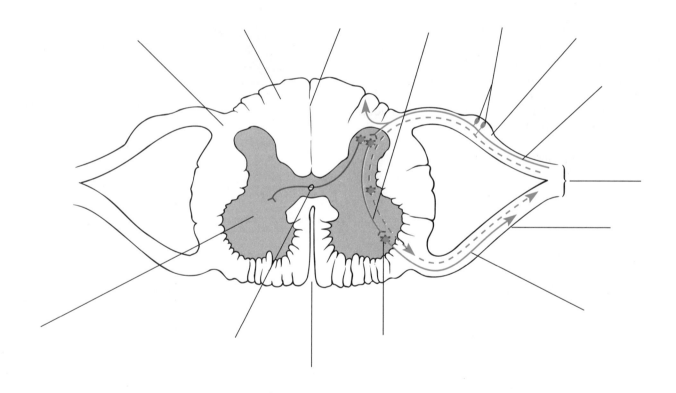

Conclusions

- What is the observable relationship between white and gray matter?

  _____

  _____

  _____

  _____

- Were you able to distinguish the tracts of the spinal cord? Explain.

  _____

  _____

  _____

  _____

 ## Quick Quiz 2

Match the numbered structure in A with the lettered phrase in B.

A	B
_____ 1. telencephalon	(a) bordered by pons, medulla and cerebellum
_____ 2. pons	
_____ 3. medulla	
_____ 4. mesencephalon	(b) corpora quadrigemina
_____ 5. 4th ventricle	
_____ 6. ventral root	(c) myelencephalon
_____ 7. rami	(d) cerebrum
_____ 8. plexus	(e) superior to medulla
	(f) network of nerve fibers
	(g) exit root from spinal cord
	(h) branch of spinal nerve just beyond the intervertebral foramen

Selected Functions of the Nervous System

OVERVIEW

In this unit you will study some of the functional aspects of the nervous system. You will be examining some simple reflexes and making observation about cutaneous sensations.

OUTLINE

The following exercises involve experimentation on living material. A commercial audio-visual tape of this subject may be substituted if appropriate.

Cutaneous Sensations

Background

One of the major functions of the nervous system is to provide a means of receiving information about the body's environments, both external and internal. Our ability to do this can affect our survival. The structures of the nervous system which receive the information, or stimuli, are the **receptors** and they are distributed throughout the body. Receptors can be classified according to location, e.g. exteroceptors, interoceptors, or classified according to the type of stimulus, e.g., mechanoreceptor, thermoreceptor.

Touch, pressure, heat, cold, and pain are the cutaneous senses. A variety of specialized receptors, the exteroceptors, are responsible for detecting these stimuli. Most of the specialized cutaneous receptors are **mechanoreceptors** (i.e., they detect mechanical deformation), but some may be classified as **thermoreceptors** (which detect changes in temperature). Still others can be called **nociceptors**, which detect damage to tissue that causes pain.

EXERCISE 1
Distribution of Cutaneous Receptors

Discussion

The distribution of the exteroceptors for cutaneous sensations over the entire body surface is unequal. Some regions such as the fingertips have a plentiful supply of receptors and are quite sensitive. Other areas, like the back, are relatively insensitive as a result of a sparse supply of receptors. Within a particular region, the distribution of the various types of receptors varies.

Materials

Blunt muscle probes; blunt end of opened paper clip; hairbrush bristles; ice water bath; hot water bath (43–45°C); ballpoint pen; solvent (to remove marks).

Procedure

Work with a lab partner.

1. Using the grids in the Observations section as models, mark off a similar area with a ballpoint pen on the forearm, back of the neck, and sole of the foot of your lab partner.
2. Have the subject look away so as not to be able to observe what is happening.
3. Touch each of the small squares of the grid on the forearm with the bristle. Ask the subject what it feels like. Record the results.
4. Record the location of the receptor with a "T" on the appropriate grid.
5. Locate the heat receptors by the same procedure, using muscle probes that have been heated in the hot water bath and quickly dried. Several probes may be needed to maintain a constant temperature. Record on the grid with an "H."
6. Repeat step 5 using muscle probes that were cooled in the ice water bath. Record the results on the grid with a "C."
7. Using a blunt point, locate the pain receptors in the same manner and record the results with a "P."
8. Carry out the entire procedure two more times, once using the grid on the back of the neck and the second time using the grid on the sole of the foot. Record the results in the appropriate grids.
9. Do not remove the grid on the forearm—other grids may be removed with solvent according to your instructor's directions.

Observations

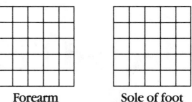

Forearm   Sole of foot   Back of neck

Conclusions

- Compare the results obtained in the three different body areas. Explain the differences.

_____

_____

_____

_____

_____

_____

EXERCISE 2

Adaptation to Stimuli: Cutaneous Receptors

Discussion

When you first put your watch on, you are aware of the band on your wrist, but this awareness diminishes quickly. Prolonged application of some stimuli results in **adaptation** of the receptors. The number of impulses initiated depends on the nature of the stimulus and the nature of the receptor. Rapidly adapting receptors such as those of touch and temperature fire impulses when the stimulus is first applied and quickly stop firing. Slowly adapting receptors such as those for some pain sensations continue to respond to repetitive stimuli. Both types of stimuli are important in cutaneous sensations.

■ **TABLE 12.1 Selected Cutaneous Exteroceptors**

RECEPTOR NAME	DESCRIPTION	MODALITY
Free nerve endings	Uncovered nerve endings	Touch, pressure, pain
Meissner's corpuscles	Encapsulated nerve endings	Light touch
Pacinian corpuscles	Encapsulated nerve endings	Deep pressure
Krause end bulbs	Multibranched encapsulated nerve endings	12–35°C temperature and at 50°C, and skin being rapidly cooled
Corpuscles of Ruffini	Network of nerve endings surrounded by a thin connective tissue capsule	25–45°C temperature and skin being warmed

Materials

A coin; muscle probe; ice cube.

Procedure

Work with a lab partner.

1. Place a coin on the grid from Exercise 1 (forearm). Note how long you are aware of its presence and record the time in the Observations section.
2. Have your lab partner place the blunt point on one of the pain receptor points found in Exercse 1 and note how long you are aware of its presence. Record the time below.
3. Place an ice cube on the grid and record the length of time you are aware of the cold sensation.
4. Remove the grid with solvent.

Observations

a. Awareness of coin _____ min

b. Awareness of blunt point _____ min

c. Awareness of cold ice _____ min

Conclusions

• Is "adaptation time" the same for these three receptors?

_____

_____

_____

• Which of these receptors is least adaptive? What is the significance of this?

_____

_____

_____

EXERCISE 3
Tactile Discrimination

Discussion

If you close your eyes and someone places a tennis ball in one hand and a golf ball in the other, you can tell what each object is. You "feel" the difference in size and texture, noting the "puckered" surface of the golf ball and "fuzzy" surface of the tennis ball. By squeezing, you can

determine that one is hard and the other is more pliable. Tactile sensations provide a great deal of information. Sensory impulses traveling over sensory nerves are relayed through the thalamus to the appropriate area of the cerebral cortex (parietal lobe) where they are interpreted.

Materials

Skeletal bones: cervical and thoracic vertebrae, radius and ulna, tibia and femur.

Procedure

Work with a lab partner.

1. Close your eyes and have your partner select several of the materials listed above.
2. Using your sense of touch, identify each item.
3. Record the results in Table A and indicate any mistakes. For example, if you gave the subject a cervical vertebra and it was identified as thoracic, indicate that mistake.

Observations

■ **TABLE A Touch Discrimination**

	IDENTIFICATION		
BONE	**CORRECT**	**WRONG**	**MISTAKE**
Cervical vertebra			
Thoracic vertebra			
Radius			
Ulna			
Tibia			
Femur			

Conclusion

• If any mistakes in identification were made, review Unit 6 and repeat this exercise.

Quick Quiz 1

Match the numbered term in A with the lettered phrase in B.

A		B
_____ 1.	interoceptor	(a) light touch
_____ 2.	nociceptor	(b) tactile discrimination
_____ 3.	Meissner's corpuscles	(c) internal receptor
		(d) thermoreceptor
_____ 4.	rapidly adapting receptor	(e) detects tissue damage
_____ 5.	determine texture	

Reflexes

Background

Reactions to stimuli which are automatic, predictable and consistent are called **reflexes**. The neural pathway by which sensory impulses from receptors reach effectors, without traveling through the conscious portions of the brain are called spinal reflex arcs. Typically, spinal reflexes are structurally described as a **reflex arc** which has a minimum of five components: (1) receptor, (2) sensory neuron, (3) synapse, (4) motor neuron, and (5) effector. The general series of events is as follows: **sensory receptors** detect various stimuli and initiate the transmission of sensory information via **sensory neurons** into the CNS; within the CNS this information is "processed" and appropriate response signals are conducted via **motor neurons** to the **effectors**, which carry out the response. The path *to* the CNS is called the **afferent pathway** and the path *away* from the CNS is called the **efferent pathway**.

The nervous system allows us to adapt to the environment. Some of the body's responses to a changing environment are conscious, deliberate actions that involve the cerebral cortex; others are "automatic" and are carried out without conscious awareness at a level "below" the cerebral cortex. For example, our responses to hot, humid weather can be at the conscious level and also below the conscious level. We can consciously choose to wear light, loose-fitting clothing and even go to the beach to "beat the heat." Simultaneously, the body is also responding subconsciously in several ways. Among the several responses, the sudoriferous glands of the skin produce more perspiration which, as it evaporates, lowers the body temperature. We do not consciously control the activities of the nervous system that stimulate the increased production of sweat.

EXERCISE 1
Cranial and Spinal Responses: The Frog

Discussion

There are both somatic and visceral reflexes. One way of classifying reflexes is according to the area through which they pass. Some reflexes pass through the brain, while others pass through the lumbar spinal cord.

The simplest reflex pathways pass only through the spinal cord. The knee jerk (patellar) reflex is an example of a **simple reflex**. More complex reflex pathways (e.g., those controlling the caliber of blood vessels in the leg) involve both pathways in the spinal cord and integration of information with the medulla. The preceding examples do not entail conscious activity because they do not involve the cerebral cortex. The knee jerk reflex is an example of a reflex that is localized in the spinal cord. In general, reflexes localized in the spinal cord are those in which the response causes withdrawal from harmful stimuli or responses to a local "irritating" stimulus. In this exercise you determine which responses of the frog travel through cranial pathways and which responses travel through the spinal cord.

Materials

A frog; pithing probe; dissecting pan; fish tank or sink and stopper; forceps; heavy textbooks; tissues.

Review safety procedures before working with fresh, live, or preserved specimens.

Procedure

This procedure is carried out three times. The first time it is done with a normal frog, the second time with a single pithed frog, and the third time with a double pithed frog. Pith the frog according to your instructor's directions.

1. Place the frog right side up on the lab table. Determine the respiratory rate by observing the nostrils. Turn the frog over on its back and hold it so that you can determine the heart rate. To do this place your index finger over the midchest region and record the number of beats for 30 sec. Record the results in Table B.
2. Place the frog on its back and record the results.
3. With the frog right side up, bang together two heavy textbooks, near the frog to observe its reaction to loud noise. Do not vibrate the surface on which the frog sits. Record the results.
4. Put the frog in the filled sink or aquarium and observe any swimming movements. Record the results.

Observations

■ **TABLE B Frog Responses**

	NORMAL FROG	SINGLE-PITHED FROG	DOUBLE-PITHED FROG
Respiratory rate			
Heart rate			
Righting action			
Noise response			
Swimming activity			
Reaction to pinching of digits			
Corneal response			

5. Hold the frog in your hand so that its hind limbs are free. Lightly pinch the digits of one of the hind limbs and record the results.
6. Gently touch the frog's cornea with the twisted end of a tissue, observe what happens in the eye, and record.

Conclusions

• Based on your observations, in which of the responses does the brain play a role? Explain.

_____

_____

_____

• Was there a difference in the respiratory rate and heart rate of the frog under each of the three conditions? Explain.

_____

_____

_____

_____

• Which of the responses does not involve the spinal cord? What evidence in this experiment supports your conclusion? Explain.

_____

_____

_____

_____

_____

• Based on your observations, what do you think the term "spinal animal" means?

_____

_____

_____

_____

EXERCISE 2
Reflexes: The Human

Discussion

Human reflexes operate in the same fashion as those of other animals. In some cases, however, behavior can be modified as the result of the overriding cerebral activity.

For example, in the simple knee jerk reflex, deliberate and conscious effort to "not respond" to the stimulus can lessen the response and even eliminate it. Unhampered reflexes are important to the neurologist because the responses can indicate normal health or disease conditions.

Materials

A rubber hammer; flashlight; soft tissue; sterile applicator sticks; blunt probe.

Procedure

Working with a lab partner, carry out the reflex tests below. One partner is the subject and the other performs the test and observes the results. The subject should attempt to relax and not override the reflex. Record the results in Table C. The tester should stand facing the subject, but to one side.

Observations

1. **Corneal Reflex** With the subject looking away, gently touch the cornea with the corner of a clean handkerchief. Observe the result and record in Table C.
2. **Nasal Reflex** Tickle the nasal mucosa with a sterile applicaor stick. Observe and record the result.
3. **Uvular Reflex** Have the subject open the mouth wide. With a sterile applicator stick, carefully touch the uvula at the back of the mouth. Use a flashlight to see the reaction, if necessary. Record the result.
4. **Achilles Tendon Reflex** Have the subject remove shoes and socks, then kneel on a lab chair so the feet hang freely over the edge. Tap the Achilles tendon and record results.
5. **Babinski Reflex** Have the subject lie on his back. Stimulate the plantar surface of the bare foot by running a blunt probe firmly up the sole of the foot. Observe the digits and record the result.
6. **Chaddock's Reflex** Have the subject sit on the lab bench and dangle his bare feet from a sitting position. Stroke the subject's lateral malleolus and observe reaction of the foot. Record the result.
7. **Gonda Reflex** Position as in step 6. Press one of the subject's smaller toes downward and release quickly. Observe the other toes and record results below.
8. **Ciliospinal Reflex** Position as in step 6. Pinch the back of the subject's neck and observe the eyes. Record the results.

■ TABLE C Human Reflexes

REFLEX	RESPONSE
Corneal reflex	
Nasal reflex	
Uvular reflex	
Achilles tendon reflex	
Babinski reflex	
Chaddock's reflex	
Gonda reflex	
Ciliospinal reflex	

Conclusions

- Were there certain reflexes in this exercise for which you observed no response? Explain.

  _____

  _____

  _____

  _____

- Complete Table D, indicating the sensory nerve, localization center, and motor nerve involved in these selected reflexes.

■ TABLE D Selected Reflexes

REFLEX	SENSORY NERVE	CENTER	MOTOR NERVE
Corneal			
Nasal			
Uvula			
Achilles tendon			
Ciliospinal			

• Based upon your knowledge, explain why an airway is not employed as an emergency measure when a patient is conscious?

_____

_____

_____

_____

_____

_____

_____

❖ **Quick Quiz 2**

Match the numbered terms in A with the lettered phrases in B.

A		B
_____ **1.** Babinski reflex	(a)	usually a withdrawal reflex
_____ **2.** Ciliospinal reflex	(b)	pass only through spinal cord
_____ **3.** Patellar reflex	(c)	large toe extends, others fan out
_____ **4.** simple reflex	(d)	pupillary-skin reflex
_____ **5.** spinal reflex	(e)	knee jerk

Special Senses

OVERVIEW

In this unit, you will study the special senses and the receptors which receive the stimuli. Many of these senses provide us with important information from the external environment. You will become familiar with the structural features of the eye and ear as well as some aspects of their functions in vision, hearing and balance. In addition, you will learn about the receptors for taste and smell and recognize how they function and are interrelated.

OUTLINE

❖ The Eye

Background

The eye houses the **receptors**, which are stimulated by light and are able to convert this energy into electrical energy. This electrical energy, in the form of a nerve impulse, travels from the eye's sensory neurons along the **optic nerve** (cranial nerve II) to the occipital lobe of the cerebrum where it is interpreted as a visual image.

As seen in Unit 12, there are many different types of receptors which are classified in several ways, i.e., location, nature of the stimulus and modality. In the case of the eye, vision is the modality, light is the stimulus, and the body surface is the location. Thus the receptors of the eye can be referred to as visual receptors, photoreceptors, or exteroceptors.

EXERCISE 1
Anatomy of the Eye

Discussion

Although it appears to be a relatively simple structure, the eye is a complex apparatus that contains not only photoreceptors, but also structures that adjust the amount of light entering the eye and structures that refract these light rays. The eye is situated in a bony orbit of the skull and is protected by the eyelids. The eye has several accessory structures associated with it externally. Among the major accessory structures are the **extrinsic eye muscles**, which connect the eye to the orbital cavity and provide for movement of the eyeball or orb. The **lacrimal apparatus** is another accessory structure that helps, through the production and distribution of tears, to keep the exposed surface of the eye moist and free of dust and microorganisms.

These structures are illustrated in Figure 13.1 a and b, respectively. Information about the innervation and action of the extrinsic muscles of the eye is presented in Table 13.1.

The eyeball is a generally spherical structure, about 2.5 cm in diameter, composed of three coats or tunics.

1. The external coat called the **sclera**, is a tough fibrous layer. This is the "white of the eye" but its most ante-

---

■ **FIGURE 13.1   The human eye**
(a) Extrinsic eye muscles. (b) Lacrimal apparatus.

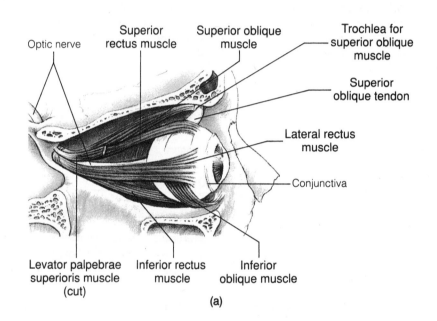

(a)

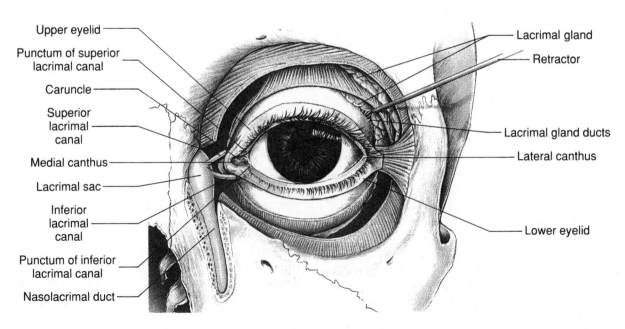

(b)

■ **TABLE 13.1 The Eye Muscles**

MUSCLE	INNERVATION AND ACTION
Superior rectus	Oculomotor nerve; moves eyeball superiorly and medially
Inferior rectus	Oculomotor nerve; moves eyeball inferiorly and medially
Lateral rectus	Abducens nerve; moves eyeball laterally
Medial rectus	Oculomotor nerve; moves eyeball medially
Superior oblique	Trochlear nerve; moves eyeball inferiorly and laterally
Inferior oblique	Oculomotor nerve; moves eyeball superiorly and laterally

rior portion is modified into a transparent window called the **cornea**.

2. The middle vascular coat or **uvea**, includes a thin vascular layer posteriorly, the **choroid**, which is darkly pigmented to prevent internal reflection of light. Anteriorly the uvea shows several structures.

   a. The round **ciliary body** is composed of the ciliary muscle and suspensory ligaments. The ligaments suspend the **crystalline lens**, which serves to bend light rays as they enter the interior of the eye.

   b. The **iris** is a round disc made of two sheets of smooth muscle, a circular sheet (the **sphincter muscle**) and radial sheet (the **dilator muscle**). Contraction of these muscles regulates the amount of light entering the eye through the opening in the center of this disc, the **pupil**. Coloration of the iris is the result of pigmentation, which is genetically determined.

3. The innermost layer of the eye is the "visual" layer or **retina**. It is the layer that contains the photoreceptors, called **rods** and **cones** because of their shape. The structure of the retina is complicated because there are several layers of cells.

   a. The retinal layer, adjacent to the choroid, is a pigmented epithelium that absorbs excess light.

   b. Next is the layer of photoreceptor cells, rods and cones.

   c. The remaining layers consist of **neuroglial** cells, and other neurons with which the rods and cones synapse directly. The first neurons that synapse with the rods and cones are called "second-order neurons." These "second-order neurons" synapse with "third-order neurons," which ultimately converge toward the middle of the retina and exist as the **optic nerve**. Where the nerve fibers leave the eye as the optic nerve, there are no rods and cones. This area is called the **blind spot** or **optic disc**. Rods are more numerous than cones except for a small spot near the center of the retina called the **macula lutea**. The small depressed area in the center of the macula lutea contains only cones and is called the **fovea centralis**. This is the area of greatest visual acuity.

The rods and cones contain photosensitive pigments that break down when exposed to light. The pigment present in the rods, **rhodopsin**, is very sensitive to light. Thus, rods function in dim light (night vision). Presumably the pigment within the cones requires bright light, and thus cones are responsible for vision in daylight and for color vision. It is thought that when light strikes the photosensitive pigments, the configuration of the molecules is altered causing them to absorb light. This ultimately results in polarity changes in the cell. This, in turn, results in the transmission of neural signals via the optic nerve to the visual cortex.

The interior of the eye is not empty; rather, it contains **refractive media** that bend the light rays entering the eye so that the rays strike the photosensitive layer. Suspended behind the iris is the biconvex crystalline **lens**, which separates the interior of the eye into anterior and posterior compartments. The anterior compartment is filled with a watery fluid called **aqueous humor**. It is produced by the ciliary body and circulates throughout the anterior compartment, providing nutrients for the lens and cornea. This fluid is continually replaced as it drains out through the canal of Schlemm. Between the retina and the lens is the posterior compartment, which is filled with a gelatinous material called the **vitreous humor**. This aids in giving shape to the eye.

Materials

Eye models; preserved sheep eye; dissecting pan; dissecting instruments.

Be sure to review safety precautions prior to handling preserved specimens.

Procedure

1. Using the pictures in your textbook, Figures 13.1 through 13.5 and the eye models, familiarize yourself with the overall structure of the eye and note the location of various parts, compartments, and refractive media.

2. Observe the external structure of the preserved sheep

eye. Identify the sclera, cornea, pupil, and optic nerve (which exits in an inferior direction when the eye is properly oriented).

3. Carefully remove the fat and connective tissue around the eyeball to reveal the extrinsic eye muscles. Identify the six muscles.

4. To identify the internal features of the eye, make an incision about 1 cm from the edge of the cornea. Cut around the periphery of the cornea leaving a "hinge" of attachment for the flap you separate by cutting. Let the aqueous humor drain. Open the "door" and fold back the anterior portion of the eye. Gently loosen the vitreous humor from the lens.

5. Examine the internal structures and identify: ciliary muscle, suspensory ligaments, and muscles of the iris, retina, and optic disc.

6. Gently remove the vitreous humor from the posterior compartment. Identify the three layers forming the wall of the eyeball: the tough outer sclera, the pigmented middle choroid and the whitish inner retina. In the sheep eye, the middle coat is irridescent. This is the **tapetum lucidum**, which is absent in humans.

Conclusions

● Is the shape of the iris muscle and pupil the same in sheep and human? Explain.

_____

_____

_____

_____

_____

● What are the differences in vitreous humor and aqueous humor with respect to their consistency?

_____

_____

_____

■ **FIGURE 13.2   The human eye, sagittal section**

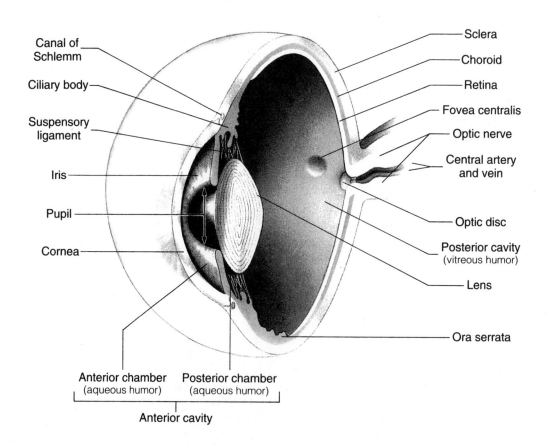

• How could a blockage in the canal of Schlemm affect the eye? What is the problem called?

_____

_____

_____

_____

_____

_____

• Look up the definition of "cataract." Where could the problem be manifest in the eye?

_____

_____

_____

_____

_____

_____

■ **FIGURE 13.3 Structure of the retina**

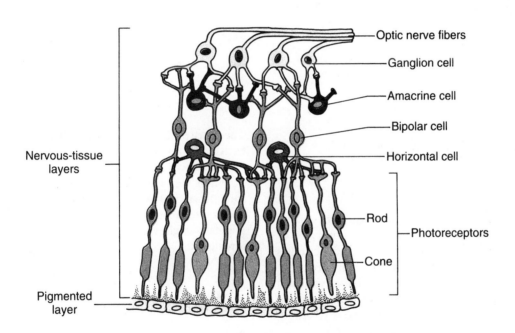

■ **FIGURE 13.4 The sheep eye — anterior view**
Refer to the color photo gallery.

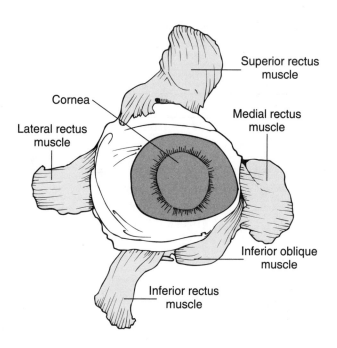

■ **FIGURE 13.5 The sheep eye — coronal section**
Refer to the color photo gallery.

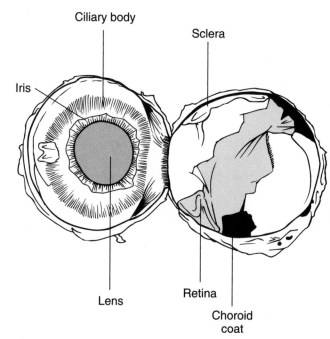

EXERCISE 2
Selected Eye Function Tests

Discussion

Clear, sharp vision involves more than light rays striking the rods and cones. It involves control of pupil size so that appropriate amounts of light enter the eye; changes in the shape of the lens to ensure that light rays are properly bent to strike receptors; **binocular vision**, whereby images from a particular object fall on corresponding points of the retinas of both eyes simultaneously; and integrity of the optic nerves and visual cortex.

Changes in the size of the pupil are reflex responses and were considered in Unit 12. Changing the shape of the lens so that light rays from different distances converge on the retina is called **accommodation**, and this process is very important in focusing as the eyes view objects at varying distances. Light rays coming from close objects are more divergent than those from distant objects and must, therefore, be bent more to bring them to focus on the retina. To achieve this, the ciliary muscles contract, pulling the ciliary body forward, thus releasing the tension on the suspensory ligaments. As a result the lens bulges and gains more refracting power.

We see one object rather than two, as a result of integration, in the occipital lobe of the cerebrum, of the input from both eyes. The eyes move so that the images of an object fall on corresponding points of both retinas.

The visual fields of each eye are slightly different because each eye looks at an object from a slightly different angle. This leads to depth perception.

When viewing close objects, the eyes turn toward each other. This is called **convergence**. The nearer the object, the greater the degree of convergence required to maintain single vision. Work in pairs to perform the following tests.

A. Blind Spot Distance

Materials

White index cards; marker pen; meter stick.

Procedure

1. On an index card print, in dark ink, a plus sign and a circle, each about 1 cm (½ in.) in diameter and about 5 cm (2 in.) apart from each other; plus sign on left, circle on right.
2. Hold the card, at eye level, 45 to 50 cm (18 to 20 in.) in front of your face. Cover your right eye.
3. Focus continually on the circle with your left eye and gradually, in small increments, move the card toward your face until the circle disappears. This distance is your **blind spot** distance.

4. Have your partner measure the distance from the card to your face and record the distance in the Observations section.
5. Repeat the procedure for the right eye, but this time focus on the plus sign.
6. Switch with your partner and repeat the procedure and record the results.

Observations

Blind spot distance:

	Student 1	Student 2
a. Left eye	_____	_____
b. Right eye	_____	_____

Conclusions

- Explain why the plus sign disappears in the blind spot test.

  _____
  _____
  _____
  _____
  _____

- Were the values obtained the same for both eyes? Explain.

  _____
  _____
  _____
  _____
  _____

- Were your values about the same as your partner? Explain.

  _____
  _____
  _____

B. Near-Point Accommodation

Materials

An index card; marker pen; meter stick.

Procedure

The closest distance at which an object appears to be in focus is called the **near point**. Determine your nearpoint of accommodation.

1. Print a distinct letter on an index card and hold it at arm's length from the body.
2. Have your lab partner hold the meter stick under your eye, parallel to the floor.
3. Gradually bring the card toward you along the meter stick until the letter becomes fuzzy.
4. Move the card away until it becomes perfectly clear. This is your near-point distance. Record the distance from the eye in the Observations section.
5. Repeat the procedure for your other eye.
6. The relation between age and near point is tabulated in the Conclusions section.

Observations

Near point:

	Student 1	Student 2
a. Left eye	_____	_____
b. Right eye	_____	_____

Conclusions

AGE (YEARS)	10	20	30	40	50	60
Near point (Inches)	2.9	3.5	4.5	6.7	20.6	32.8
(Centimeters)	7.4	8.9	11.4	17.0	52.3	83.3

- Should the values you obtained be considered normal? Discuss.

  _____
  _____

- Why do you think the near point changes gradually during the five decades of life, and change so drastically during the 50–60 age range?

  _____
  _____
  _____
  _____
  _____
  _____
  _____

C. Accommodation

Materials

A pencil.

Procedure

1. Holding the pencil at arm's length in front of the body, focus on the eraser and have your lab partner observe your eyes as you move the pencil toward yourself.
2. Again hold the pencil at arm's length and focus on the eraser, but this time gently press one eyeball out of line while focusing on the pencil.
3. Record what happens in the Observations section.

Observations

a. Activity of eyeballs as pencil is moved closer

_____

b. Vision without pressure

_____

c. Vision with pressure

_____

Conclusions

• What is near-point accommodation?

_____

• Is the near-point accommodation the same for both eyes? Discuss.

_____

_____

_____

• What happened to your vision when you pushed one eye out of line? Why did this occur?

_____

_____

_____

_____

D. Visual Acuity

Materials

A Snellen eye chart; index card.

Procedure

1. Stand (6.1 m) (20 ft) from the eye chart while your lab partner stands next to the chart to check your accuracy.
2. Visual acuity numbers are determined at the end of the line of print. Record the numbers next to the line of smallest print you can read.
3. Hold the index card over the right eye to block vision in that eye. (Keep the covered eye open.)
4. Read the lines of the Snellen chart as indicated by your lab partner. Note the distance markings at the end of the line.
5. Repeat the process for the other eye and record your visual acuity in the Observations section.
6. Read the chart with both eyes open. Record visual acuity.

Observations

a. Vision in right eye _____

b. Vision in left eye _____

c. Vision in both eyes _____

Conclusions

• Based on a 20/40 reading from the Snellen chart, what does 20/200 mean for "vision in the left eye"?

_____

_____

_____

• Does the Snellen chart indicate what "normal" vision is? Explain.

_____

_____

- As the lens loses its ability to change shape, the near point of vision recedes and ability to see close objects becomes impaired. Explain why hypermetropia (far sightedness) often occurs as an individual ages.

_____

_____

_____

E. Astigmatism

Materials

An astigmatism chart; index card.

Procedure

1. Stand about (3 m) (10 ft) away from the chart and cover one eye as in Part D of this exercise.
2. Observe the lines of the wheel and note if any are out of sharp focus or of different intensity.
3. Repeat the procedure for the other eye and record the results in the Observations section.
4. Observe the wheel with both eyes open. Record the results.

Observations

a. Appearance of wheel with left eye open

_____

b. Appearance of wheel with right eye open

_____

c. Appearance of wheel with both eyes open

_____

Conclusions

- Astigmatism results when there is unequal curvature of portions of the refractive media of the eye. List, in sequence, the refractive media of the eye starting with the medium the light rays strike first.

_____

_____

_____

F. Color Blindness

Materials

Ishihara color charts.

Procedure

1. Have your lab partner hold the plates about 3 ft (1 m) in front of you.
2. Pick out the "colored" figure in each plate within a few seconds of observing it.
3. Record the results in the Observations section.

Observations

a. Number of mistakes _____

b. Description of mistakes _____

_____

_____

Conclusions

- Color blindness results from a deficiency or absence of one or more of the cone types. Based upon your understanding of how the visual pigments work, explain this disorder.

_____

_____

_____

_____

❖ Quick Quiz 1

Match the numbered items in A with the most appropriate lettered phrase in B.

A	B
_____ 1. ciliary body	(a) fibrous tunic
_____ 2. transparent "window"	(b) vascular tunic
	(c) retina
_____ 3. uvea	(d) posterior chamber
_____ 4. fovea centralis	(e) anterior chamber
_____ 5. tapetum lucidum	(f) tears
_____ 6. vitreous humor	
_____ 7. lacrimal apparatus	
_____ 8. rhodopsin	

❖ The Ear

Background

The ear contains receptors for sensing phenomena of two different types. One group plays a role in hearing and the second group in maintaining equilibrium or balance. The receptors for hearing are stimulated by mechanical energy (sound waves) and are able to convert this into the electrical energy of nerve impulses. The nerve impulses travel over sensory neurons in the statacoustic (VIII) cranial nerve to the temporal lobe of the cerebrum, where they are interpreted as meaningful sounds. The receptors of the ear are thus auditory receptors (modality), mechanoreceptors (type of stimulus), or exteroceptors (surface location).

In the role of maintaining equilibrium, the ear has receptors that are stimulated by changes in the position of the head. Such changes stimulate receptors that transmit impulses over cranial nerve VIII to the cerebellum, which in turn initiates impulses to maintain the body's equilibrium. Thus, these receptors are balance receptors, mechanoreceptors, and exteroceptors.

EXERCISE 1
Anatomy of the Ear

Discussion

The **external ear** is an air-filled area that basically acts as a passageway for sound waves; the **middle ear** is an air-filled chamber that transmits the sound waves to the inner ear; the **inner ear** is a series of fluid-filled compartments that contain the receptors for hearing and equilibrium. Parts of the external ear, the middle ear, and inner ear are situated within the **temporal bone**.

Consisting of an elastic cartilage framework covered by skin, the **auricle** or **pinna** is the most prominent feature of the external ear. It serves to gather sound waves and direct them to the **external auditory canal**, which tunnels through the temporal bone. This canal is lined with skin and has special glands that produce a waxy material called **cerumen**, which serves a protective function. The auditory canal directs sound waves toward the **tympanic membrane**, the boundary between the external and middle ear. This membrane vibrates and transmits the sound waves to the middle ear, which in turn transmits them to the inner ear. The middle ear is a small air-filled cavity within the temporal bone (**petrous portion**) and is lined with a **mucous membrane**. Extending across the cavity of the middle ear are three bones or **ossicles**: the **malleus**, the **incus**, and the **stapes** which transfer sound energy from the air of the external ear to the fluid of the inner ear. The "foot plate" of the stapes fits into the **oval window**, one of two openings in the thin, bony wall that separates the middle and inner ear. The second opening is called the **round window**.

Equalization of the pressure on the internal side of the tympanic membrane is accomplished by the **Eustachian tube**. The opening of this tube connects the cavity of the middle ear with the **nasopharynx**, and the mucous membrane lining of both is continuous.

A system of intercommunicating bony channels within the temporal bone forms the inner ear and is referred to as the **bony labyrinth**. Suspended within the bony labyrinth is a series of fluid-filled membranous sacs referred to as the **membranous labyrinth**. The fluid in the sacs is called **endolymph**. The areas between the membranous labyrinths also contains a fluid called **perilymph**. These labyrinths are divided into three areas: the cochlea, the vestibule, and the semicircular canals. The **cochlea** is that part of the inner ear concerned with hearing; the **vestibular apparatus** and the **semicircular canals** have the receptors for static and dynamic equilibrium, respectively.

Resembling a snail's shell, the cochlea consists of a bony spiral canal around a central bone core called the **modiolus**. Inside, the cochlea is divided into three **canals**: the **scala vestibuli** and **scala tympani**, which are part of the bony labyrinth, and the **cochlear duct**, which is the membranous labyrinth running through the center of the cochlea and filled with endolymph. The **basilar membrane** forms the floor of the cochlear duct and houses the receptors for hearing, the **organ of Corti**. The organ of Corti is made of a series of epithelial cells extending through the cochlear duct. It consists of **hair cells**, which are the receptors for auditory stimuli, and supporting cells. Forming a canopy over the hair cells is a gelatinous membrane called the **tectorial membrane**. As sound waves strike the tympanic membrane, it starts to vibrate. This results in the vibration of the malleus which, in turn, is transmitted to the incus and ultimately to the stapes. Movement of the foot of the stapes against the oval window sets up waves of motion in the perilymph of the scala vestibuli. Movement of the perilymph sets the **vestibular membrane** in motion which, in turn, creates waves in the endolymph. This causes the basilar membrane to vibrate, which in turn causes displacement of the hair cells of the organ of Corti against the tectorial membrane. This mechanical event initiates the development of an action potential (in the receptor cells) that is transmitted over the cochlear division of the **eighth cranial nerve** to the cerebrum.

Within the vestibular apparatus are two small membranous sacs, the **saccule** and the **utricle**, which contains hair cells that project into the inner region of the sac. Covering the hair cells is a gelatinous material containing crystals of calcium carbonate referred to as **otoliths**. When the head is pulled down, the otoliths descend and brush against the hair cells, generating sensory impulses over the **vestibular branch** of the eighth cranial nerve. These sensory impulses provide information concerning the position of the head in relation to the pull of gravity, or **static equlibrium**. The semicircular canals are three bony structures that house the receptors for **dynamic equilibrium**. The **superior**, **posterior**, and **lateral canals** are situated at right angles to each other and are thus able to detect *any change* in the movement of the head. At the base of each canal is an enlargement, the **ampulla**, which contains the hair cells covered by a gelatinous mass called the **cupula**. Movement of the head sets the fluid inside the canals in motion; this bends the hairs of the hair cells and initiates impulses, which are in turn relayed to the brain via the vestibular branch of the eighth cranial nerve.

Materials

Models and diagrams of human ear; human skull.

Procedure

1. Using Figures 13.6 through 13.8 and any additional references available as guides, examine the human ear model. Locate the three divisions: external, middle, and inner.
2. Identify the pinna, external auditory meatus, and tympanic membrane.
3. Identify the malleus, incus, stapes, oval window, round window, and Eustachian tube.
4. Identify the cochlea, vestibule, and semicircular canals, and the ampullae of canals and both branches of cranial nerve VIII.
5. Review the structures of the temporal bone of the skull that relate to the ear: petrous portion, external auditory meatus and internal auditory meatus, and mastoid air cells.
6. Identify on the organ of Corti model the scala vestibuli and scala tympani, cochlear duct, basilar membrane, tectorial membrane, vestibular membrane, perilymph, endolymph, hair cells, and supporting cells.
7. On the accompanying diagram label: hair cells, tectorial membrane, nerve fibers, support cells, and basilar membrane.

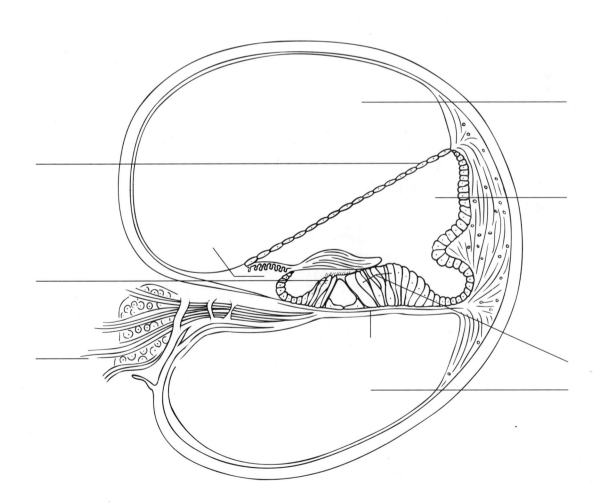

■ **FIGURE 13.6 The human ear, three divisions**

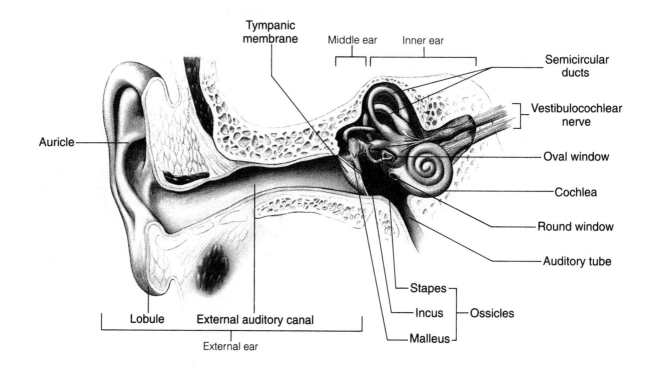

■ **FIGURE 13.7 The inner ear**

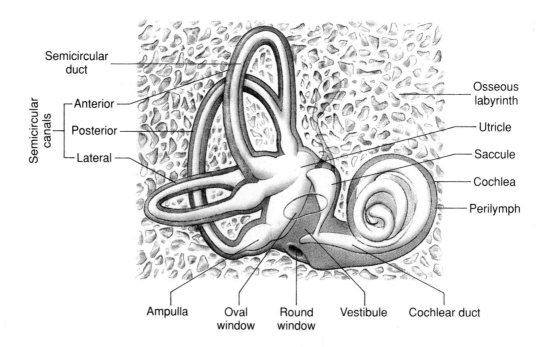

■ **FIGURE 13.8 The cochlea**

(a) Section through cochlea. (b) Magnified cross section of one turn of cochlea.

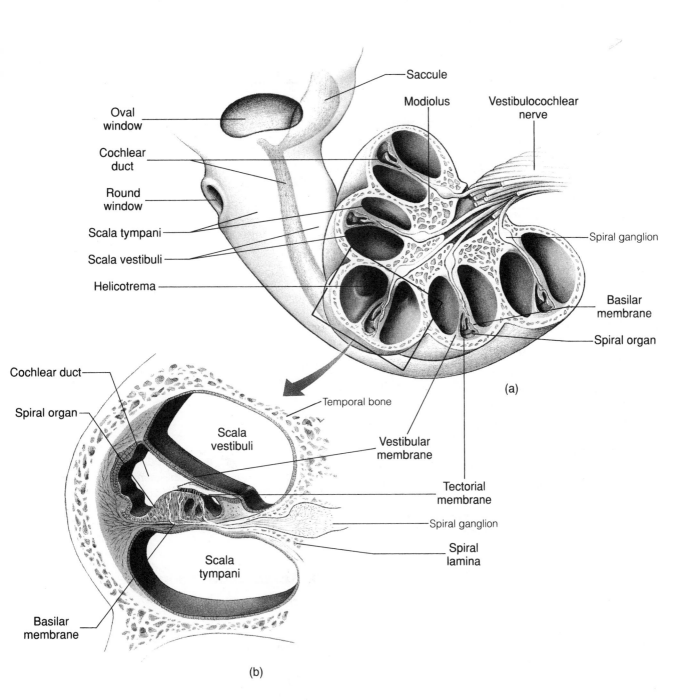

Conclusions

- Based upon your knowledge of the anatomy of the ear, what symptoms would be anticipated in an individual suffering from a middle ear infection? Explain.

  _____

  _____

  _____

  _____

- Why do throat infections often result in an ear infection?

  _____

  _____

  _____

EXERCISE 2
Selected Ear Function Tests

Discussion

Loss of hearing can be caused by many different conditions, among which are conduction deafness, nerve deafness, and central deafness. In **conduction deafness** there is interference with the transmission of sound vibrations to the inner ear. The reason can be as simple as having wax impacted on the eardrum or a more serious condition such as fusion of the ossicles. When damage to the cochlea or eighth cranial nerve has resulted in loss of hearing, the condition is called **nerve deafness**. **Central deafness** involves hearing centers of the brain.

A. Transmission of Sound

Materials

A tuning fork.

Procedure

1. Holding the tuning fork by the handle (not the prongs) rap the prongs on the heel of your hand to set the fork vibrating.
2. Perform the **Weber test** for hearing loss by placing the handle of the vibrating tuning fork in the center of your forehead. Record how you perceive the direction of sound to be in the Observations section.
3. Perform the **Rinne test** by holding the handle of the tuning fork on the mastoid process. When you no

longer hear the sound, move the fork so it is in front of the auditory meatus. Record the results.

Observations

a. Hearing perception in the Weber test:

  _____

  _____

b. Hearing perception in the Rinne test:

  Mastoid process _____

  _____

  In front of meatus _____

  _____

  _____

Conclusions

- Mr. X can hear the vibrating tuning fork when it is held on the mastoid process but not when it is held in front of the auditory meatus. What type of deafness does Mr. X have?

  _____

- Mrs. Y's hearing is profoundly impaired. She cannot hear any sound when the stem of the tuning fork is held between her teeth. What kind of deafness is this?

  _____

B. Localization of Sound

Materials

A tuning fork.

Procedure

In this procedure, work with your lab partner as a subject and then reverse roles.

1. Have your partner, with eyes closed, hold the vibrating tuning fork in different positions around the head (as listed in the Observations section) at a distance of at least 12.7 cm (5 in.).
2. Note how your partner is able to localize the sound. Ask where the sound is coming from.

Observations

a. In front of head

_____

b. Behind head

_____

c. To left of head

_____

d. To right of head

_____

Conclusions

● Explain your results in light of the data obtained.

_____

_____

_____

_____

_____

C. Balance

Materials

Masking tape.

Procedure

1. Put a straight line about 3 m (10 ft) long on the floor, using masking tape.
2. Walk this straight line heel to toe, arms at your side. Record the results in Table A.
3. Walk the line again, this time close your eyes. Record the results.
4. Now have your lab partner spin you around for 30 sec and then try to walk the line again, eyes open. Have your lab partner observe the movements of your eyes and body movements and record in Table A.

Observations

■ **TABLE A Balance**

CONDITION	ABILITY TO WALK LINE		
	GOOD	FAIR	POOR
Arms at side			
Eyes closed			
After rotation			

Conclusions

● Why was your ability to walk the line different in each of the three conditions? Explain.

_____

_____

_____

● The jerking movement of the eyes after rotation is called **nystagmus**. Based upon your understanding of the sense of balance, how can you account for the movement?

_____

_____

_____

❖ Quick Quiz 2

Match the items in column A with the terms in column B.

	A		B
_____	**1.** Eustachian tube opening	(a)	outer ear
		(b)	membranous labyrinth
_____	**2.** Organ of Corti	(c)	middle ear
_____	**3.** endolymph	(d)	basilar membrane
_____	**4.** cupula	(e)	gelatinous mass
_____	**5.** external auditory canal		

Gustatory and Olfactory Sensation

Background

Specialized **gustatory receptors** are called **taste buds** and are located in raised areas of the tongue called the **papillae**. **Olfactory receptors** are neurons located in the mucosal epithelium lining the upper part of the nasal cavity. Both types of receptors are stimulated by certain kinds of chemicals. For the sense of taste, the chemicals must be in solution, and for the sense of smell they must be in a vapor or gaseous state. The receptors for taste and smell are referred to as gustatory and olfactory receptors (modality), chemoreceptors, and exteroceptors.

EXERCISE 1
Taste Sensation

Discussion

Although some taste receptors are distributed on the palate and upper throat, the majority are found on the tongue. Each taste bud consists of specialized neuroepithelial cells that have hairlike projections protruding from the free border into the mouth through a **taste pore**. Chemicals in solution stimulate these cells and impulses are conducted over cranial nerves V, VII, IX and X to the brain. The area of the brain responsible for interpretation of taste is in the cerebral cortex, at the base of the postcentral sulcus near the lateral fissure.

Most of the foods we eat are mixtures of taste-stimulating compounds. The various "tastes" we can differentiate are combinations of four **primary tastes**: sweet, sour, salty, and bitter. It is thought that certain taste buds respond to a particular primary taste, but to some extent are capable of being stimulated by the other primary tastes. There appears to be a pattern of distribution on the tongue of the different taste buds. It is generally recognized that certain types of compounds "produce" characteristic taste sensations. The sweet taste is produced by nonionized organic compounds and the salty taste is associated with inorganic salts. Acids produce the sour taste, while the bitter taste is elicited by compounds like alkaloids and long-chain organic substances.

A. Identifying Primary Tastes

Materials

Sterile cotton applicator sticks; paper cups; solutions of glycerol (10% by volume); quinine (commercial preparaion of quinine water); sodium nitrate (10%); vitamin C (one 500-mg table dissolved in 25 ml of water), and an unknown solution supplied by the instructor.

Procedure

Work with a lab partner.

1. Moisten a sterile applicator stick with the glycerol solution.
2. Dab the moist stick to your partner's tongue at the apex; repeat on the sides and middle posterior portions of the tongue.
3. Note which taste is elicited and where the taste is the "strongest."
4. Repeat the procedure for each of the remaining solutions, being sure to rinse the mouth thoroughly between applications.
5. Record in the Observations section what taste is elicited and where it is strongest. On the diagrams to the right, record the area of the tongue affected.
6. Test the unknown; note where it is tasted and which of the four categories it falls under. Record your results and check your conclusion with the instructor.

Conclusion

● Which of the solutions produced a sour taste?

_____

_____

_____

_____

● Is it "easy" to distinguish sour from bitter?

_____

_____

B. Adaptation

Materials

A sterile cotton applicator stick; paper cups; saccharin solution (one tablet dissolved in 250 ml of water).

Procedure

1. Moisten a sterile applicator stick with the saccharin solution.
2. Dab the stick on the regions of the tongue as in the Procedure for Part A of this exercise.

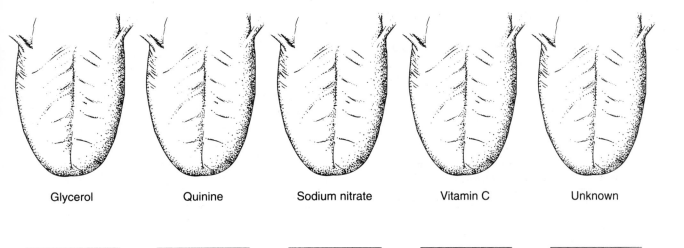

| Glycerol | Quinine | Sodium nitrate | Vitamin C | Unknown |

_____ _____ _____ _____ _____

Taste sensation

3. Note the taste sensation, and the region where it is the strongest.
4. Wait for a few minutes and note the taste sensation and region where it is strongest.

Observations

a. Initial taste

_____

b. Region of concentration

_____

c. Taste after few minutes

_____

d. Region of concentration

_____

Conclusions

• Explain the results obtained in Part B of this exercise with respect to adaptation time.

_____

_____

_____

• Do you think that saliva plays a role in taste? Explain.

_____

_____

_____

_____

EXERCISE 2
Olfactory Adaptation

Discussion

The **olfactory cleft** is an area between the nasal septum and the superior turbinate, which contains the receptors for smell. The receptor cells are small bipolar neurons with hairlike processes projecting toward the nasal cavity. These processes respond to substances inhaled into the nose. Classification of odors has not been successful, probably because of the subjective nature of the attempts.

Olfactory receptors are quite sensitive—they can be stimulated by very weak odors. They are also very adaptable—persistent odors soon go unnoticed, yet when a different odor is introduced into the environment, it is readily detected.

Materials

A clove of garlic; (mild) perfume; solid air freshener or room deodorizer; petri dishes; watch.

Procedure

1. Cut a garlic clove in half and place the pieces in a petri dish; hold the dish about 15 cm (6 in.) below the nostrils.
2. Breathe in and out normally and time how long it takes for the odor of garlic to "disappear." Record the results in the Observations section.
3. Repeat the procedure using a small amount of perfume in another petri dish.
4. Freshen the air with air freshner.

5. Place the two petri dishes on the desk and seat yourself in front of them.
6. Time how long it takes for the odor of the garlic and/or perfume to become noticeable. Which odor appears first?

Observations

■ **TABLE A Olfactory Adaptation**

	DISAPPEARANCE (min)	REAPPEARANCE (min)
Garlic odor		
Perfume odor		

Conclusions

• What do you think the term olfactory fatigue means?

_____

_____

_____

• How do you think air fresheners work?

_____

_____

_____

EXERCISE 3
Olfactory and Gustatory Interaction

Discussion

Many of the flavors we recognize are really a combination of taste and olfactory stimulation: the sense of taste is greatly influenced by the sense of smell.

Materials

Equally thin slices of apple and onion; equal sized strips of carrot and celery; paper cups.

Procedure

Work with a lab partner.

1. Close your eyes, pinch your nose shut and have your lab partner place a slice of apple or onion on your tongue. Close your mouth. Record your ability to identify the food in the Observations section.
2. Rinse out your mouth.
3. Close your eyes and this time without closing off the nose, repeat the procedure. Record the results.
4. Working now with the carrot or celery, repeat steps 1 though 3. Record the results.

Observations

■ **TABLE B Olfactory and Gustatory Interaction**

	COULD	/	COULD NOT
Ability to recognize onion/apple Nose closed	_____	/	_____
Ability to recognize onion/apple Nose open	_____	/	_____
Ability to recognize celery/carrot Nose closed	_____	/	_____
Ability to recognize celery/carrot Nose open	_____	/	_____

Conclusions

• Explain the results obtained in this exercise.

_____

_____

_____

• Why are many foods "tasteless" when an individual has a cold?

_____

_____

_____

Blood

OVERVIEW

In this unit you will examine blood under the microscope and perform several hematological tests, so that you will become thoroughly familiar with the characteristics of blood and be able to identify its many components.

OUTLINE

See the safety precautions for rules dealing with the use of body fluids.

All Background and Discussion material is based on human blood. Exercises use horse or domesticated dog blood.

 ## Blood Cell Counts

Background

Blood is a fluid connective tissue that provides a medium for circulating substances throughout the body. These substances include respiratory gases, electrolytes, nutrients, hormones, and metabolic wastes. In addition to its function as a transport medium, blood plays a role in the body's defense mechanisms, in the maintenance of an optimum pH of body fluids and in the distribution of body heat.

Numerous components of blood can be identified but a simple view divides these components into two aspects: the liquid medium called **plasma** and the cells and cell fragments referred to as **formed elements**. See Figure 14.1. The combination of these components provides blood with certain characteristics. Normally, whole blood has a viscosity about four times greater than water, and its specific gravity (SG) is 1.055 to 1.065. The average pH is 7.4 and the color ranges from bright scarlet to dark red, depending on the level of oxygenation. Venous blood, because of its poor oxygenation, is dark red, whereas arterial blood is bright red.

■ **FIGURE 14.1 Blood: Formed elements**

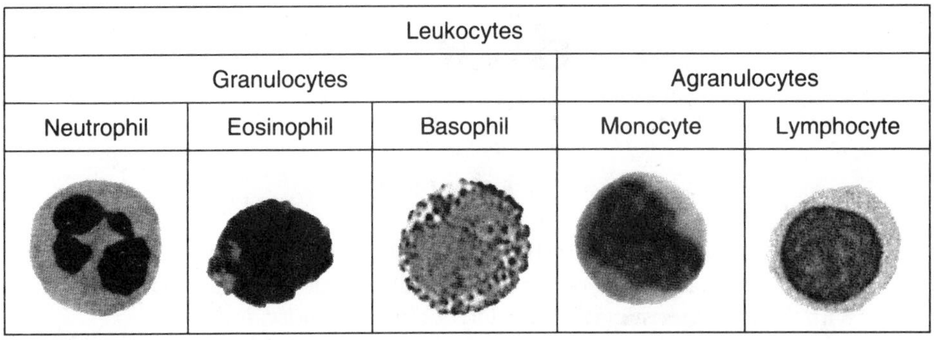

Leukocytes				
Granulocytes			Agranulocytes	
Neutrophil	Eosinophil	Basophil	Monocyte	Lymphocyte

EXERCISE 1
Total Red Cell Count

The following blood experiments are performed on horse blood.

Discussion

One of the formed elements of blood is the red blood cell (RBC), or **erythrocyte**. It is the smallest of the whole cells, with a diameter ranging from 6 to 9 μm. Erythrocytes have no nuclei and are shaped like biconcave discs (think of a ping-pong ball with the center recessed on both sides). These are the most numerous of the formed elements, the average number being less for women than for men. The average erythrocyte count for adult women is 4.6 million ($\pm$500,000) per cubic millimeter, while for adult men the average is 5.4 million ($\pm$600,000) per cubic millimeter. In children, depending on the age, the range is 4.5 to 5.1 million per cubic millimeter. **Polycythemia** is an abnormal increase in erythrocytes by more than 2 to 3 million per cubic millimeter. **Anemia** is a decrease in the oxygen-carrying capability of blood, which can be caused, among other things, by a reduction in erythrocyte number.

To determine cell counts a special counting chamber called a **hemocytometer** is used. Engraved on the hemocytometer is a grid (Figure 14.2). After diluting the blood sample with special reagents, a small quantity of blood is put in the **counting chamber**. After counting cells in the appropriate square, a mathematical calculation is done to determine the total cell count of the sample. This calculation is needed because (1) the sample of blood is diluted, and (2) the sample used is less than a cubic millimeter. For red cells, the formula is:

total cells per millimeter =
    cells counted in 5 small squares $\times$ 50 $\times$ 200

The "50" factor is needed to bring the total volume of 0.2 mm³ used up to 1 mm. The "200" factor is needed to adjust for the amount by which the solution has been diluted.

Materials

A hemocytometer and cover slip; RBC diluting pipette and tubing; RBC diluting fluid (isotonic saline); clean soft cloth; hand counter; pipette shaker; gauze pad; microscope; blood sample.

Be sure to review safety precautions before handling fresh or preserved specimens.

Procedure

1. Be sure that the hemocytometer and cover slip are clean and dry. Use distilled water and soft cloth for this.
2. Practice steps 3 through 7 with distilled water until you are comfortable performing these techniques.
3. Draw a drop of blood up into the RBC diluting pipette to the 0.5 mark. If you exceed this mark, touch the tip of the pipette to a clean gauze pad to draw the blood back to the 0.5 mark.
4. Dilute the sample with isotonic saline fluid by drawing the fluid into the pipette to the 101 mark. Be careful not to draw the solution past this mark.
5. Place the pipette in a shaker and mix for 2 min. If manual mixing is performed, hold the pipette with your finger on one end and thumb on the other end and rotate the pipette back and forth for 2–3 min.
6. After the blood and diluting fluid have been thoroughly mixed, discharge about one third of the fluid from the bulb by touching the tip of the pipette to a gauze pad.

7. Place the cover slip on the hemocytometer counting chamber and "charge" the chamber. To do this, place a drop of the pipette fluid in the chamber by touching the tip of the pipette to the edge of the cover slip. DO NOT OVERFILL THE CHAMBER.

8. Place the hemocytometer on the microscope stage centered under the low-power objective.

9. Focus with low power and reduce the light so that the cells and lines of the grid are in focus.

10. Using high power, count the red blood cells in the areas of the grid marked in Figure 14.2a with "R." To ensure consistency and accuracy and to avoid duplicate counting, *count the cells that touch the lines on the left and top of each square and those within the squre.* A total of five squares should be counted as indicated in Figure 14.2. Record the results in the Observations section.

11. Multiply the total number of cells in the five square by 10,000 to determine the total number of erythro-cytes per cubic millimeter. As previously noted in the Discussion section, $10,000 = 50 \times 200$. Record the results.

Observations

Square 1   _____

Square 2   _____

Square 3   _____

Square 4   _____

  Total   _____

_____ $\times$ 10,000 = _____ RBC/mm³

Note: Horse blood has a high (7–11 million/mm³) red count, but the cells are microcytic.

■ **FIGURE 14.2 The hemocytometer**
(a) Grid. (b) "Charging" the hemocytometer.

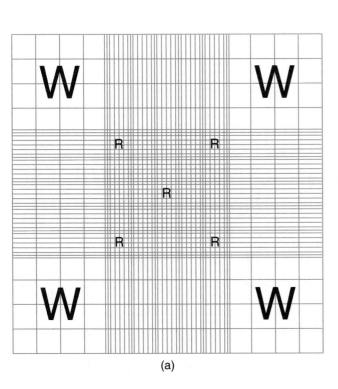

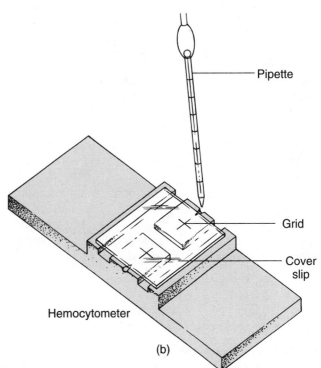

(a)   (b)

Conclusions

• Would an RBC count of 6 million for an adult human female be normal? Why?

_____

_____

_____

_____

_____

• The mathematical calculations of cell number times 10,000 is to correct for certain factors. What two factors need to be "corrected" in the calculation?

_____ and _____

• What changes in RBC count would be expected for the following people?

   a. Big Al from Denver _____·____

   b. Roseann, a runner _____

   c. Joe, a leukemic patient _____

   d. Lily, who has a heavy menstrual flow

_____

EXERCISE 2
Total White Cell Count

Discussion

White blood cells, (WBCs) or **leukocytes** are the least numerous and most varied of the formed elements. There are actually several different leukocytes, which are distinguished in the next exercise. Although each type has its own characteristics and functions, the group as a whole are **nucleate**, as well as being less numerous and much larger than red blood cells. Because they can leave blood and function in the tissues against foreign material, there may be variations in the blood cell count between 5000 and 10,000 per cubic millimeter. A count increased above 10,000 per cubic millimeter is referred to as **leukocytosis** and is a normal response to an infection. Leukocyte counts below 5000 are abnormal and are referred to as **leukopenia**. Changes in the count are characteristic of certain acute and chronic diseases. To determine cell counts, a chamber called a hemocytometer is used.

Materials

A hemocytometer set; lens paper; WBC diluting fluid, see Appendix C; WBC diluting pipette and tubing; hand counter; pipette shaker; gauze pad; microscope; animal blood sample; 70% alcohol swabs; lancet.

   Be sure to review safety precautions before handling fresh or preserved specimens.

Procedure

The procedure in this exercise is similar to that for Exercise 1.

1. Follow steps 1 through 4 for the procedure in Exercise 1, being sure to use a WBC pipette.
2. Dilute the sample with WBC diluting fluid by drawing the fluid into the pipette to the 11.0 mark.
3. Mix the contents of the pipette, then discard about one third of the fluid in the bulb by touching the tip of the pipette to a gauze pad. Proceed with filling the hemocytometer chamber as in Exercise 1.
4. Using low power, count the number of cells in the four squares of the grid marked with a "W." See Figure 14.2a. Employ the same pattern as for the red blood cell count. Record numbers in the Observations section.
5. Multiply the total number of cells from these four squares by 50 to determine the total number of leukocytes per cubic millimeter. One "W" square contains $0.1 \ mm^3$ of blood. Four squares contain $0.4 \ mm^3$ of blood. A correction factor of 50 is used to bring the volume up to 1 ml ($\times$ 2.5) and to account for the dilution ($\times$ 20). Record.

Observations

Square 1   _____

Square 2   _____

Square 3   _____

Square 4   _____

  Total   _____

_____ $\times$ 50 = _____ WBC/mm^3

Note: Horse and dog blood have WBC values similar to human.

Conclusions

• Why is the correction factor (50) less for the white cell count than for the red cell count?

_____

_____

_____

• What would a human leukocyte count of 4500/mm³ be referred to as?

_____

• What does the WBC diluting fluid do to the blood sample? Why is this necessary?

_____

_____

_____

_____

_____

EXERCISE 3
Differential White Cell Count

Discussion

As indicated previously, there are several different types of leukocyte. The two major groupings are based on the granulation of the cytoplasm: **granulocytes**, which have a granular cytoplasm, and **agranulocytes**, which do not. Granulocytes occur in three types: **neutrophils, basophils**, and **eosinophils**, which are distinguished on the basis of the cell's response to stains. The agranulocytes are **monocytes** and **lymphocytes**. A cubic millimeter of blood contains between 5000 and 10,000 of these five types of leukocytes, and normally there is a definite proportion of each type within that total number. The most predominant leukocyte is the neutrophil, followed by the lymphocyte, the monocyte, the eosinophil, and the basophil. Fluctuations in the ratio of the various leukocytes is indicative of certain abnormal conditions. For example, an increase in the number of eosinophils is observed in allergic reactions, whereas increased levels of certain hormones like adrenocorticoids can depress the level of eosinophils.

Materials

Prepared human blood smears.

Procedure

1. Examine the smear with oil-immersion magnification.
2. Observe the differences between the erythrocytes and a typical leukocyte. Note the difference in size, shape, staining characteristics, and nucleus. Sketch these cell types in space provided in Table A.
3. Use a definite pattern (e.g., scan across the slide and back) in observing the slide and count the different types of leukocytes. Be sure to count to at least 100 total leukocytes.
4. Record the results of the differential count in the Observations section.

■ **TABLE 14.1 White Human Blood Cells**

CELL TYPE	FUNCTION	ADULT SOURCE	DESCRIPTION	STAINING CHARACTERISTICS
Neutrophil (60–70%)	Phagocytosis	Bone marrow	12–14 μm diameter; multilobed nucleus (usually 2–5 lobes) fine granular cytoplasm	Nucleus, dark blue; cytoplasm, pink; granules, liliac
Eosinophil (1–3%)	Can phagocytize and release antihistamines	Bone marrow	9 μm diameter; bilobed nucleus; cytoplasm contains large uniform granules	Nucleus, dark blue; cytoplasm, pink; granules, reddish orange
Basophil (0.5–1%)	Releases heparin, histamine, and serotonin	Bone marrow	12 μm diameter; "continuous" S-shaped nucleus; cytoplasm contains heavy, irregular granules	Nucleus, blue; cytoplasm, pink; granules, dark blue-black
Lymphocyte (20–25%)	Antibody formation; may be phagocytic	Lymphoid tissue (lymph nodes, spleen, thymus)	Large cells about 12 μm diameter; small cells about 8 μm diameter; large round nucleus; cytoplasm clear	Nucleus, dark blue; cytoplasm, light blue
Monocyte (3–8%)	Phagocytosis	Lymphoid tissue and bone marrow	20 μm diameter; oval- to kidney-bean-shaped nucleus; cytoplasm clear	Nucleus, dark blue; cytoplasm, light blue

Observations

■ **TABLE A   Blood Cells**

RBC	Neutrophil	Eosinophil
Basophil	Lymphocyte	Monocyte

Neutrophils   _____ _____ %

Eosinophils   _____ _____ %

Basophils   _____ _____ %

Lymphocytes   _____ _____ %

Monocytes   _____ _____ %

    Total cells counted   _____ _____ %

Conclusions

• Look up "band cells" and define.

    _____

    _____

    _____

• What do band cells look like?

    _____

    _____

    _____

• Were any of these forms observed in the smears you looked at?

    _____

• What would account for this?

    _____

    _____

    _____

• What were the tiny, pale blue, irregularly shaped structures with purple granules on the stained preparation?

    _____

    _____

• Are these structures complete cells?

    _____

• What is the normal number of these formed elements in blood?

    _____

 Hematocrit, Sedimentation Rate, and Hemoglobin Content

Background

In a given sample of blood, there should be not only a certain number of formed elements but also a certain amount of plasma. As indicated earlier, plasma is the fluid portion of blood of which 8 to 10% is solutes and colloids and 90 to 92% is water. Although the solutes and colloids play important roles in homeostasis, the major function of the plasma portion is to act as a solvent and suspending medium for the numerous substances found in blood, including the formed elements.

If blood is prevented from clotting and is allowed to stand in a tube, after awhile the formed elements will settle to the bottom of the tube, leaving the clear plasma in the upper region. Unless the erythrocytes hemolyze and release the hemoglobin, the plasma will appear as a clear, straw-colored fluid.

EXERCISE 1
Microhematocrit Determination

Discussion

A useful estimate of the oxygen-carrying capacity of blood is the **hematocrit** or **packed cell volume** (PCV), since the PCV is a measurement of the percentage of red blood cells in a given volume of blood. When a sample of blood is centrifuged, the red blood cells are forced to the bottom of the tube where they are packed together, leaving the "lighter" plasma in the upper portion of the tube. Often a thin "whitish" layer can be observed between the clear plasma and red cell mass. This represents the leukocyte fraction and is called the **buffy coat**.

Normal human hematocrit values generally average around 45%, with adult male values of 47.0 ± 7.0%, and adult female values of 42.0 ± 5.0%. There is an apparent correlation between the hematocrit and hemoglobin content, and usually the PCV is about three times the grams of hemoglobin per 100 ml of blood.

Materials

Heparinized capillary tubes; capillary tube sealant (e.g., Seal-Ease); microhematocrit centrifuge and reader; blood sample; beaker.

Procedure

1. Drop between 1–5 drops of normal saline in your sample of blood and mix by inverting the tube, to make it "anemic."
2. Fill a capillary tube with blood.
3. Holding your index finger over one end of the tube, seal the opposite end with sealant.
4. Place the tube into the centrifuge with the sealed end against the outer circumference of the wheel. (If a tube is spun singly, balance the centrifuge by putting an empty capillary tube opposite the filled tube. If other students are centrifuging at the same time, distribute the tubes evenly around the centrifuge.)
5. Secure the inside cover and fasten the outside cover.
6. Set the time for 4 min and allow to spin.
7. Remove the tube from the centrifuge and determine the hematocrit by placing the tube in the microhematocrit reader. Instructions for operating the reader are located on the instrument.
8. Record your results.

Observations

a. What was the hematocrit value? _____

b. Was a "buffy" layer of leukocytes visible?

_____

c. Was the plasma clear? _____

Note: Since horse RBC size is small, the red count will *not* be in concordance with the hematocrit. Dog blood *will* be in concordance.

Conclusions

● What would account for a plasma that was pink rather than clear?

_____

_____

● What conditions would produce an elevated hematocrit?

_____

_____

_____

_____

_____

● What conditions could result in a decreased hematocrit?

_____

_____

_____

_____

EXERCISE 2
Sedimentation Rate

Discussion

In looking at the blood slides under the microscope, you may have noticed that the erythrocytes often appeared "stuck" together at their broad surfaces like a stack of coins. These stacks, or **rouleaux**, are a natural occurrence in slowly flowing blood. The tendency to form rouleaux contributes to the **sedimentation rate**, which is the rate at which red blood cells settle to the bottom of a tube.

A sample of **citrated blood** (blood that will not clot) is allowed to stand in a tube for an hour. The distance the erythrocytes fall in that time is the sedimentation rate. Several procedures are available and sedimentation rate differs depending on the method employed. Using the **Wintrobe method**, the rate for males is 0–6.5 mm/hr and for females 0–15 mm/hr. The rate for the **Westergren method** for males is 0–15 mm/hr and for females 0-20 mm/hr.

A. Wintrobe Method

Materials

Wintrobe tubes and stand; dry Heller-Paul-Wintrobe mixture (see Appendix C); test tubes; Pasteur pipettes; clock; balance; blood sample.

   Be sure to review safety precautions before handling fresh or preserved specimens.

Procedure

1. Place dry Heller-Paul-Wintrobe mixture in a test tube, 2 mg for every milliliter of blood.
2. Add 1–5 ml blood.
3. Mix well by inverting gently.
4. Draw 1 ml of the mixture into a Pasteur pipette and fill a Wintrobe tube from the bottom (to exclude air bubbles) up to the 100–mm mark.
5. Place the tube in a vertical position in the Wintrobe stand. Note the time.
6. Read the length of the plasma column after 1 hr. This is the sedimentation rate. Record in Table B.

B. Westergren Method

Materials

Westergren tubes and stand; 3.8% citrate solution (see Appendix C); test tubes; 5-ml pipettes; blood sample.

Be sure to review safety precautions before handling fresh or preserved specimens.

Procedure

1. Place 0.6 ml of the citrate solution in a test tube.
2. Add 2.4 ml of blood to the tube and mix.
3. Draw this citrated sample of blood up to the 200-mm mark of the Westergren tube.
4. Place the tube in a vertical position in the Westergren stand.
5. Read the length of the plasma column after 1 hr. This is the sedimentation rate. Record in Table B.

Observations

■ **TABLE B   Wintrobe versus Westergren Method**

	SEDIMENTATION RATE (mm/hr)
Wintrobe	
Westergren	

Conclusions

• What does citrate do to a sample of blood? (See Figure 14.3)

_____

_____

• Why is it used for sedimentation rate procedures?

_____

_____

EXERCISE 3
Determination of Hemoglobin Content

Discussion

Erythrocytes consist of an inner framework of lipids and proteins called the **stroma**, bounded by an outer cell membrane. Also inside the red cells, held to the stroma, is the conjugated protein **hemoglobin**, consisting of the iron-containing pigment **heme** and a protein portion called **globin**. Since hemoglobin carries more than 90% of the oxygen transported by blood, the level of hemoglobin can be of clinical significance. The hemoglobin content is measured in grams per 100 ml of blood and the values differ for men, women, and children. Normal values for males are 16.0 ± 2.0 g/100 ml, and for females the values are 14.0 ± 2.0 g/100 ml. The range for children is 11.2–16.5 g/100 ml, with younger children having higher vales.

Reduction in the amount of hemoglobin or number of RBC's is referred to as **anemia**. Obviously if the number of erythrocytes is reduced, hemoglobin levels will be below normal. This type of anemia can be caused by different conditions. If the cause is a reduction in blood cell production the condition is called **aplastic anemia**; if it is caused by a deficiency in **intrinsic factor** (a protein in the stomach lining that promotes absorption of vitamin B_{12}) the condition is referred to as **pernicious anemia**. If the production of erythrocytes is faulty, resulting in fragile cells that rupture, the condition is called **hemolytic anemia**. If the cells are very small, the condition is called **microcytic anemia**. Still another cause may be reduced iron supplies, producing **iron-deficiency anemia (hypochromic anemia)**. An excess production of RBC's is called **polycythemia** with red cell counts between 8–11 cells/mm^3.

A. Hemoglobinometer/Colorimeter

Materials

A hemoglobinometer (Hb-meter) kit or colorimeter; blood sample.

Procedure

1. Read the directions for operating the hemoglobinometer or colorimeter that is available. (There are sev-

eral models, and you should become familiar with the instrument you will use.)

2. Once you have learned how to operate the instrument, obtain a sample of blood.
3. Following the directions for use of the instrument, determine the grams of hemoglobin per 100 ml of blood and record the results.

Observations

_____ g Hb/100 ml

Note: Horse and dog hemoglobin content will be similar to human.

Conclusions

• Why would hemoglobin values differ for men and women?

_____

_____

_____

_____

_____

• If each gram of hemoglobin carries 1.34 ml of oxygen, what is the total oxygen-carrying capacity of hemoglobin in average adult males?

_____

_____

_____

B. Tallquist Method

Materials

Tallquist paper and scale; blood sample.

Procedure

1. Obtain a section of Tallquist paper from the booklet.
2. Obtain a drop of blood.
3. Place the blood on the Tallquist paper.
4. When the drop of blood loses is glossy appearance, place the sample under the color comparison chart in the back of the booklet. Match the sample with the closest color.
5. Record the results.

Observations

_____ g Hb/100 ml

Conclusions

• Which of the two procedures (hemoglobinometer or Tallquist method) of this exercise is the more accurate? Explain.

_____

_____

_____

 ## Coagulation Time and Bleeding Time

Background

Blood is a vital fluid, and excessive loss of this fluid can disrupt homoestasis and even be fatal. There are three aspects to preventing the loss of blood. They are the three stages of **hemostasis**. The first is **vascular spasm** where the vessel constricts and minimizes blood loss. The second is the formation of a **platelet plug** where the fragments called platelets adhere to a torn vessel and plug it up. The third is the actual formation of the clot; these stages are diagrammed in Figure 14.3.

The **clotting process** is a protective mechanism that prevents the loss of blood when a vessel is injured. It is a series of chemical reactions that take place very quickly in a specific sequence, since blood contains all the materials that are needed for clotting to occur. This mechanism works in conjunction with the platelet plug and constriction of the blood vessel.

The three general stages in the clotting process are outlined in Figure 14.3. The diagram illustrates in simplified form the sequence of events in the clotting mechanism. It is important to remember, however, that several other factors have been identified and play a role in the process. These factors are summarized in Table 14.2. Assessing the time it takes for a clot to form (the **coagulation time**), and the time it takes for a small wound to stop bleeding (the **bleeding time**), can be of clinical significance in preparation for surgery.

The overall process of coagulation or clot formation involves chemical interactions that ultimately convert the soluble plasma protein **fibrinogen** to the insoluble form **fibrin**. The time it takes to form a clot can vary depend-

■ **FIGURE 14.3 Clotting process**

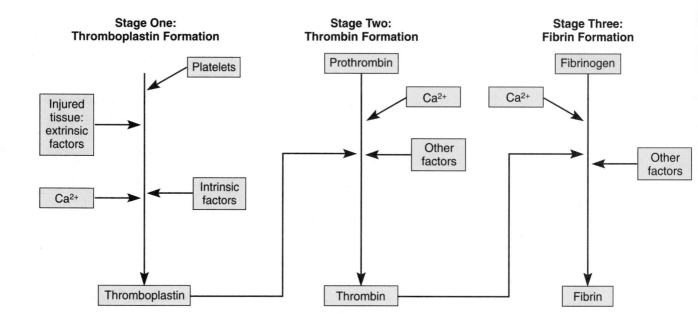

Stage One: **Thromboplastin Formation**   **Stage Two:** **Thrombin Formation**   **Stage Three:** **Fibrin Formation**

ing on factors such as size of the wound and level of clotting factors in blood. When assessing clotting time by laboratory procedures, the time also varies depending on the method used.

EXERCISE 1
Coagulation Time (Clotting Time)

Discussion

Among the methods used to assess clotting time are the **tube method** and **slide method**. If a blood sample is placed in a test tube and incubated in a 37°C water bath, the normal clotting time ranges from 5 to 8 min. When the capillary tube method is carried out, the normal range is 2 to 6 min. Normal clotting time, using the slide method, ranges from 2 to 8 min. This test roughly measures the capability of the intrinsic clotting mechanism. This mechanism involves factors XII, XI, IX, VII, Ca^{2+}, and the platelet factor. They initiate the activation of factors X and V, which in turn, along with calcium ions, convert prothrombin into thrombin and the subsequent events outlined in Figure 14.3.

These tests cannot be performed on shipped blood which contains an anticoagulant. However, the Discussion and Procedures are presented so that you may see what steps are involved.

A. Capillary Tube Method: Clotting Time

Procedure

1. Obtain free-flowing blood.
2. Note the time the blood first appears. This is the "beginning time."
3. Holding the capillary tube at a slight downward slant, place one end of the tube in the drop of blood and let capillary action draw the blood through, completely filling the tube.
4. Sixty seconds after the first appearance of blood on your finger, break off a small section of the tube (less than 1 cm) and observe whether strands of coagulated blood are visible between the two sections of glass.
5. Continue to break off small sections of the tube at 30-sec intervals until coagulated threads are visible.
6. If you were actually performing this procedure, you would record the total coagulation time.

B. Slide Method: Clotting Time

Procedure

1. Obtain a blood sample and note the "beginning time" of the experiment as in Part A.

■ **TABLE 14.2 Human Clotting Factors**

FACTOR NUMBER	NAME	TYPE OF FACTOR AND ORIGIN
I	Fibrinogen	A plasma protein produced in the liver
II	Prothrombin	A plasma protein produced in the liver
III	Tissue thromboplastin	A complex mixture of lipoproteins containing one or more phospholipid substances; released from damaged tissues
IV	Calcium ions	An ion in the plasma that is acquired from the diet and from bones
V	Proaccelerin (labile factor, accelerator globulin)	A plasma protein produced in the liver
VI	Not utilized	
VII	Serum prothrombin conversion accelerator (stable factor, proconvertin)	A plasma protein produced in the liver
VIII	Antihemophilic factor (antihemophilic globulin)	A plasma protein produced in the liver
IX	Plasma thromboplastin component (Christmas factor)	A plasma protein produced in the liver
X	Stuart factor (Stuart-Prower factor)	A plasma protein produced in the liver
XI	Plasma thromboplastin	A plasma protein produced in the liver
XII	Hageman factor	A plasma protein
XIII	Fibrin stabilizing factor	A protein present in plasma and in platelets

2. Place the drop of blood on a glass slide.
3. At 30-sec intervals draw the dissecting needle or pin through the drop of blood. Gently draw the instrument up away from the blood and observe when coagulated strands of blood adhere to the point.
4. If you were actually performing this procedure, you would record the total coagulation time.

Conclusions

• Where are all the clotting factors made in the body?

_____

_____

_____

_____

_____

EXERCISE 2
Bleeding Time

Discussion

When a small blood vessel is cut, thrombocytes (platelets) accumulate at the injured site, forming a temporary "plug," which helps arrest the bleeding. Bleeding time is the result of the formation of this **platelet plug**, the **constriction** of capillaries, and a working **clotting cascade**. Clotting occurs when both intrinsic factors (described in Exercise 1 of this section) and extrinsic factor operate. Extrinsic factor is released from damaged tissue and activates factor VII. Active factor VII initiates the activation of factors X and V, and the process continues as for the intrinsic clotting mechanism. Although the size of the vessel and severity of the injury will affect bleeding time, the normal range is 3 to 6 min.

Procedure

1. Clean the shaved portion of a laboratory animal's skin with an alcohol swab and let it air dry.
2. Puncture the skin with a sterile lancet and note the time the first drop of blood appears.
3. Blot this drop with a piece of filter paper and continue to blot the drops of blood that form every 30 seconds until the wound does not stain the paper.
4. Calculate the bleeding time.
5. Record the results.

Observations

Bleeding time _____

Conclusions

• How does the concept of coagulation time differ from that of bleeding time? Which is the longer of these times and by how much?

_____

_____

_____

_____

• Based on your knowledge of the clotting mechanism, name several disease conditions that might contribute to an increased bleeding time and indicate why.

_____

_____

_____

_____

_____

_____

• Look up the definition of hemophilia and relate it to the clotting mechanism.

_____

_____

_____

_____

_____

_____

◆ Blood Type

Background

Erythrocytes may contain various protein-carbohydrate complexes on the surface of the cell membrane. Which of the various complexes are present on the red blood cell membrane is determined genetically and is of clinical significance when an individual requires a transfusion. If an individual receives the "wrong type" of blood, **agglutination** or clumping of erythrocytes will result because of a reaction between the complexes on the red cells (which are foreign to the recipient's body) and protein antibodies found in the plasma. The complexes on the cell membrane are called **agglutinogens** and the antibodies in plasma are called **agglutinins**. The reaction between agglutinogens and agglutinins is similar to the reaction between antigens and antibodies in that it is an attempt to counteract introduction of foreign molecules into the body. The agglutinogen-agglutinin interaction is highly specific. Many of the various antigenic factors that may be on the red cell membrane have been classified into groups. The two primary groups for which typing is routinely done are the ABO system and the Rh system.

NOTE: Exercises 1 and 2 may be carried out simultaneously.

EXERCISE 1
The ABO System

Discussion

Blood type is determined by the agglutinogen present on the erythrocyte membrane. Within the ABO system two agglutinogens have been identified, A and B. Depending on genetic composition, an individual may have both agglutinogens, one of the agglutinogens, or neither agglutinogen. Those who have the A agglutinogen have **type A** blood; those who have the B agglutinogen have **type B** blood; those who have both A and B agglutinogens are **type AB** blood, and those who lack both agglutinogens have **type O** blood. Normally, cells do not agglutinate within an individual's own bloodstream because the plasma does not contain the agglutinin *corresponding* to the agglutinogen; rather, it contains the *opposite* agglutinin. In the ABO system the agglutinins are **anti-A** and **anti-B**. Thus, individuals with type A blood have anti-B in their plasma; those with type B blood have anti-A in their plasma, and those with type AB blood have neither anti-A nor anti-B. Since individuals with type O blood lack both agglutinogens, they have both agglutinins (anti-A and anti-B) in their plasma. Table 14.3 summarizes the characteristics of each blood type.

In most transfusion reactions, the agglutinins in the recipient's plasma agglutinate the donor's cells. Thus, if a person who is type A receives blood from a type B individual, the anti-B of the recipient will agglutinate the type B cells received in the transfusion. The danger of a severe **transfusion reaction** comes when the agglutinated cells **hemolyze** (i.e., release hemoglobin), which can accumulate in and ultimately block kidney tubules, causing renal shutdown. The major blood groups in which severe transfusion reactions occur are the ABO and Rh groups.

Although typing will not be done on student blood, the Discussion and Procedure sections are presented so you may know what is involved.

Materials

Blood sample; glass slide; wax pencil; mixing sticks; anti-A, anti-B, and group O sera.

■ TABLE 14.3 Human ABO Blood Types

BLOOD TYPE	AGGLUTINOGEN ON RBC	AGGLUTININ IN PLASMA
A	A	Anti-B
B	B	Anti-A
AB	A and B	Neither anti-A nor Anti-B
O	Neither A nor B	Anti-A and anti-B

Procedure

1. Draw a vertical line down the center of a clean glass slide, separating it into two sections. In the upper left corner mark an A, and in the upper right corner mark a B.
2. Place two drops of anti-A serum on the left side of the slide and two drops of anti-B serum on the right side.
3. Obtain blood.
4. Add a drop of blood to the anti-A serum and to the anti-B serum by placing the drops of blood next to the antiserum and then mixing them together with sticks. *Use separate sticks for mixing.*
5. Gently rotate the slide for about 1 min and observe whether agglutination occurs.
6. If no agglutination occurs, check results by adding a drop of blood to group O antiserum on a separate slide.
7. If you were actually performing this procedure, you would determine and record the number of type A, type B, type AB and type O.

Conclusions

• What type of cell produces the agglutinins?

_____

_____

EXERCISE 2

The Rh System

Discussion

Another system of agglutinogens that may be present on the red blood cell membrane is the Rh system. It includes agglutinogens of at least eight different types, which are referred to as **Rh factors**. Certain of these are more antigenic than others and are likely to cause transfusion reactions.

In the Rh system the anti-Rh agglutinins are not present in the plasma until after an individual has received massive exposure to one or more of the antigenic factors, as in a transfusion. This is unlike the ABO system in which individuals are continually producing anti-A and/or anti-B because of continued exposure to the antigens from the environment. People who have the potent Rh agglutinogen are classified as Rh positive. Those who lack these antigenic factors are said to be Rh negative.

Materials

Blood sample; anti-Rh serum; warming box; glass slide; mixing stick.

Procedure

1. Place two drops of anti-Rh serum on a clean glass slide.
2. Add a drop of blood to the antiserum.
3. Place the slide on the lighted warming box and rotate the slide back and forth for about 2 min.
4. Check the slide for agglutination and record the results in the Observations section.
5. If in doubt about agglutination, examine the slide under the microscope.
6. Record the number of Rh+ and Rh- students in the class.

Conclusions

• What agglutinins are present in your blood for the Rh system? Explain.

_____

_____

_____

_____

• Statistics show that 85% of the general population is Rh positive. Does the number of Rh positive students in the class agree with the distribution for the general population?

_____

_____

_____

_____

- Look up the definition of RhoGam. How is it used?

_____

_____

_____

_____

- Look up the causes and symptoms of **erythroblastosis fetalis**. Why is the infant transfused?

_____

_____

_____

_____

_____

_____

The Anatomy of the Heart

OVERVIEW

In this unit you will have a chance to examine the anatomy of the sheep heart, which is quite similar to the human heart. You will be able to look at the microscopic and macroscopic features which make it a self-excitable, one-way pump in the body.

OUTLINE

Anatomy of the Heart

Background

The heart can be thought of as a four-chambered pump that plays the key role in propelling blood through the body. The circulation of the blood is achieved through the coordination of the activities of cells that are specialized for contraction, self-excitation, and impulse conduction.

Functionally the heart can be thought of as consisting of the **right heart** (carrying poorly oxygenated blood) and the **left heart** carrying well-oxygenated blood), with the lung circuit on the route between these two. Blood enters the heart by way of vessels called **veins** and exits by way of **arteries**. See Figure 15.1.

The right heart receives blood from tissues of the body through the superior vena cava and the inferior vena cava, and pumps the blood to the lungs via the pulmonary artery. The left heart receives blood from the lungs through the pulmonary veins and pumps blood to the body through the aorta. As blood flows into each side of the heart, it enters a receiving chamber, the **atrium**, and flows on into a pumping chamber, the (right or left) **ventricle**, before leaving through the outflow tracts.

EXERCISE 1
Microscopic Examination of Cardiac Tissue

Discussion

The muscle tissue of the heart is heterogeneous. About 90% is contractile tissue with striations, central nuclei, and a highly branched network of cells. As you will remember from Unit 4, it is characterized by **intercalated discs**, which are areas of fusion of the membranes between cells. This fusion allows the contractile tissue of the heart to act as though it were one enormous cell; it acts as though the individual cell membranes were not there. This is possible because of **gap junctions** which allow for communication from one heart cell to another. In addition, the areas of fusion between cells act to facilitate the transmission of tension from cell to cell.

There are two upper chambers, the atria, comprising one of these "enormous cell areas;" the two lower chambers, the ventricles, comprise the other. They are completely separated by connective tissue, part of what is known as the **cardiac skeleton**. The remaining 10% of the muscle tissue of the heart is specialized muscle that is readily self-excitable (although under abnormal circum-

■ **FIGURE 15.1   The heart**

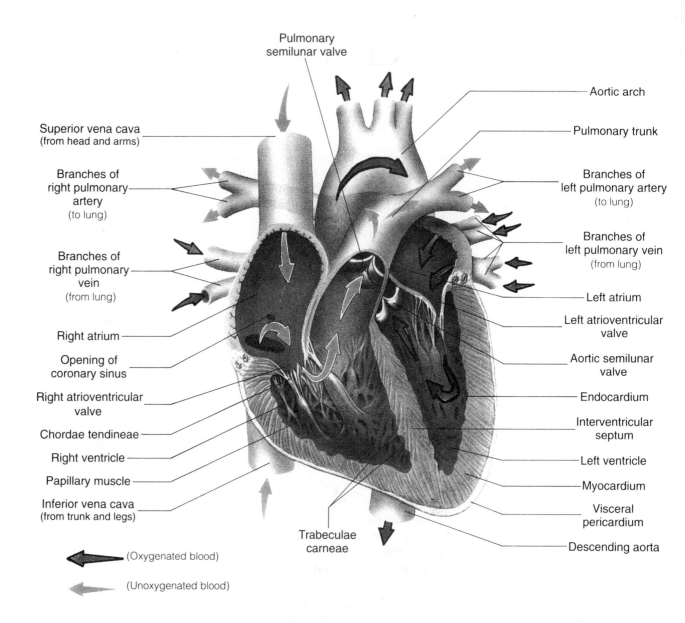

Pulmonary
semilunar valve

Aortic arch

Superior vena cava
(from head and arms)

Pulmonary trunk

Branches of
right pulmonary
artery
(to lung)

Branches of
left pulmonary artery
(to lung)

Branches of
left pulmonary vein
(from lung)

Branches of
right pulmonary
vein
(from lung)

Left atrium

Right atrium

Left atrioventricular
valve

Opening of
coronary sinus

Aortic semilunar
valve

Right atrioventricular
valve

Endocardium

Chordae tendineae

Interventricular
septum

Right ventricle

Left ventricle

Papillary muscle

Myocardium

Inferior vena cava
(from trunk and legs)

Visceral
pericardium

Trabeculae
carneae

Descending aorta

⬅ (Oxygenated blood)

⬅ (Unoxygenated blood)

stances any area of the heart can be self-excitable). It does not contract; rather, it initiates the impulse and conducts it rapidly throughout the heart. You will study the excitatory tissue of the heart in Unit 16.

Materials

A microscope; slides of cardiac tissue (c.s. and l.s.).

Procedure

1. Review the material of Unit 4.

2. Note intercalated discs, position of nucleus, and striations, and compare them to Figure 15.2.

Conclusions

• Look up the rate at which cardiac cells undergo mitosis in the adult human. What can you conclude about the repair of damaged heart tissue?

_____

_____

_____

A. External Anatomy

Major external features of the heart include the epicardium, the auricles, the grooves and the coronary vessels. See Figures 15.4 and 15.5.

The great vessels are the **superior vena cava**, which drains the head, neck, chest and arms, and opens into the right atrium; the **inferior vena cava**, which drains the lower body and opens into the right atrium; the **pulmonary artery**, which carries blood to the lungs from the right ventricle; the **pulmonary veins**, which return blood from the lungs to the left atrium; and the **aorta**, which carries blood out to the body from the left ventricle.

The **auricles** lie on the superior ventral surfaces of the right and left sides of the heart. They are pocket-like structures that most likely act as overflow structures for the atria, since the space inside the auricles is continuous with that within the atria.

The external grooves are markings that denote the separation of chambers within. The **anterior interventricular groove** runs along the ventral surface and marks the separation of right and left ventricles within, by the **interventricular septum**. The **atrioventricular** (or **coronary**) **groove** is a demarcation of atrial and ventricular separation within. Both grooves are embedded with adipose tissue and carry the coronary blood vessels.

EXERCISE 2
The Sheep Heart

Discussion

The sheep heart you are to examine is very similar to the human heart. It is a triangular mass that consists mainly of muscle tissue called **myocardium**. In the body, the base of the triangle is located superiorly, and the apex points inferiorly toward the left leg.

There is a thin layer of **epicardium** on the surface of the muscle mass. This is the same serous membrane that is described as part of the **pericardial** membranes in Exercise 3 (see Figure 15.3). There is also a layer of epithelium, termed **endocardium**, lining the inner surface of the chambers of the heart. The endocardium is continuous with the **endothelium** of the blood vessels. When this tissue is redoubled on itself and anchored into the connective tissue of the cardiac skeleton (see Exercise 1), it functions as **valves** for the heart.

■ **FIGURE 15.2 Cardiac muscle tissue**
(a) Photo. (b) Diagram.

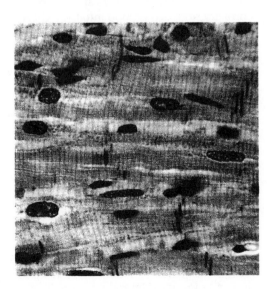

(a)

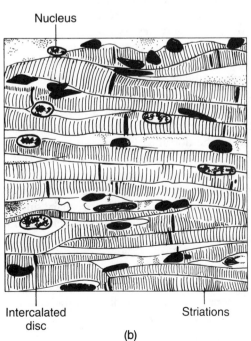

Nucleus

Intercalated disc   Striations

(b)

■ **FIGURE 15.3**   **The pericardial membranes**

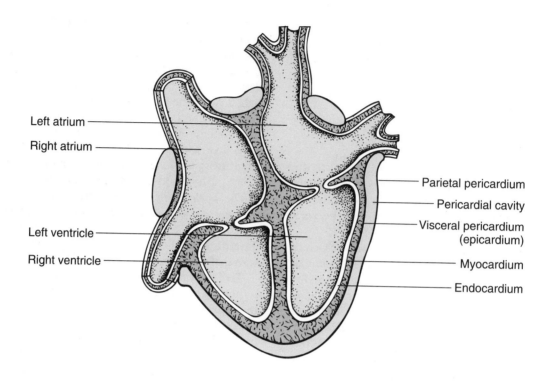

The **coronary blood vessels** are visible running on the external surface of the heart. The muscle of the heart has a nutritive demand as does any muscle tissue in the body. The **coronary arteries and veins** feed and drain the myocardium. There are two main coronary arteries branching off either side of the aorta just as the aorta leaves the heart: the **right and left coronary arteries**. The left coronary artery is a short vessel that divides into the **anterior descending artery** (which runs in the interventricular groove, on the ventral side, toward the apex), and the **circumflex branch** (which runs along the inferior border of the left auricle around to the dorsal side of the heart). The right coronary artery branches off the aorta just opposite the left coronary artery. This single vessel runs along the inferior border of the right auricle and around to the dorsal surface of the heart. The **cardiac veins** run fairly parallel to the arteries. The **great cardiac vein** runs in the anterior interventricular groove and continues toward the dorsal side in the coronary groove as the **coronary sinus**. The coronary sinus receives blood from many smaller cardiac veins and drains into the right atrium. A small amount of blood may flow into the heart directly. A remnant of fetal circulation called the **ductus arteriosus** runs between pulmonary semilunar valve and aorta. This was once a bypass of the pulmonary circuit. It remains as the closed structure, the **ligamentum arteriosum**, in the adult heart.

B.   Internal Anatomy

The study of the heart is best understood by following the path of blood through it. Follow Figure 15.6 as the interior of the heart is discussed.

From the tissues of the body, blood poor in oxygen enters the right side of the heart. The vena cavae and the coronary sinus carry blood into the right receiving chamber of the heart, the right atrium. Overflow of blood may be accommodated by the auricle whose chamber is continuous with that of the atrium.

The **interatrial septum** separates right and left atria and contains another remnant of fetal circulatory anatomy. What was in the fetus a direct opening from right and left atrium, the **foramen ovale**, is a shallow depression, the **fossa ovale**, in the adult heart.

Most of the blood flows (because of a pressure gradient) into the lower right chamber of the heart, the right ventricle. (Only a small amount of blood is *pumped* into the ventricle when the atrium contracts.) At the juncture between atrium and ventricle is the **right atrioventricular valve (right A-V valve)** or the **tricuspid valve**. When the right ventricle contracts, it forces blood up against this valve and shuts it, thus preventing regurgitation of blood into the right atrium. The valves are prevented from being pushed into the atrium by stringlike attachments to the ventricular wall. These are called the

chordae tendineae and insert on the ventricular myocardium at small projections of muscle called **papillary muscles**. The band of muscle running across the right ventricle is called the **moderator band**. This structure may play a role in electrical conductivity through the heart, or in preventing overdistention of the ventricle.

Blood is pumped out of the right ventricle to the lungs through the **pulmonary artery**. Backflow of blood is prevented in the pulmonary artery by the cusps of the **pulmonary semilunar valve**. The pulmonary artery divides into the right and left branches ultimately to a capillary bed in the lung where gas exchange occurs. It returns from the lungs as two double-branched vessels, the **pulmonary veins**, to the left atrium of the heart. Blood in the pulmonary veins and left heart is well oxygenated as a result of this circuit through the lungs. As with the right heart, most of the blood *flows* immediately into the

left ventricle below (by a pressure gradient). A small percentage of the blood is *pumped* into the ventricle from the atrium. At the juncture of the left atrium and left ventricle is the **mitral valve**, or **left atrioventricular valve** (left A-V valve). This is also called the **bicuspid** valve. It functions for the left side of the heart, as the tricuspid does for the right side. See Figure 15.7.

When the left ventricle contracts and pumps blood out of the body, it sends blood into the aorta. Backflow of blood down the aorta toward the ventricle is prevented by the cusps of the **aortic semilunar valve**. The aortic arches give off the paired coronary arteries to the heart and three large vessels to the head and neck in the human: the **brachiocephalic** (innominate) artery, the **left common carotid artery**, and one left **subclavian** artery. The artery continues posteriorly and inferiorly as the descending aorta.

■ **FIGURE 15.4 The sheep heart: exterior views of anterior portion**
Refer to the color photo gallery.

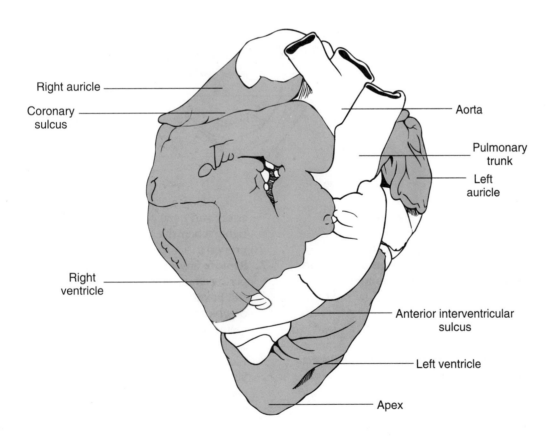

■ **FIGURE 15.5 The sheep heart: exterior views of posterior portion**
Refer to the color photo gallery.

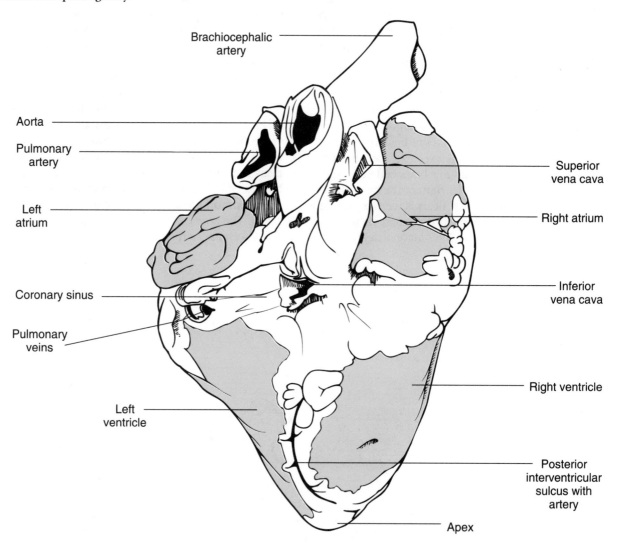

Brachiocephalic artery

Aorta

Pulmonary artery

Left atrium

Coronary sinus

Pulmonary veins

Left ventricle

Superior vena cava

Right atrium

Inferior vena cava

Right ventricle

Posterior interventricular sulcus with artery

Apex

Materials

A sheep heart; dissecting tray and instruments.
  Be sure to review safety precautions before handling either fresh or preserved specimens.

Procedure

Remember that the *right* side of the heart in the dissecting tray is on your *left* side and vice versa.

A. External Anatomy

1. Determine superior (base) and inferior (apical) ends of the heart, right and left sides, and dorsal and ventral orientation before you begin the detailed examination. The interventricular groove and the pulmonary artery run parallel to each other toward your upper right, when the ventral side of the heart faces you properly.

2. Remove any excess fat and identify the layers of the pericardium. Some of the fibrous pericardium may still be attached. The glistening visceral pericardium or epicardium will adhere closely to the myocardium. You will have to peel it off in a small section to see it.

3. Identify the auricles, grooves, and coronary blood vessels. Note the thin-walled coronary sinus in the coronary groove and follow its route into the right atrium.

4. Identify the two thin-walled venae cavae entering the right atrium, and the pulmonary artery exiting from the right ventricle diagonally toward your upper right; the pulmonary veins and their two small paired openings in the left atrium, and the thick-walled aorta exiting from the left ventricle behind the pulmonary

artery. The aortic branches, the brachiocephalic artery, the left common carotid artery, and the left subclavian artery may be seen and should be identified.

5. Note the ligamentum arteriosum betwen the pulmonary artery and aorta. It branches from the pulmonary artery at the level where the artery divides into right and left vessels.

B. Internal Anatomy

Dissect the heart, following the path of blood flow.

6. Turn the heart so that the dorsal side faces you. The large, thin-walled venae cavae are visible, although if these vessels have been cut close to the heart, only their openings into the right atrium will be seen. The superior vena cava runs in a longitudinal direc-

tion, while the inferior vena cava runs transversely in the sheep heart. Insert two dull probes into the venae cavae: they should meet in the right atrium. Carefully cut along the direction of the longitudinally oriented probe; this should give you entrance into the right atrium.

7. Open the walls of the superior vena cava and atrium and observe the smooth lining of the right atrium. Note the entrance of the coronary sinus and observe the fossa ovale in the interatrial wall. The lining of the right atrium is smooth, whereas that of the auricle is ridged. This comblike muscle is called the pectinate muscle.

8. Turn the heart back around so that the ventral side faces you. Make an incision through the pulmonary artery along its length, ending down in the right ventricle. Pull the walls of the right ventricle apart, being careful to observe the **moderator band** running between the right ventricular wall and the in-

---

■ **FIGURE 15.6 The sheep heart: interior view**
Refer to the color photo gallery.

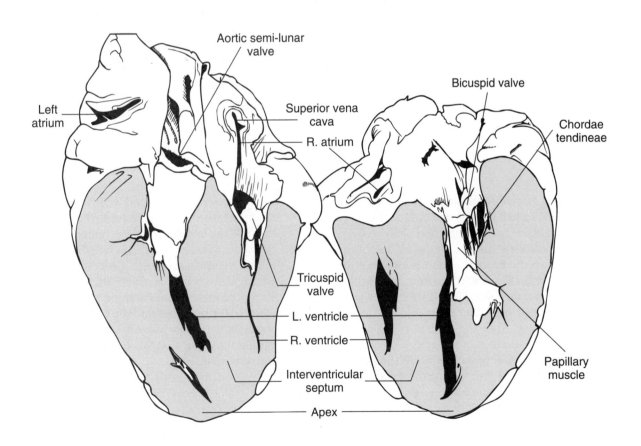

■ **FIGURE 15.7 The sheep heart—superior views**
(a) Valves open. (b) Valves closed.

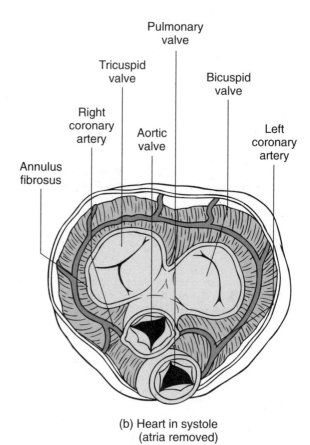

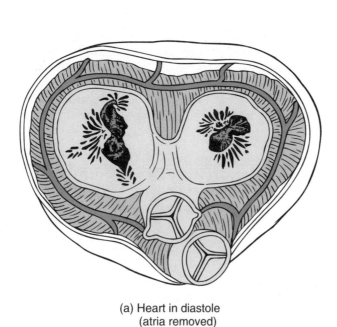

(a) Heart in diastole
(atria removed)

(b) Heart in systole
(atria removed)

terventricular septum separating right and left ventricles. Cut the band if it limits observation. Note the ridges on the ventricle wall, the trabeculae carnae.

9. Observe the tricuspid valve at the upper end of the ventricle. (Turn the heart over and notice that the valve lies between the atrium and ventricle. Return the organ to its ventral side.) Identify the three separate leaflets of this valve, and the stringlike chordae tendineae running from the leaflets to nipplelike papillary muscles on the ventricle wall.

10. Carefully pass the dull probe up inside the pulmonary artery and cut the artery along the length of the probe. Open the artery and observe the pulmonary semilunar valve. Note the three cuplike leaflets on the wall of the artery.

11. Enter the left atrium of the heart by passing a probe into one of the small openings of the pulmonary veins. Cut the lateral wall of the atrium and continue the cut to the apex. Gently open the left heart and observe the two leaves of the bicuspid or mitral valve between atrium and ventricle, the chordae tendineae, the papillary muscles, the trabeculae carnae, and the interventricular septum inside.

12. Now pass a dull probe up through the outlet tract of the left ventricle, the aorta. Observe the aortic semilunar valves, which are identical to the pulmonary semilunar valves from inferior and superior aspects.

13. Observe the openings into the coronary arteries just above the semilunar valve in the aorta. This requires cutting the aorta along its length, close to where it exits from the heart.

Conclusions

• From an external, ventral view, which chambers of the heart occupy most of the area of the surface?

_____

_____

• Is there an abundance of collateral circulation to the myocardium by the arteries and arterioles serving it? Explain.

_____

_____

_____

_____

• Based upon your knowledge of the structure and circulation of blood in the heart, what are the consequences to the myocardium if blood flow in a coronary artery is suddenly blocked?

_____

_____

_____

_____

• What would happen to the flow of blood if the chordae tendineae ruptured?

_____

_____

_____

_____

EXERCISE 3
The Heart *in Situ:* The Fetal Pig

Discussion

The heart is located in the **mediastinal space** between the lungs in the thoracic cavity. It is somewhat obscured by the lobes of the lung, the pericardial sac that encloses it, and the thymus gland, which overlies the superior end of the heart. Figure 15.8 shows the heart in the chest cavity, and it allows an understanding of the relationship of the heart to the major vessels and other thoracic structures.

As illustrated in Figure 15.3 the pericardial sac consists of three layers: the outer fibrous pericardium, the parietal pericardium (fused to the fibrous layer), and the visceral pericardium (the **epicardium**). The parietal and visceral layers form the serous pericardium, which secretes slippery, lubricating serous fluid into the pericardial cavity between them.

Materials

A preserved and injected fetal pig; dissecting tray and instruments.

Be sure to review safety precautions before handling fresh or preserved specimens.

Procedure

1. Cut into the thoracic cavity by making a shallow incision from just below the bottom of the sternum to the top of the sternum. The sternum is cartilaginous so be careful that you do not cut too deep and damage the viscera or vessels below. Be especially cautious at the superior end where the major vessels enter and leave the heart.
2. At the inferior end of the incision, make two lateral cuts. You should be just above the diaphragm. Make similar cuts at the superior end of the longitudinal incision. By gently bending the body wall laterally, you should have two doors that allow entrance into the thoracic cavity.
3. Observe the **lungs, mediastinal cavity**, and **heart** inside the **pericardial sac**. The **thymus gland**, which is large in the fetus, may be seen covering the upper part of the heart. The gland can be observed and removed.
4. Note the **pericardium** around the heart, observe its attachments to the diaphragm below and to the sternum ventrally, and its reflection on itself on the great vessels at the base of the heart. See Figure 15.8.
5. Locate the **base** and **apex** of the heart. Note the **auricles**. (It may require slight lateral rotation of the heart to observe the left auricle.)
6. Identify the **interventricular groove, coronary blood vessels**, and **left and right ventricles** of the heart.
7. Identify the **superior** and **inferior vena cavae**. Trace them each a short distance *away* from their point of entrance into the heart.
8. Locate the **pulmonary trunk** leading from the right ventricle. Trace its pathway to see where it divides into the two **pulmonary arteries** going to the lungs. Also note the short connection between the pulmonary trunk and the aorta. This is the **ductus arteriosus** which allows blood to bypass the lungs in fetal circulation.
9. The **pulmonary vein** can usually be seen from its tributaries in the lungs to its entrance into the left atrium.

10. Identify the **aorta** as it leaves the heart. Locate the arch, major branches to neck and arms, and the beginning of the dorsal aorta.
11. Label the accompanying diagram.

Conclusions

• Did you observe any differences between the heart and accompanying vessels of the sheep and fetal pig? Explain.

_____

_____

_____

_____

The heart: frontal section

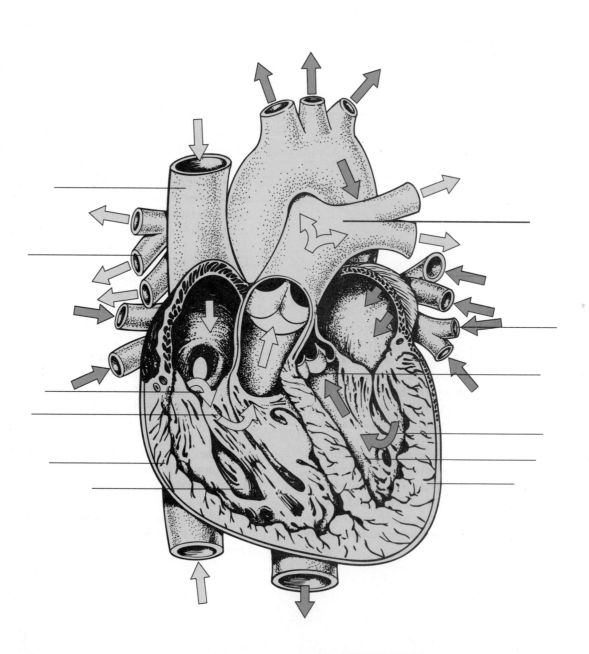

FIGURE 15.8 The thoracic cavity
Refer to the color photo gallery.

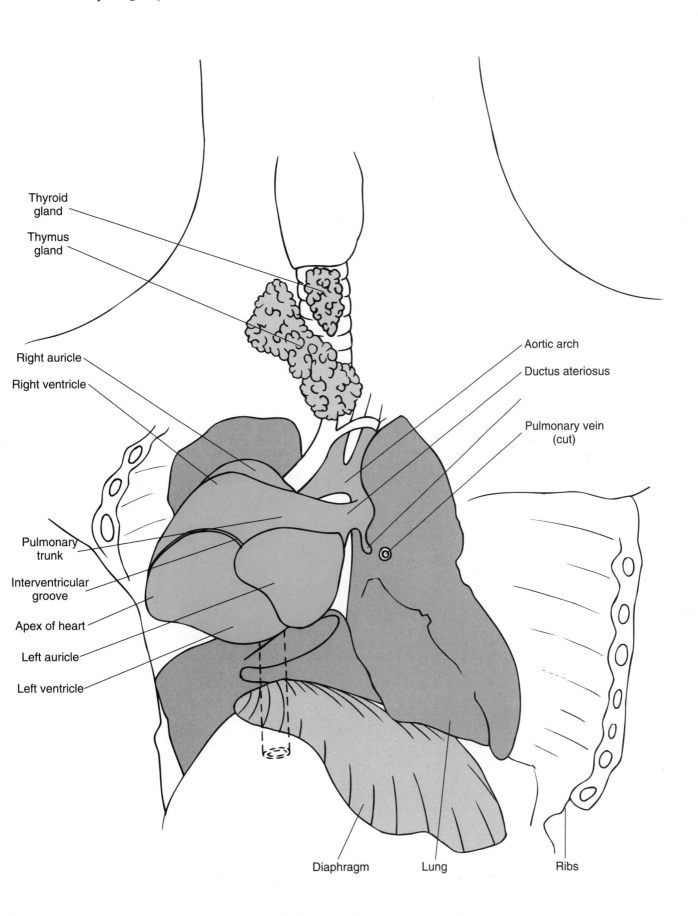

UNIT 16

The Physiology of Selected Cardiovascular Functions

OVERVIEW

You are now ready to examine the regulation of the activity of the heart. In these exercises you will examine the factors which affect the excitatory and contractile tissue of the heart. To complete your picture of heart function you will look at some electrocardiograms, as well as take pulse and blood pressure.

OUTLINE

Physiology of the Cardiovascular System

Exercise 1   Heart Activity: The Frog

Exercise 2   The Electrocardiogram (ECG): The Human

Exercise 3   Blood Pressure and Pulse: The Human

These experiments involve the use of living specimens. Commercial audio-visual materials may be substituted, if appropriate.

 Physiology of the Cardiovascular System

Background

Muscle cells of the heart contract as they perform their function: pumping blood. There are a number of ways to examine this activity. One way is to look at the rate and strength of contraction *in situ*, by monitoring the actual movements of the ventricle itself. Another is to look at the electrical activity generated by the heart each time it beats. Lastly, one can observe the effects of heart beat in the body: pulse and blood pressure.

EXERCISE 1
Heart Activity: The Frog

Discussion

Review the material "Anatomy of the Heart" in the previous unit before going on with this exercise and discussion. The myocardium consists of both non-conducting, excitatory and contractile tissue. The *rate* of heartbeat is determined by how frequently the excitatory tissue depolarizes. The *strength* of heartbeat is determined by how forcefully the contractile tissue responds.

Excitatory Tissue

Just above the entrance to the superior vena cava in the right atrium is a special area of myocardium, the sino-atrial node (SA node). This tissue excites the heart to contract. It is known as the **pacemaker tissue** of the heart. It is part of the excitatory tissue of the heart, which consists of the sinoatrial node as well as a pathway which carries the impulse throughout the entire heart. This pathway is called the **conduction system** of the heart. See Figure 16.1. The whole excitation system of the myocardium makes up less than 1–10% of the myocardium, whereas the contractile tissue makes up the rest of the myocardium.

The special aspect of this tissue is that it is **self-excitatory**. The tissue **spontaneously depolarizes** to threshold level and once threshold is reached, calcium ions flow into the cardiac muscle cells, which contributes to the action potentials developed subsequently by cardiac cells. The special excitatory cells depolarize spontaneously about 70–80 times a minute. Each time they do, an action potential is generated. See Figure 16.2. Once

threshold is reached, the impulse spreads rapidly throughout the entire atrial myocardium, and is transmitted into the ventricular myocardium; contraction now occurs.

On the graph of Figure 16.2, the junction of slope "a" and "b" is when threshold is reached; "c" denotes repolarization.

Contractile Tissue

The strength of cardiac contractile activity is determined by factors which affect the interaction of actin and myosin. Factors that alter the amount of Ca^{2+} that flows into the cell in turn will affect the number of actin-myosin cross-bridges that can form intracellularly, and therefore affect the strength of contraction. Unlike skeletal muscle, cardiac muscle requires an external source of calcium ions to flow into the cell.

Any factor which alters ion flow across the sarcolemma can alter the rate or the strength of contraction. For example, if the potassium flow were increased then

---

■ **FIGURE 16.1 The conducting system of the heart**

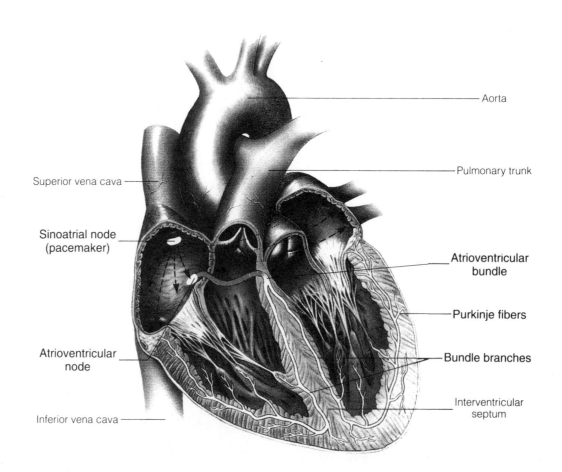

Aorta

Pulmonary trunk

Superior vena cava

Sinoatrial node
(pacemaker)

Atrioventricular
bundle

Purkinje fibers

Atrioventricular
node

Bundle branches

Inferior vena cava

Interventricular
septum

▣ FIGURE 16.2 Electrical activity of excitatory tissue of the heart

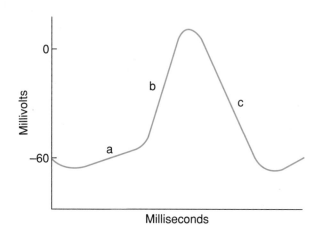

the rate might speed up; if the influx of calcium were increased, then more actin-myosin cross bridges would form and the strength of contraction would be increased.

Materials

A frog (more than one specimen may be necessary); dissecting equipment: instruments, pins, and tray; eye droppers; hypodermic syringes (1 cc); hypodermic needles (22 gauge); very thin cording (or strong thread); 5 small (100-ml) beakers; thermometers; heating tray; paper towels; amphibian Ringer's solution; Ringer's with excess K^+; Ringer's with excess Ca^+; adrenalin, 1–10,000; acetylcholine, 1–10,000; cholinergic blocker; β-adrenergic blocker; various cardiac drugs (see Appendix C) physiograph with myograph B; pencil.

NOTE: If the contractile activity of the heart diminishes greatly in any one step of the procedures that follow, prepare an identical setup with a new frog. If only one specimen is used throughout, follow the order established in the Procedure that follows.

Be sure to review safety precautions before handling fresh or preserved specimens.

Procedure

1. Double pith the frog according to your instructor's directions.
2. Pin the frog on a dissecting tray, ventral side up. Gently lifting only the skin with a forceps, make a longitudinal cut up the midline in the skin; with two transverse cuts, fold the skin back, making sure it does not touch the viscera, which you will soon expose. Then,

being careful not to penetrate the underlying organs, cut up the ventral musculature in a similar manner and pin it back.

3. Note the attachment of the pericardium to the heart and body wall. The frog heart has three chambers: two atria and one ventricle (where blood mixes). The atria are separated from the ventricle by a transverse groove. Tie the cord around the apex of the heart and hook it with a curved pin, and loop the cord onto the arm of the myograph unit which is plugged into the channel amplifier of the physiograph (see Figure 16.3). Adjust the position of the myograph arm making the thread taut so that when the amplifier is turned on there is a distinct recording of the ventricular contraction. Bathe the heart frequently in amphibian Ringer's solution, blotting excess fluid or removing it with a dropper.

4. Adjust the "sensitivity" knob of the physiograph to the highest value, and adjust the "position" knob so that the baseline of the graph falls on a line of the physiograph paper. Adjust the paper speed, and set the timer marker to "1 per second."

5. Monitor each of the following readings for about 15 sec, unless stated otherwise. Label different events directly on the graph with a pencil.

   a. Note the rate and strength of heart beat with room temperature Ringer's. Rate is determined horizontally on the paper and strength is determined vertically. Record in the Observations section.
   b. Drip Ringer's 10°C warmer than room temperature. Blot excess liquid. Note results.
   c. Drip Ringer's 10°C cooler than room temperature. Blot excess liquid. Note results.
   d. Inject 0.05 ml of the adrenalin solution into the liver very slowly. (This allows the drug to enter the portal circulation.) Count rate and assess strength. Note results for 3 to 5 min and record.
   e. With another syringe, inject 0.05 ml of the acetylcholine solution into the liver. Observe for 3 to 5 min and record results.
   f. Repeat step d. Observe for 3 min and record.
   g. Locate the vagus nerve, which runs along the external jugular vein. Tie a loop of thread around the nerve for accessibility, and place electrode tips on the nerve (wrapping wires around the nerve if necessary). Stimulate the nerve, using a 200-mV stimulus for several 5-sec intervals. Record the results.
   h. After normal heart activity has been reestablished, repeat step d, substituting a cholinergic blocking drug such as atropine. Observe for 3 to 5 min and record the results.
   i. Repeat step d, substituting β-adrenergic-blocking drug such as propanolol. Observe 3 to 5 min and record results.

j.   Repeat steps a and d with a freshly prepared frog, substituting Ringer's with Ca^{2+} solution. Observe for 3 to 5 min and record results.

k.   Repeat step d, substituting Ringer's with K^+ solution. Observe for 3 to 5 min and record results.

l.   Remove the heart from the frog and place in a petri dish with Ringer's solution. Note heart rate. Carefully snip the heart along the transverse groove and record any activity.

Observations

	Rate	Strength
a. Heart activity at room temperature	_____/min	_____ squares
b. Heart activity at 10°C above room temperature	_____/min	_____ squares
c. Heart activity at 10°C below room temperature	_____/min	_____ squares
d. Heart activity with adrenalin	_____/min	_____ squares
e. Heart activity with acetylcholine	_____/min	_____ squares
f. Heart activity with adrenalin, after acetylcholine	_____/min	_____ squares
g. Heart activity with vagal stimulation	_____/min	_____ squares
h. Heart activity with cholinergic blocker	_____/min	_____ squares
i. Heart activity with β-adrenergic blocker	_____/min	_____ squares
j. Heart activity with Ringer's and Ca^{2+}	_____/min	_____ squares
k. Heart activity with Ringer's and K^+	_____/min	_____ squares
l. Atria alone	_____/min	_____ squares
m. Ventricle alone	_____/min	_____ squares

Conclusions

• Compare cardiac activity at temperature in a, b, and c. Explain.

_____

_____

_____

_____

_____

• What effect is fever likely to have on the pulse?

_____

_____

• Epinephrine (adrenalin) has the effect of slowing K^+ outflow from the cell. How would this affect the slope of depolarization? Why?

_____

_____

_____

_____

• What does this do to the heart rate?

_____

• How is the length of the period before repolarization affected?

_____

_____

_____

_____

• How does this affect strength? Explain.

_____

_____

_____

_____

• Which autonomic nervous response releases epinephrine into the blood? From what organ, and how is it controlled?

_____

_____

_____

_____

• Acetylcholine is the neurotransmitter of the vagus nerve. What is the effect of acetylcholine on the heart?

_____

_____

_____

• How would acetylcholine affect K^+ outflow?

_____

_____

_____

• Compare the effects of acetylcholine and vagal stimulation to the heart.

_____

_____

_____

■ FIGURE 16.3 Setup for monitoring cardiac activity of the frog

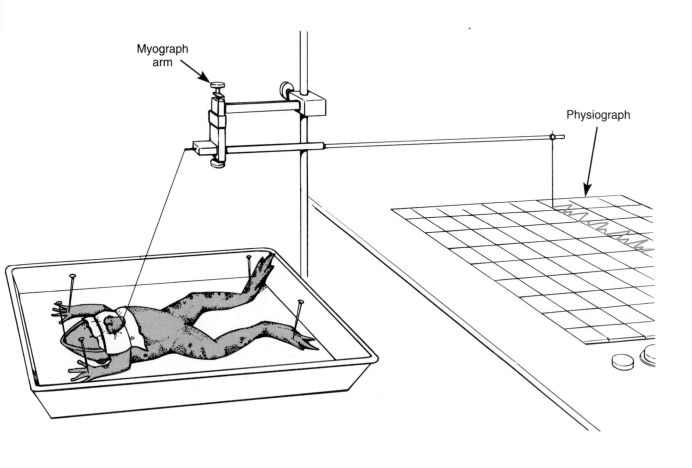

- What could an injection of atropine do to the heart rate? Explain.

  _____

  _____

  _____

  _____

- Compare the effects of propanolol (a β-adrenergic blocker) and isoproterenol (a sympathomimetic drug*) on the heart.

  _____

  _____

  _____

  _____

- How and why is cardiac activity altered with an increase in Ca^{2+} concentration?

  _____

  _____

  _____

  _____

  _____

- In separating the atria and ventricle, we are not allowing the impulse to pass in the normal fashion from atria to ventricle. In other words, the ventricle is no longer stimulated by the normal pacemaker. What happens to the ventricle?

  _____

  _____

  _____

  _____

*One that mimics the sympathetic nervous system.

- Look up what is meant by "ventricular escape"?

_____

_____

_____

EXERCISE 2
The Electrocardiogram (ECG): The Human

Discussion

Before it can contract, a muscle must exhibit an electrical event called *depolarization*. In the heart we normally see depolarization starting at the **pacemaker** in the right atrium, the **sinoatrial (S-A) node**. The impulse rapidly spreads throughout both atria (which then contract) and down toward the ventricles. The impulse is delayed at the **atrioventricular (A-V) node** so that the atria may contract before the ventricles. From the A-V node the impulse speeds down the **A-V bundle** and **bundle branches** to the terminal **Purkinje fibers** embedded in the ventricular myocardial wall. The ventricular muscle depolarizes and the ventricles then contract. See Figure 16.1.

Each depolarization is followed by a *repolarization* along essentially the same path. The electrical changes that occur in the heart are transmitted through the surrounding tissue fluid to the surface of the body where they can be "picked up." By placing recording electrodes at the three "corners" of the heart, a triangle is formed. As one section of the heart depolarizes (or repolarizes) before another, a difference in the electrical potential can be recorded; this is called the **electrocardiogram**, or **ECG**.

There are three standard "waves" or patterns on the ECG. The **P wave** represents **atrial depolarization**; the **QRS wave** represents **ventricular depolarization**; the **T wave** represents **ventricular repolarization**. See Figures 16.4 and 16.5.

ECGs are normally taken with 12 different leads.

Leads I, II, III	Standard limb leads
Leads IV, V, VI	Augmented limb leads
Leads VII-XII	Precordial leads (taken at various places on the ventral chest wall)
Leads II and III	Produce upward deflections in all waves

Materials

A physiograph with AC-DC coupler for electrocardiography; three ECG electrodes; 70% alcohol swabs; electrode paste; scrub pad; ice water and cup; standard ECG tracings of normal and abnormal heart patterns (from a local hospital).

■ **FIGURE 16.4   The three corners of the heart**

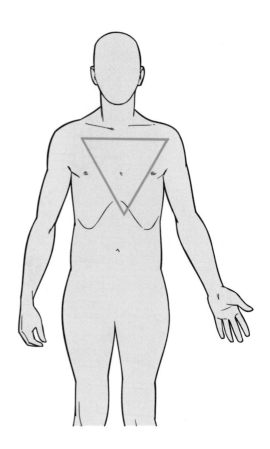

Procedure

1. Have one of the students lie down and relax. Clean with alcohol the areas to which electrodes will be attached. Apply the ECG paste to the skin of the left inside wrist, right inside wrist, and left ankle (medial side). Attach the ECG leads to these areas, and secure the straps firmly, ensuring good skin-electrode contact. The leads should be plugged into the AC-DC coupler channel amplifier, and the channel turned on. Bring the pen-line graph to a set baseline by adjusting the "position" and "balance" knobs. Calibrate the deflections of the recording so that a 1-mV deflection equals a given number of squares on the physiograph paper. Turn the "sensitivity" knob to a low setting, to obtain a clearly defined pattern. Obtain a tracing for about 30 sec. Turn off the channel amplifier and record a typical pattern in the Observations section.

2. Switch the leads on the two wrists and obtain another 30-sec tracing. Turn off the amplifier and record the pattern.

3. Obtain a tracing of the ECG right after drinking *ice cold* (not just cold) water. Record the results.

4. Examine the ECG tracings provided by your instructor and describe the various abnormalities illustrated in them.

■ **FIGURE 16.5 The ECG**

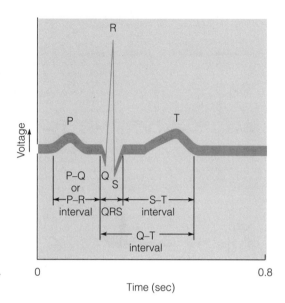

Time (sec)

Observations

Tracing of step 1 ECG

Tracing of step 2 ECG

Tracing of step 3 ECG

Conclusions

- Label the waves on the three ECGs you have recorded. Are all the waves in the same direction? Explain.

- What is the rate of heartbeat in your first tracing?

- Why is the QRS wave often used to count rate?

- How does the third tracing compare to the first? Explain.

- When does atrial repolarization occur?

- Look up the difference between "sinus tachycardia" and "ventricular tachycardia."

- What is the difference in the problem?

- Label and describe the ECGs shown on the next page.

  a. _____
  b. _____
  c. _____
  d. _____
  e. _____
  f. _____
  g. _____

EXERCISE 3

Blood Pressure and Pulse: The Human

Discussion

As the heart beats it forces blood through the vascular tree. A wave of pressure, known as **pulse**, is transmitted

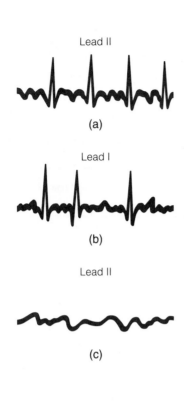

Lead II

(a)

Lead I

(b)

Lead II

(c)

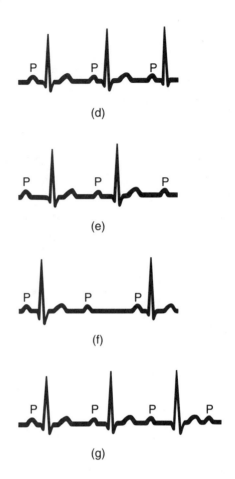

(d)

(e)

(f)

(g)

through the blood in the vessels. When measured, it reflects how frequently the heart beats. Pulse can be measured in a number of places on the body where the vessel is close to the surface and can be compressed against firm tissue (like a bone or muscle). Most conveniently, pulse is measured at the wrist in the distal portion of the radial artery. The carotid artery in the neck and the femoral artery are other areas where pulse can be measured.

In addition to rate of heart beat, the **pressure** generated by the heart and the vessels can be measured. As the heart contracts and pumps in systole, a pressure known as **systolic blood pressure** (normally about 120 mm Hg) is generated in the vessel. When the heart relaxes in diastole, the blood is under pressure from the walls of the vessel only. This is known as **diastolic blood pressure** (normally about 80 mm Hg). Various factors such as increased fluid in the vascular bed and changes in the elasticity of the vessel wall may affect blood pressure, as can age, weight, sex, and level of activity. The difference between systolic and diastolic pressure is called **pulse pressure** and usually measures about 40 mm Hg.

Clinically, blood pressure can be measured with a **blood pressure cuff** (sphygmomanometer) and a **stethoscope**, as illustrated in Figure 16.6. When sufficient pressure is applied to a blood vessel by increasing the pressure of the sphygmomanometer, the vessel collapses. As the pressure on the vessel is released, eventually the

coursing blood begins to rush through the vessel under the pressure developed by the heart in systole. By listening to the vessel with a stethoscope, the sound of this moving blood can be heard and read on the pressure gauge as systolic blood pressure. When these sounds cease, diastolic pressure is read on the pressure gauge.

Materials

A sphygmomanometer and pressure gauge; stethoscope; clock; physiograph with blood pressure cuff and bulb, sphygmomanometer, pressure coupler transducer, and pulse transducer with pulse "pickup" unit.

Procedure

Work in pairs or groups as necessary.

A. Pulse Measurements

1. Locate the radial pulse on your wrist by gently pressing the index and middle fingers into the "groove" just medial to the radius at the wrist.

2. Have your partner locate your carotid pulse at the "hollow," lateral to the larynx below the jaw.
3. Monitor the pulse simultaneously at each place for 30 sec. Record results in Table A.

B. Manual Blood Pressure Reading

1. Make sure you are able to operate the blood pressure bulb with one hand (increasing and decreasing pressure, tightening and loosening knob).
2. Secure the blood pressure cuff of the manually operated unit snugly around the lower biceps brachii muscle of your lab partner's arm. The arrow should point toward the palm of the hand.
3. Palpate the antecubital area for pulsations in the brachial artery.
4. Once you have located the brachial pulse, pump the pressure up to about 160 mm Hg.
5. With the stethoscope in place, loosen the pressure bulb screw so that the pressure drops *slowly*, and listen for the *first regular* beating sounds. Remember this as systolic blood pressure.
6. Continue dropping pressure and listening with the stethoscope until the sounds *cease or change* in nature. This is the point at which diastolic pressure is read.
7. Open the pressure bulb up and remove the cuff. Do not repeat before a recovery time of a few minutes has elapsed.
8. Record both pressures in Table A.

■ **FIGURE 16.6 Manual blood pressure measurement**

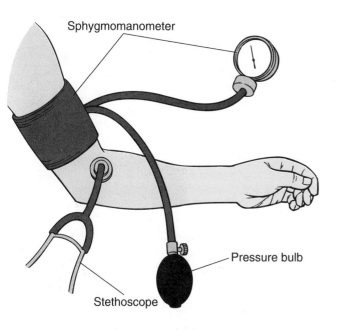

Sphygmomanometer

Pressure bulb

Stethoscope

C. Variations in Pulse and Blood Pressure

Monitor and record below pulse and blood pressure as in Procedures A and B in the following circumstances.

1. Student runs (with blood pressure cuff in place) for a few minutes or does some equivalent vigorous exercise. Take pressure immediately after exercise and again 5 min later. Record in Table A.
2. Student sits quietly and relaxes for about 15 minutes in order to lower both pulse and pressure.
3. Student lies down. Take pressure and record in Table A.

D. Pulse and Blood Pressure Reading on the Physiograph

1. Plug the blood pressure cuff and bulb in the physiograph, and secure the cuff on a student volunteer.
2. Adjust the pressure dial for the appropriate pressure (0–100, 0–200, 0–300), position the recorder arm, and calibrate the recording by pressing the "100 mm Hg" button and noting the deflection produced in the recording.
3. On the index finger of the same arm, attach the pulse "pickup" and plug the transducer cable into the physiograph.
4. Pump the pressure about 100 mm Hg above the level where the blood pressure reading loses pulsations and flattens out.
5. Note what is happening to finger pulse.
6. Start dropping the pressure slowly until pulsations once again appear in the blood pressure recording. Note the finger pulse recording.
7. Open the pressure bulb up and remove the cuff.

Observations

The instructor may want to make a "class" table comparing certain male/female values, age, physical condition, and other variables.

Conclusions

- How do radial, carotid, and physiograph pulses compare in terms of strength?

  _____

  _____

- Why are the radial and carotid areas considered "pulse" areas?

  _____

  _____

■ **TABLE A Blood Pressure and Pulse**

	REST (SITTING)	AFTER EXERCISE		AFTER QUIET PERIOD	LYING DOWN
		IMMEDIATE	5 MIN		
Pulse (radial)					
Pulse (carotid)					
Pulse (physiograph)					
Blood pressure (manual)					
Blood pressure (physiograph)					

● What are some other areas at which to take pulse?

_____

_____

_____

● Explain what happens to pulse beats when pressure generated by the sphygmomanometer in the brachial artery rises above systolic value.

_____

_____

_____

● Where does the pressure come from in diastolic pressure?

_____

_____

_____

● How would blood pressure taken at the ankle of a standing man differ from "standard" blood pressure? Explain.

_____

_____

_____

● Explain the variation in pulse and blood pressure compared to your standard after:

a. Exercise _____

_____

_____

b. Quiet rest _____

_____

_____

c. Lying down _____

_____

_____

d. Blood pressure vary during the day? Explain.

_____

_____

_____

Blood Vessels and Fetal Circulation

O V E R V I E W

In this unit you will begin by examining the microscopic structure of the arteries, veins and capillaries. Then you will locate some of the major blood vessels of the fetal pig, and do some comparisons with the human. At last, you will look at the beginning—the fetus, and examine the circulatory plan.

O U T L I N E

The following experiments involve the use of living specimens. Commercial audio-visual materials may be substituted, if appropriate.

Blood Vessels

Background

With the heart serving as the pump, blood is circulated through a closed system of blood vessels that serves every organ of the body. The three major types of vessels —arteries, veins, and capillaries—are organized into sev-

eral circulatory circuits. These circuits are the pulmonary, systemic and hepatic portal circuits. The **pulmonary circuit** incorporates the vessels that transport poorly oxygenated blood from the heart to the lungs and well-oxygenated blood back to the heart. The **systemic circuit** has the vessels that transport oxygenated blood from the left side of the heart to all the systems and organs of the body and back to the right side of the heart. The **hepatic portal circuit** involves the vessels that drain blood from the stomach, small and large intestines, pancreas, and spleen into the liver and from there to the inferior vena cava and back to the heart.

EXERCISE 1
Microscopic Examination of Blood Vessels

Discussion

The three major blood vessel types, arteries, veins, and capillaries, can be distinguished both structurally and functionally. **Arteries** carry blood away from the heart, and are considered mainly to **conduct** the flow of blood in the body (conductance vessels), while **veins** transport blood back toward the heart. Since most of the blood at rest flows in the veins, they have the greatest **capacity** (capacitance vessels). In between the arteries and veins are the **capillaries**, which serve as the sites of exchange between the blood and the tissues. These are called exchange vessels. See Figure 17.1. It is possible to make the *generalization* that arteries carry well-oxygenated blood and veins carry poorly oxygenated blood, provided the exceptions are noted, namely in the pulmonary and fetal circuits. In pulmonary circulation the pulmonary artery is carrying deoxygenated blood to the lungs and the veins carry oxygenated blood from the lungs back to the heart. The same is true in fetal circulation: the umbilical vein carries oxygenated blood from the placenta to the fetus while the umbilical arteries carry deoxygenated blood from the fetus to the placenta (see Exercise under "Fetal Circulation: The Human").

If the general pattern of circulation is considered, the structural differences between arteries, veins, and capillaries are easy to understand. For example, where the pressure in the system is highest, strong walls are necessary. Where pressure is low and blood flow is slowed, the system possesses thinner walled vessels with the means of preventing backflow, such as valves. Certainly the efficiency of nutrient and waste exchange between vessels and tissues is enhanced by thin semipermeable vessels. Table 17.1 summarizes the structural differences in the major blood vessel types. Note that both arteries and veins consist of three basic layers, **tunica externa, tunica media**, and **tunica interna**, whereas the capillaries are only one layer thick. This layer is a continuation of the epithelial lining of the arteries and veins, and is known as endothelium. See Figure 17.2 and Table 17.1.

Materials

A microscope; prepared slide of c.s. artery, vein, and capillary.

Procedure

1. Using a low-power microscope, examine the slide of an artery and vein.
2. Distinguish the different layers, illustrated in Figure 17.2, of each and compare their relative thicknesses.
3. Observe the capillary, noting size and makeup of the wall.
4. Sketch the artery, vein, and capillary in cross section in the Observations section.

Observations

Artery

Vein

Capillary

Conclusions

- From your observations, which blood vessel has a larger

   a. Inner diameter? _____

   b. Outer diameter? _____

■ **FIGURE 17.1 General circulatory pathways**

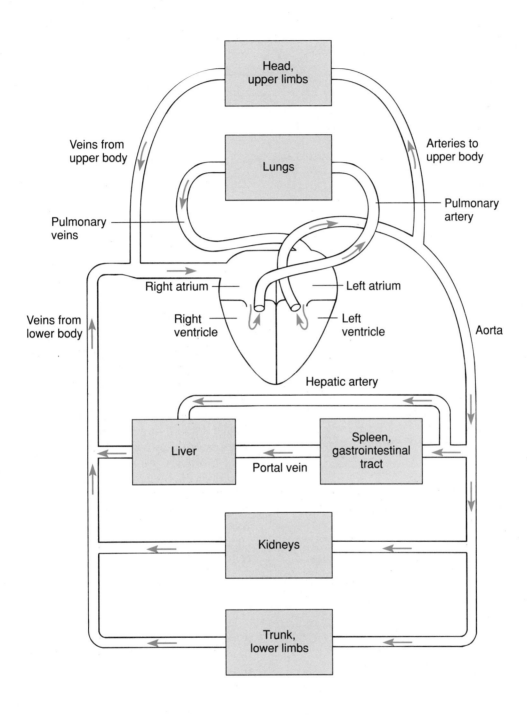

- Which layer is thickest in a vein? _____
- Why do veins collapse when the blood pressure in them drops?

  _____

  _____

  _____

- Look up and describe varicosities in veins.

  _____

  _____

  _____

  _____

■ **TABLE 17.1   Structural Comparison of Blood Vessel Types**

		COATS		
VESSEL	**FUNCTION**	**TUNICA ADVENTITIA**	**TUNICA MEDIA**	**TUNICA INTERNA**
Artery	Conductance vessels	Loosely arranged fibrous connective tissue	Bulk of arterial wall consisting of circular smooth muscle and/or elastic tissue	Simple squamous epithelium resting on elastic membrane
Vein	Capacitance vessels	Loosely arranged fibrous connective tissue, which may contain some longitudinal smooth muscle	Circular smooth muscle with few elastic fibers	Simple epithelium tissue with valves that are flaps of endothelium strengthened by some connective tissue
Capillary	Exchange vessels	Absent	Absent	Simple layer of squamous epithelium

- In which area of the body would the veins be likely to have the greatest concentration of valves? Explain.

  _____

  _____

  _____

- Why is the tunica media of arteries much thicker than in the corresponding layer of veins?

  _____

  _____

  _____

- What makes up the wall of a capillary?

  _____

  _____

- Look up and define the term "aneurysm." Why is it serious?

  _____

  _____

  _____

EXERCISE 2
Microcirculation

Discussion

A view into the small arterioles and capillaries of an organism allows us to see circulation at the level where exchange with the tissues occurs. The arteriole is from 30 to 50 μm in diameter and has a continuous coat of mul-

tiunit **smooth muscle** encircling it. This muscle receives a dense sympathetic nerve supply, and is the major control over resistance to blood flow and therefore, pressure. The vessels leading off the arteriole are the **thoroughfare channels**, which have smooth muscle along the first third of their length, but with less innervation than the arterioles. This portion is called the **metarteriole**. Capillaries (3–8 μm diameter) branch off the metarterioles. At the entrance to the capillaries is a ring of unitary smooth muscle with only a sparse nerve supply. The smooth muscles of the metarteriole and **precapillary sphincter** are responsible for regulating the local flow of the blood into a particular area of the body. These sphincters are sensitive to levels of local metabolites such as oxygen and carbon dioxide, as well as temperature. Blood drains into **venules**, which have a collagenous connective tissue wrapping and a few smooth muscle cells. See Figure 17.3.

Materials

A frog or goldfish; frog board with cut-out circle; string; dissecting pins; rubber bands; cloth; epinephrine solution, 1:1000; histamine solution, 1:10,000; microscope; amphibian Ringer's solution.

Be sure to review safety precautions before handling fresh or preserved specimens.

Procedure

1. Wrap the frog (fish) in a cloth and secure it with string to the frog board so that the webbing of the foot (or tail fin) lies over the circular hole in the board.
2. Stretch the webbing (or fin) and pin it to the board so it is taut.
3. Keep the area moistened with saline.

■ **FIGURE 17.2 Veins, arteries, and capillaries**
(a–d) Diagrams. (e–g) Photomicrographs.

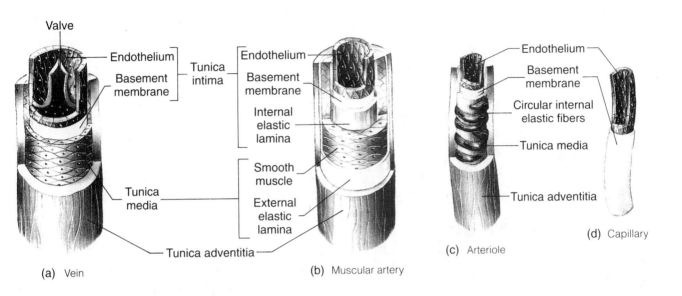

(a) Vein

(b) Muscular artery

(c) Arteriole

(d) Capillary

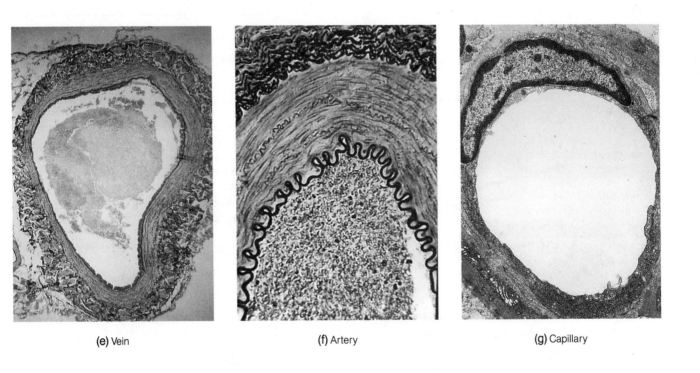

(e) Vein

(f) Artery

(g) Capillary

■ **FIGURE 17.3   Capillary bed**

The entrance of blood into a capillary bed is controlled by a smooth muscle ring at the entrance called the sphincter. Otherwise, blood will flow from arteriole to venule by way of the thoroughfare channel.

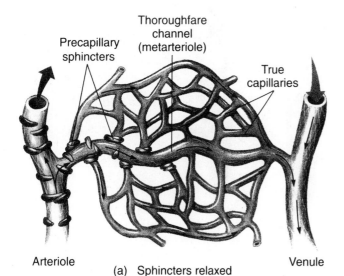

Arteriole

(a)   Sphincters relaxed

Venule

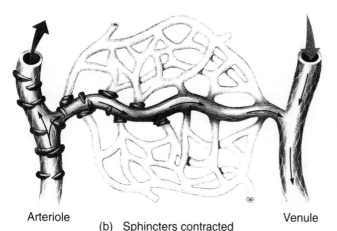

Arteriole

(b)   Sphincters contracted

Venule

4. Using a 10× objective, observe the blood vessels in the webbing (or fin) under the microscope. Locate an arteriole, a capillary, and a venule.

5. Bathe the area with a few drops of the histamine solution and observe any changes in blood flow. Record in the Observations section.

6. Rinse the area well with Ringer's solution and blot dry.

7. Bathe the area with a few drops of the epinephrine solution. Observe the blood flow and record the results.

8. Rinse the area again and return the specimen to the aquarium.

Observations

a. Type of blood flow in arteriole _____

b. Type of blood flow in capillary _____

c. Type of blood flow after bathing with histamine

_____

d. Type of blood flow after bathing with epinephrine

_____

Conclusions

● Under normal conditions, what structure controls the characteristic blood flow into capillaries?

_____

_____

_____

● What is the effect and purpose of histamine being released in the inflammatory response?

_____

_____

_____

● What does epinephrine do to the smooth muscle of an arteriole?

_____

_____

_____

● Epinephrine was formerly used as an asthma medication, but it was learned that blood pressure was affected by this hormone. How does epinephrine affect blood pressure, and why?

_____

_____

◆ **Quick Quiz 1**

1. What appears to be the function of the structures called valves in the veins?
   (a) allowing for greater volume of blood
   (b) helping accept a greater pressure in the vessel
   (c) preventing backflow
   (d) all of these

2. When you observe the structure of a capillary, what do you see as far as a "match" between structure and function?
   (a) a strong, reinforced epithelium to withstand pressure
   (b) a thin epithelium to allow diffusion
   (c) a layer of smooth muscle to allow stretching
   (d) none of these

3. Why is the tunica intima "wavey"?
   (a) it sits on an elastic membrane
   (b) it has contracted
   (c) the smooth muscle throws it into folds
   (d) none of these

4. Epinephrine, released in a sympathetic response would do what to the blood pressure?
   (a) decrease it
   (b) increase it
   (c) cause no change
   (d) cause a slow, gradual drop

5. When the precapillary sphincter is shut, where does most of the blood flow in a capillary bed?
   (a) into the thoroughfare channel
   (b) into the metarteriole
   (c) into the venule
   (d) all of these

EXERCISE 3
Dissection of Circulatory System: The Fetal Pig (Chest, Neck, and Upper Extremity)

Discussion

The circulatory system of the fetal pig has been injected with a latex material to aid in identification of the blood vessels. Veins are more superficial and contain a *blue* latex; arteries are deeper and contain a *red* latex. Often the arteries appear pink because the thick walls do not permit the color to show through easily. For the most part, the system is **symmetrical**: that is, the major systemic vessels, the aorta and the vena cavae, give off paired branches. There are some exceptions, notably in the pectoral region. The branches of the aorta to the head and arms are not symmetrical because the first branch, the brachiocephalic artery, sends blood to both sides of the head and to the right arm, while the second branch, the left subclavian artery, sends blood to the left arm. However, the symmetry is generally such that you can trace the arteries and veins on one side of the neck, shoulder, and arm knowing that the other side is the same, except

■ **TABLE 17.2 Veins of the Anterior Region of the Fetal Pig**

VESSEL	DESCRIPTION
Anterior vena cava (Cranial vena cava or precaval vein)	Unpaired vessel that enters the right atrium of heart, formed by junction of brachiocephalic veins; drains head, neck and upper extremity
Hemiazygos (Azygos)	Unpaired vessel on left side of midline that enters the anterior vena cava near heart; drains dorsal body wall
Intercostals	Paired vessels found between the ribs and enter the hemiazygos
Internal mammary (Sternal)	Unpaired vessel that empties into the anterior vena cava below the junction of the brachiocephalic vein; drains ventral thoracic wall
Costocervical trunk	Vein that empties into the anterior vena cava usually just above or below the entrance of the internal mammary; drains the cervical vertebrae, muscles of neck and back and intercostal muscles
Internal thoracic	Paired veins that empty into ventral surface of the anterior vena cava just above its origin; drains chest wall
Brachiocephalic (Innominate)	Paired vessels formed by junction of jugulars and subclavian veins; drains blood from head, neck and upper extremity
External jugular	Paired vessels located along each side of the neck and are more lateral and superficial than internal jugulars; drains head and neck
Internal jugular	Paired vessels located along each side of neck medial to the external jugulars; drains brain and spinal cord
Maxillary	Paired vessels that lead into external jugular and lie near mandibular gland
Linguofacial	Paired vessels that lead into external jugular and cross mandibular gland
Subclavian	Paired vessels that are a continuation of axillary; drain blood from upper extremity and shoulder
Axillary	Paired vessels in axillary region that are a continuation of the brachial; drain blood from the arm
Brachial	Paired vessels in upper arm that are a continuation of the axillary; drain blood from the arm
Radial	Paired vessels found distal to the elbow; drain blood from the lower arm
Subscapular	Paired vessels that empty into the subclavian or axillary near the base of the external jugular; drain blood from the shoulder region
Cephalic	Paired veins on the lateral surface of the foreleg, extending from the shoulder area to the base of the external jugular

■ **TABLE 17.3   Arteries of the Anterior Region of the Fetal Pig**

VESSEL	DESCRIPTION
Aorta	Largest systemic artery; leaves left ventricle, arches toward the left (aortic arch) and continues through the thoracic and abdominal cavities
Brachiocephalic	First systemic branch (after the coronary arteries) of the aortic arch; extends anteriorly and divides into the common carotids and right subclavian; supplies blood to the right side of the head and neck and the right arm
Left subclavian	Second systemic branch of the aortic arch; supplies the left arm and left ventral wall
Common carotid	Paired arteries of the brachiocephalic artery; lie on either side of the trachea; supply head and neck
Right subclavian	Branch of the brachiocephalic; supplies the right arm and the left ventral wall
External carotid	Paired branches of the common carotid near the base of the head; supply the external parts of the head
Internal carotid	Paired branches of the common carotid near the base of the head; supply the internal parts of the head
Axillary	Paired arteries in the axillary region that are a continuation of the subclavian; supply the arms
Brachial	Paired arteries in the arms that are a continuation of the axillary; supply the arms
Radial	Paired arteries in the lower arms that are a continuation of the brachial; supply the lower arms
Costocervical trunk	Paired arteries that branch from the subclavian; supply the deep muscles of the neck
Thyrocervical	Paired arteries that branch from the subclavian usually opposite the costocervical; supply deep muscles of the neck and shoulder
Internal mammary (thoracic)	Paired arteries that branch from the subclavian; supply the ventral thoracic wall
Pulmonary trunk	Artery that leaves the right ventricle and branches into right and left pulmonary arteries that supply the lungs
Ductus arteriosus	Fetal vessel that brings blood from the pulmonary trunk into the aorta
Intercostal arteries	Paired arteries that branch from the thoracic aorta and run between the ribs; supply the intercostal muscles

where indicated. Tables 17.2 and 17.3 describe the major veins and arteries of the anterior region of the fetal pig.

A. Veins of Anterior Region of the Fetal Pig

Materials

A preserved and injected fetal pig; dissecting tray and instruments.

Be sure to review safety precautions before handling preserved or fresh specimens.

Procedure

Use Table 17.2 and Figure 17.4 as guides in tracing the veins.

1. Cutting through the muscles, carefully extend the incision previously made in the thorax up into the neck region.
2. Remove the thymus gland and uncover the trachea and larynx. Using a probe or dissecting needle, gently tease the fat and connective tissue to expose the blood vessels. Once the blood vessels can be clearly seen, begin to identify them using the heart as the starting point. However, keep in mind that the veins are returning blood to the heart.
3. Locate the **anterior (cranial) vena cava** and trace it anteriorly.
4. Push the left lung and heart medially and look for a vein running next to the aorta on the left side of the dorsal body wall. This unpaired vein is the **hemiazygos (azygos)**. In the human the azygos is on the right side of the body.
5. Located between the ribs, observe the paired **intercostal veins** that drain into the azygos.
6. Push the heart and lung back to its normal position. Return to the anterior vena cava and locate the **internal mammary vein** on its ventral surface just above the heart.
7. Somewhat variable in location, but usually near the entrance of the internal mammary, is the **costocervical trunk**.

■ **FIGURE 17.4 Veins of the anterior region of the fetal pig**
Refer to the color photo gallery.

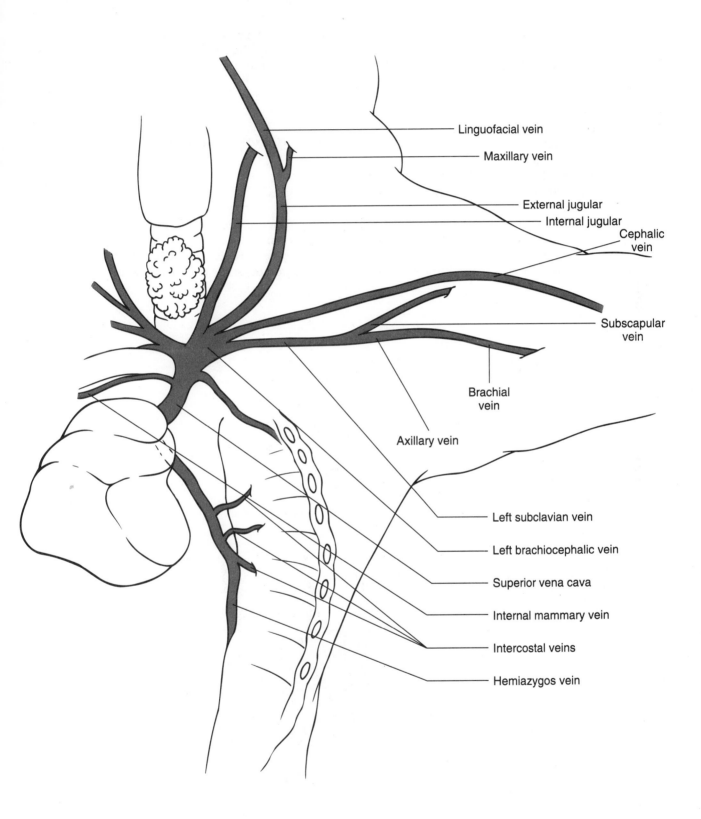

8. If your specimen is well injected, you may be able to observe a pair of **internal thoracic veins** entering the vena cava just above the juncture of the costo-cervical trunk.

9. Continue to trace the anterior vena cava anteriorly and observe the V-shaped junction of the right and left **brachiocephalic (innominate) veins**. The vessels leading into the brachiocephalic veins are symmetrical.

10. Trace either the right or left brachiocephalic vein and note the juncture of the jugulars and subclavian vein. The **external and internal jugulars** lie along each side of the neck.

11. Trace the external jugular anteriorly and observe the **maxillary vein**. The internal jugular continues as the **linguofacial vein**.

12. The **subclavian vein** goes out toward the arm where it becomes the **axillary vein** in the axillary region, continuing into the arm as the **brachial vein**. It may be necessary to cut through the skin

and muscles to trace the vessel. Below the elbow it becomes the **radial vein**.

13. Observe the **subscapular vein**, usually entering the subclavian or axillary vein close to the base of the external jugular. A **cephalic vein** may also be seen entering the base of the external jugular.

14. Label the diagram in the Observations section below.

B. Arteries of Anterior Region of the Fetal Pig

Materials

A preserved and injected fetal pig; dissecting tray and instruments.

Be sure to review safety precautions before handling preserved or fresh specimens.

Observations

Anterior veins of the fetal pig

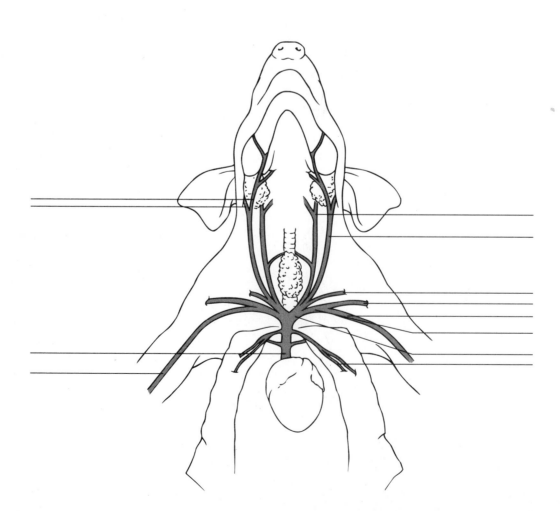

Procedure

Use Table 17.3 and Figure 17.5 as guides in tracing the arteries.

1. Note the large **aorta** leading out of the left ventricle and arching toward the left side of the body.
2. Locate the two arteries branching off the aortic arch, the **brachiocephalic artery** and the left **subclavian artery**.
3. Follow the brachiocephalic artery anteriorly to where it branches into a pair of **common carotid arteries** and the **right subclavian artery**. Take notice that the right subclavian is a branch off the brachiocephalic while the left subclavian is a branch directly off the aorta.
4. Follow the common carotid arteries running along side the internal jugulars. Note that the common carotid divides near the base of the head into the **external and internal carotid arteries**. The internal carotid is usually smaller than the external carotid.
5. Trace the subclavian artery as it becomes the **axillary, brachial, and radial arteries**.
6. Several arteries leading off the subclavian include the **costocervical trunk**, the **thyrocervical trunk** and the **internal mammary (sternal) arteries**.
7. As observed when the heart was studied in situ, the **pulmonary trunk** leads out from the right ventricle and divides into two **pulmonary arteries** going to the lungs. Remember the specimen is a fetus so that you can observe the **ductus arteriosus** continuing from the pulmonary trunk and entering the aortic arch.
8. Observe the aorta and trace it as it curves posteriorly and descends along the dorsal thoracic wall. Observe the many pairs of **intercostal arteries** that branch from the aorta.
9. Label the diagram in the Observations section below.

Observations

Anterior arteries of the fetal pig

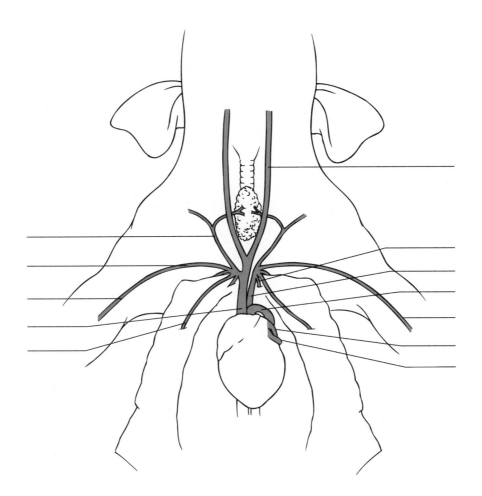

■ **FIGURE 17.5 Arteries of the anterior region of the fetal pig**
Refer to the color photo gallery.

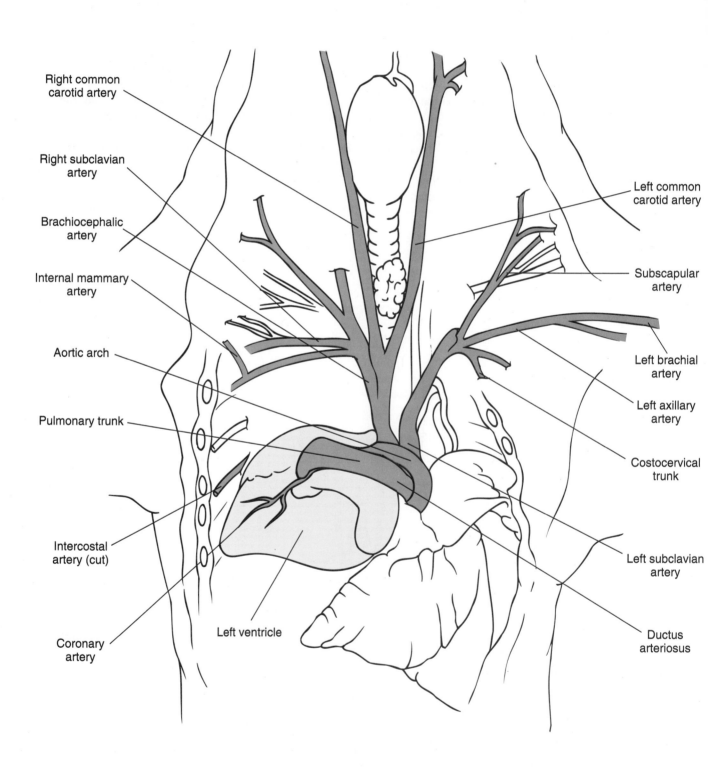

Right common carotid artery

Right subclavian artery

Brachiocephalic artery

Internal mammary artery

Aortic arch

Pulmonary trunk

Intercostal artery (cut)

Coronary artery

Left ventricle

Left common carotid artery

Subscapular artery

Left brachial artery

Left axillary artery

Costocervical trunk

Left subclavian artery

Ductus arteriosus

EXERCISE 4

Dissection of Circulatory System: The Fetal Pig (Abdomen and Lower Extremity)

Discussion

The general pattern of the circulatory system below the diaphragm is such that the aorta serves as the major arterial supply to the viscera and utlimately divides into two vessels, each of which feeds a lower extremity. Correspondingly, except for the hepatic portal system, the drainage of blood from the leg is from a major vein that unites with the vein from the opposite leg to form the posterior vena cava, which then receives blood from the abdominal organs.

An outstanding exception to this pattern is the hepatic portal system, which includes the portal vein and its tributaries from the organs of the digestive tract. The **hepatic portal vein** enters the liver and divides into numerous branches which eventually empty into the posterior vena cava. The significance of this system is that blood from digestive organs is transported directly to the liver, where absorbed nutrients can be processed more quickly than would be possible if they made the complete circuit of the systemic circulation.

A. Veins of the Abdomen and Lower Extremity of the Fetal Pig

Materials

A preserved and injected fetal pig; dissecting tray and instruments.

Be sure to review safety precautions before handling preserved or fresh specimens.

Procedure

Use Table 17.4 and Figure 17.6 as guides.

1. Extend the incision from the inferior border of the thorax posteriorly to the umbilical cord. Make an incision on both sides of the umbilical cord and continue the incisions to the legs. This results in a flap of tissue containing the umbilical cord and urogenital organs.
2. Gently pull the umbilical cord and observe the **umbilical vein** extending to the liver. In order to expose the body cavity, cut the umbilical vein close to the cord so that it can be reidentified later. If there is

■ **TABLE 17.4 Selected Veins of the Posterior Region of the Fetal Pig**

VESSEL	DESCRIPTION
Umbilical	Single fetal vessel extending from the umbilical cord to the liver
Posterior vena cava	Unpaired vessel; enters right atrium; formed by union of common iliac veins in pelvic region; drains blood from abdomen, pelvis and lower extremity
Hepatic	Vessels that drain liver; enter into posterior vena cava
Adrenolumbar	Paired vessels; enter posterior vena cava above kidney; drain adrenal glands and dorsal body wall
Renal	Paired vessels (however, may find two on the right side); drain kidneys and enter posterior vena cava
Genital	Paired vessels; either testicular or ovarian depending on the sex; right vessel drains into posterior vena cava and left vessel empties into left renal vein
Iliolumbar	Paired vessels below the kidney; drain dorsal body wall
Common iliac	Paired vessels; formed by union of internal and external iliac; drain legs
Internal iliac	Paired vessels; enter common iliacs medially; drain blood from pelvic and gluteal regions
External iliac	Paired vessels; unite with internal iliacs to form common iliacs
Femoral	Paired vessels that lead into external iliacs; drain legs
Deep femoral	Paired vessels; enter femoral veins medially; drain deep thigh muscles and external genital area
Saphenous	Paired vessels; enter into femoral veins from medial calf and thigh; drain legs
Hepatic portal	Unpaired vessels located in lesser omentum and enters the liver; drains digestive organs
Gastrosplenic	Unpaired vessel formed by union of veins from stomach and spleen; empties into hepatic portal vein
Mesenteric	Unpaired vessel; drains large intestine and pancreas; empties into hepatic portal vein

■ **FIGURE 17.6 Selected veins of the posterior region of the fetal pig**
Refer to the color photo gallery.

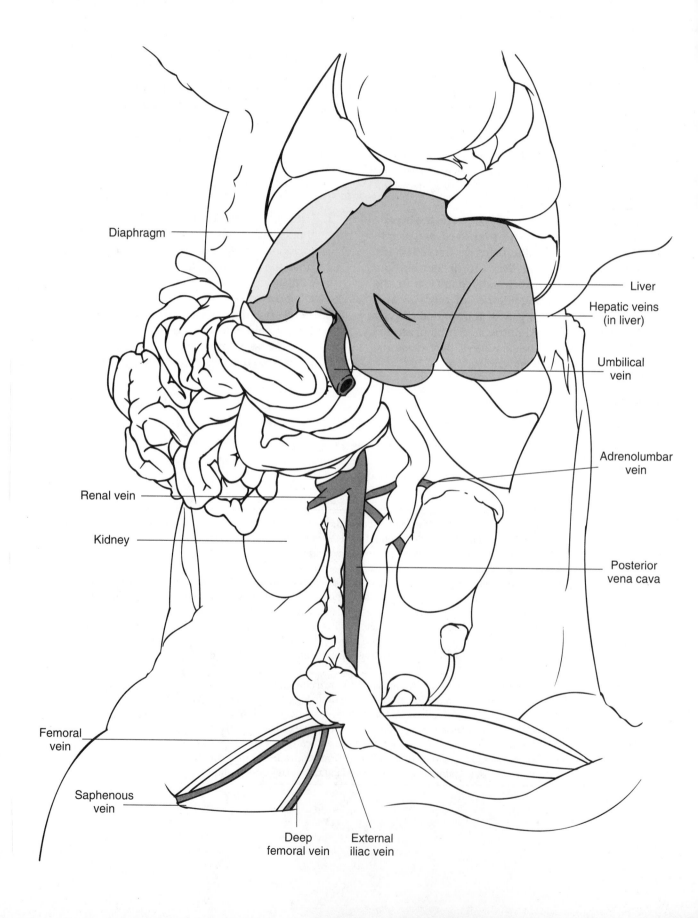

Diaphragm

Liver

Hepatic veins
(in liver)

Umbilical
vein

Adrenolumbar
vein

Renal vein

Kidney

Posterior
vena cava

Femoral
vein

Saphenous
vein

Deep
femoral vein

External
iliac vein

a brown fluid in the body cavity you may wish to flush it out with water.

3. To further expose the body cavity, make an incision in both sides of the body wall just posterior to the diaphragm.

4. Using the heart as a starting point, locate the **posterior vena cava** leading into the right atrium. Trace it posteriorly as it passes through the diaphragm into the abdominal cavity.

5. Scrape away some of the liver tissue and note several **hepatic veins**. The umbilical vein enters the liver and connects with one of the larger hepatic veins.

6. Gently push the abdominal organs to the left side of the body cavity and note the vessels draining into the posterior vena cava.

7. The **adrenolumbar vein** can be seen passing from the body wall and adrenal gland just anterior to the kidney.

8. Observe the **renal vein** draining the kidney. You may notice two renal veins on the right side.

9. If your specimen is well injected you may see the small **genital veins** (testicular or ovarian).

10. **Iliolumbar veins** can be seen draining the back muscles.

11. Draining each leg is the **common iliac vein**. These two veins unite to form the posterior vena cava.

12. Trace the common iliac vein posteriorly. Each common iliac is formed by the uniting of the **internal iliac vein** and the **external iliac vein**.

13. The **femoral vein** and **deep femoral vein** can be seen leading into the external iliac vein. The **great saphenous vein** in the leg drains into the femoral vein.

14. In most specimens it is difficult to trace the portal system since it is usually poorly injected. If possible, return to the abdominal region and locate the **hepatic portal vein** located in the lesser omentum, a membrane extending between the stomach and duodenum to the liver. Veins from the spleen and stomach form the **gastrosplenic vein** which joins with the **mesenteric vein** from the intestine. These unite to form the hepatic portal vein which enters the liver.

15. Label the diagram in the Observations section below.

Observations

Posterior veins of the fetal pig

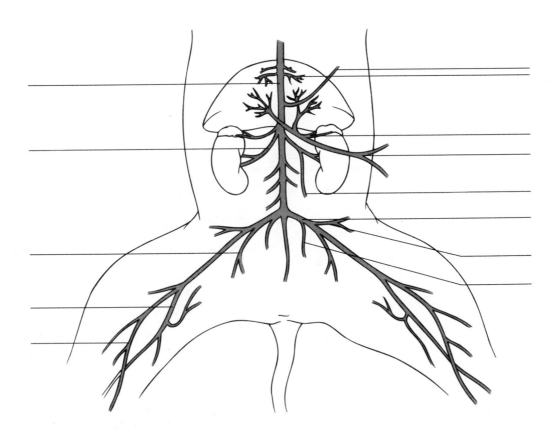

B. Arteries of the Abdomen and Lower Extremity of the Fetal Pig

Materials

A preserved and injected fetal pig; dissecting tray and instruments.

Be sure to review safety precautions before handling preserved or fresh specimens.

Procedure

Use Table 17.5 and Figure 17.7 as guides.

1. Using the heart as a starting point, locate the **aorta** and trace it posteriorly into the thoracic cavity, passing through the diaphragm into the abdominal cavity.
2. Just below the diaphragm, gently tease away the peritoneum and observe the unpaired **celiac trunk** branching off the aorta.
3. Trace the celiac artery as it gives rise to the **splenic artery** and the **gastrohepatic artery**.
4. A short distance posterior to the celiac trunk is the **anterior (superior) mesenteric artery** seen as a branch off the aorta and going to the small intestine.
5. Observe the pair of **adrenolumbar arteries** branching from the aorta posterior to the superior mesenteric artery.
6. Note the **renal arteries** supplying the kidneys. These may sometimes branch before entering the kidneys.
7. Posterior to the renal arteries locate a pair of long, thin **genital arteries (ovarian or testicular)**.
8. Posterior to the genital arteries is an unpaired **posterior (inferior) mesenteric artery** supplying the large intestine.
9. Locate the paired **iliolumbar arteries** branching from the aorta below the posterior mesenteric and supplying the dorsal body wall.
10. Trace the aorta posteriorly and observe the two **external iliac arteries** extending into each leg.
11. Follow the external iliac artery in the leg to where it becomes the **femoral artery**. Note the **deep femoral artery** as a medial branch of the femoral artery.
12. In the region of the knee the femoral artery becomes the **saphenous artery**.
13. Below the branching of the aorta into the external iliac arteries, locate the pair of **internal iliac arteries**.
14. Locate the pair of **umbilical arteries** that run along each side of the bladder and continues into the umbilical cord.
15. Label the diagram in the Observations section on page 224.

■ **TABLE 17.5 Selected Arteries of the Posterior Region of the Fetal Pig**

VESSEL	DESCRIPTION
Aorta	Large unpaired vessel leaving left ventricle and continuing through thoracic and abdominal cavities to pelvic region where it divides; supplies blood to all organs
Celiac trunk	Unpaired vessel; first branch of aorta below diaphragm; has branches that supply stomach, liver, pancreas, spleen and duodenum
Splenic	Unpaired vessel; branch of celiac that supplies spleen
Gastrohepatic	Unpaired vessel; branch of celiac that supplies stomach and liver
Anterior (superior) mesenteric	Unpaired vessel; second major branch of aorta located just below celiac trunk; supplies small intestine and a portion of large intestine
Adrenolumbar	Paired vessels near origin of anterior mesenteric; supply adrenal gland and dorsal body wall
Renal	Paired vessels arising from aorta just posterior to adrenolumbars; may branch as they approach kidney; supply blood to kidneys
Posterior (inferior) mesenteric	Unpaired vessel; originates from ventral surface of aorta
Iliolumbar	Paired vessels off aorta below posterior mesenteric; supply blood to dorsal body wall
External iliac	Paired vessels formed by bifurcation of aorta in lower pelvic region; supply blood to thigh

■ FIGURE 17.7   Selected arteries of the posterior region of the fetal pig
Refer to the color photo gallery.

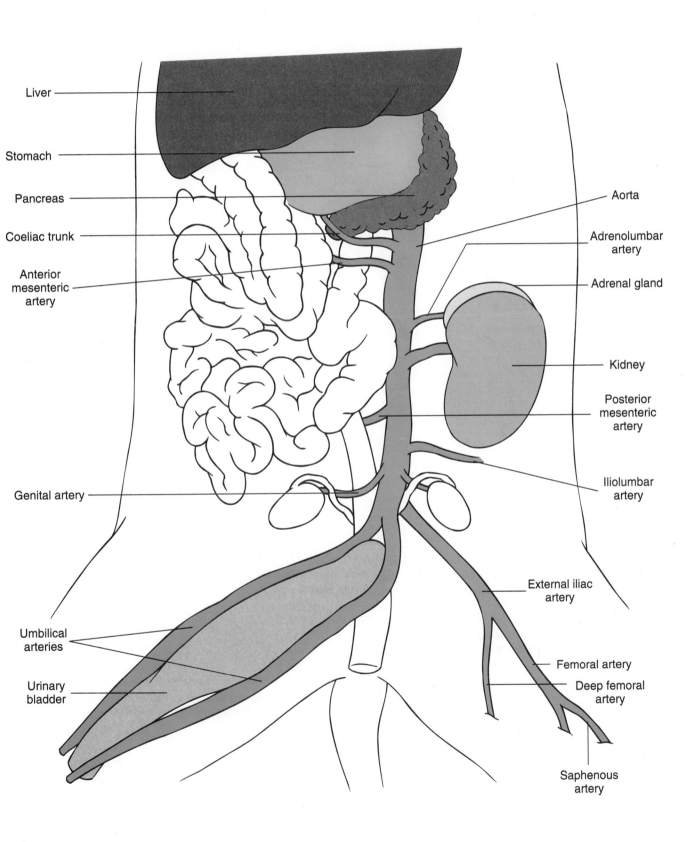

Observations

Posterior arteries of the fetal pig

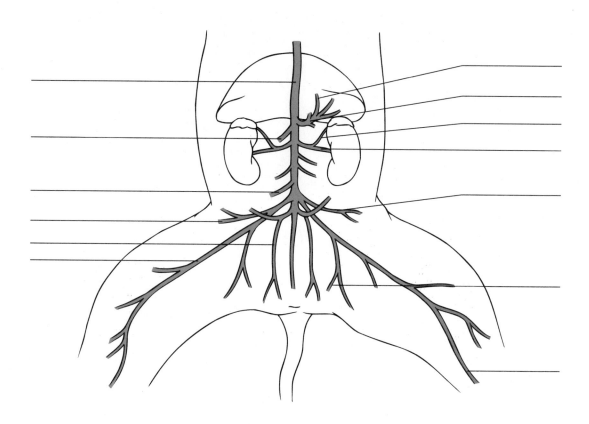

Conclusions

• What differences exist in the major blood vessels associated with the arterial supply of blood to, and venous drainage of blood from, the areas of the fetal pig listed in Table A?

■ **TABLE A Arterial Supply and Venous Drainage Differences in the Fetal Pig**

	ARTERIES	VEINS
Hind limb		
Digestive tract		
Arm		
Head		

EXERCISE 5

Comparison of the Fetal Pig and Human Circulatory Systems

Discussion

The pig may be used to illustrate the circulatory system because it is very similar to the human in that respect. However, differences between the two systems exist. A good example can be seen in the branches of the arch of the aorta. In the pig there are two branches, the brachiocephalic and left subclavian. In the human there are three branches, the brachiocephalic, left common carotid and left subclavian. Both the pig and human have a pair of internal and external jugulars. In the pig the internal jugular is very small while in the human the internal jugular is larger than the external jugular. A difference can also be noted in the venous supply to the thoracic region. In the pig the hemiazygos, after receiving the intercostals, enters the right atrium of the heart. In the human,

■ FIGURE 17.8 The human circulatory system
(a) Arteries. (b) Veins.

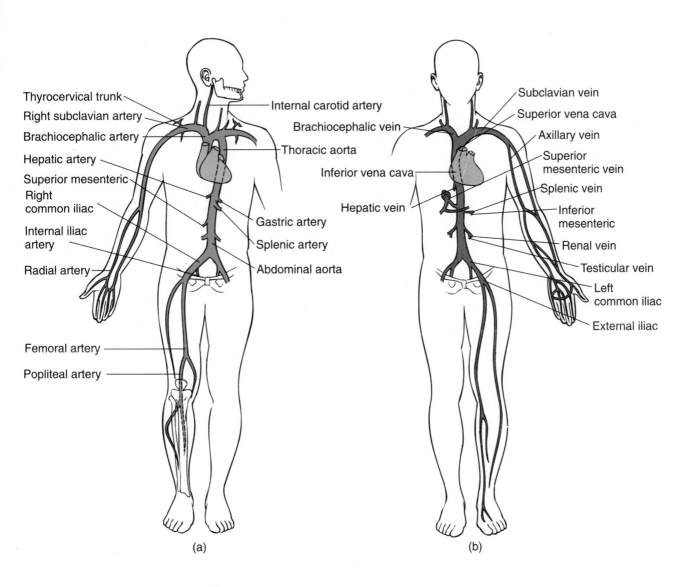

Thyrocervical trunk
Right subclavian artery
Brachiocephalic artery
Hepatic artery
Superior mesenteric
Right common iliac
Internal iliac artery
Radial artery
Femoral artery
Popliteal artery

Internal carotid artery
Brachiocephalic vein
Thoracic aorta
Gastric artery
Splenic artery
Abdominal aorta

(a)

Subclavian vein
Superior vena cava
Axillary vein
Superior mesenteric vein
Splenic vein
Inferior mesenteric
Renal vein
Testicular vein
Left common iliac
External iliac

Inferior vena cava
Hepatic vein

(b)

the corresponding azygos enters the posterior vena cava. See Figure 17.8.

Materials

Charts and models of pig and human circulatory systems.

Procedure

1. Trace the circulatory system of the pig by diagram as in the preceding exercises.
2. Compare the major blood vessels of the pig and human, noting the differences between the two.
3. Record the differences between the two circulatory systems in Table B.

 Quick Quiz 2

1. The most distal artery of the vessels below is:
   (a) subclavian
   (b) axillary
   (c) brachial
   (d) brachiocephalic
2. Which name is common to both arterial and venous systems?
   (a) brachiocephalic
   (b) iliac
   (c) subclavian
   (d) all of these

■ **TABLE B   Differences in Human and Fetal Pig Circulatory Systems**

VESSEL GROUP	FETAL PIG	HUMAN
Carotid artery		
Brachiocephalic artery		
Subclavian artery		
Brachial and radial arteries		
Ulnar arteries		
Azygos vein		
Subclavian vein, jugular group		
Adrenolumbar artery		
Iliolumbar artery		
Iliac arteries		
Femoral, saphenous vein		

3. Which is the first branch of the descending aorta under the diaphragm which branches to the liver, spleen and stomach?
   (a) superior mesenteric
   (b) inferior mesenteric
   (c) celiac
   (d) hepatic
4. The hepatic portal vein drains:
   (a) the liver
   (b) the spleen, large and small intestines
   (c) the posterior abdominal wall
   (d) all of these
5. The most superior vein draining into the inferior vena cava in the abdominal cavity is the:
   (a) iliolumbar vein
   (b) internal iliac
   (c) caudal
   (d) adrenolumbar

 # Fetal Circulation

Background

During fetal development the placenta supplies the fetus with nutrients and oxygen and removes wastes and carbon dioxide. Blood rich in nutrients and oxygen is brought to the fetus by the **umbilical vein**. The umbilical vein passes through the umbilical cord to the liver. Much of the blood goes through the liver by the **ductus venosus**. It then passes into the posterior vena cava which continues to the right atrium of the heart. A very small portion of the blood continues to the lungs by the pulmonary trunk. However, most of the fetal blood bypasses the lungs. This can be accomplished in two ways: (1) Some blood passes from the right atrium into the left atrium through an opening in the interatrial septum called the **foramen ovale**. (2) Some blood passes from the right ventricle to the pulmonary trunk and directly into the aorta through the **ductus arteriosus**. The aorta carries blood throughout the body and eventually into two **umbilical arteries** to the placenta. See Figure 17.9.

At birth or shortly after, several things occur that allow for changes in the circulatory pathway. These include:

1. The umbilical vein becomes the **round ligament** of the liver.
2. The ductus venosus becomes a fibrous band, the **ligamentum venosum**.
3. The foramen ovale closes and becomes a depression, the **fossa ovale**.
4. The ductus arteriosus becomes connective tissue, the **ligamentum arteriosum**.
5. The umbilical cord dries and is lost several days after birth, leaving the remnant, the **umbilicus** or naval.

EXERCISE 1
Tracing Fetal Circulation Using the Fetal Pig

Materials

A preserved and injected fetal pig; dissecting tray and instruments.

Be sure to review safety precautions before handling preserved or fresh specimens.

Procedure

1. Observe the umbilical cord and within it, one **umbilical vein** and two **umbilical arteries**. The umbilical vein extends from the umbilical cord to the liver. Observe the umbilical arteries running along both sides of the urinary bladder.

■ **FIGURE 17.9 Fetal circulation**

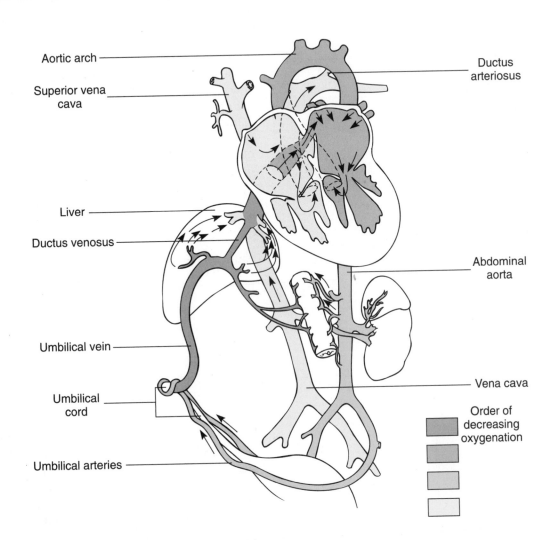

Aortic arch
Superior vena cava
Liver
Ductus venosus
Umbilical vein
Umbilical cord
Umbilical arteries
Ductus arteriosus
Abdominal aorta
Vena cava
Order of decreasing oxygenation

2. The **ductus venosus** is within the liver, connecting the umbilical vein to the posterior vena cava. This may be difficult to observe.
3. Locate the **ductus anteriosus** connecting the pulmonary trunk to the aorta.
4. Trace the aorta posteriorly and note that it eventually enters the two umbilical arteries.
5. It is difficult to see the foramen ovale in the fetal pig heart.

Conclusions

• Compare the blood in the right ventricle of the fetus and adult regarding degree of oxygenation.

_____

_____

_____

• What would be the consequence in the neonate if the ductus arteriosus failed to close?

_____

_____

_____

_____

 Quick Quiz 3

Select the best answer for each question; all are about fetal circulation.

1. Which chamber received the blood *first* from the placenta?
   (a) right atrium
   (b) right ventricle
   (c) left atrium
   (d) left ventricle
2. The blood in the right atrium *shunts* into the:
   (a) right ventricle
   (b) left atrium
   (c) left ventricle
   (d) none of these
3. What vessel brings fetal blood *to* the placenta?
   (a) aorta
   (b) vena cava
   (c) umbilical vein
   (d) ductus arteriosus
4. How does blood get from the right ventricle into the aorta?
   (a) through the foramen ovale
   (b) through the fossa ovale
   (c) through the ductus arteriosus
   (d) through the ligamentum arteriosum
5. How does oxygenated blood from the placenta bypass the liver?
   (a) through the abdominal aorta
   (b) through the ductus ateriosus
   (c) through the foramen ovale
   (d) through the ductus venosus

The Lymphatic System

OVERVIEW

In this unit you will learn about the structure and functioning of the organs of the lymphatic system through a microscopic and macroscopic examination of the organs and system as a whole. You will also explore the role of the lymphatic system in the body's defenses.

OUTLINE

The exercises in this unit involve the use of living specimens. Commercial audio-visual materials may be substituted if appropriate.

❖ Anatomy of the Lymphatic System

Background

The lymphatic system encompasses a variety of tissues and structures distributed throughout the body. These include lymphatic vessels, tissues, and organs. **Lymphatic vessels** consist of a one-way system of veinlike structures and capillaries that transport lymph, a watery, plasmalike fluid. Lymph is drawn from the tissue spaces into the vessels and ultimately returned to blood. **Lymphatic tissue** consists of a reticular connective tissue network containing lymphocytes, macrophages, and plasma cells. It may be distributed diffusely as in the **lamina propria** of the digestive and respiratory tracts or as small aggregates called nodes. Large aggregates of lymphatic tissue are found in lymphatic organs such as the spleen, thymus, and tonsils.

EXERCISE 1
Microscopic Anatomy of the Selected Lymphatic Structures

Discussion

Lymph is transported back toward the heart in vessels that are structurally similar to veins. Although the walls are thinner, larger lymphatic vessels have a three-layered wall and valves to prevent the backflow of lymph fluid. Unlike veins, however, lymph vessels have lymph nodes along the course of the vessels, which filter lymph. The fluid that is transported by the lymph vessels is retrieved from interstitial fluid by the **lymphatic capillaries**. Like blood capillaries, lymph capillaries are thin-walled tubes of endothelial cells that form a network in the tissues. Unlike blood capillaries however, lymph capillaries begin as blind-ended tubes in the tissues and have one-way permeability. Lymph is thus transported from the capillaries to larger lymphatic vessels, which ultimately drain into one of two large lymphatic ducts, the **thoracic duct** and the **right lymphatic duct**. These vessels empty into the **subclavian veins**. See Figure 18.1.

As the fluid passes through the lymphatic vessels, it is filtered at the lymph nodes. These small, oval structures contain lymphatic tissue. A fibrous connective tissue capsule encloses the node and sends extensions, the **trabeculae**, into the interior dividing the node into **sinuses**. As Figure 18.2 illustrates, the sinuses communicate freely and contain lymphatic tissue. When lymph passes through the sinuses, foreign material is phagocytized by macrophages and lymphocytes and antibodies may be added to the fluid.

Large lymphatic organs such as the **spleen** also consist of a core of lymphatic tissue. The spleen is encapsulated by a connective tissue that permeates the interior of the organ (as trabeculae) dividing it into compartments. The interior tissue, called pulp, exists in two types: red pulp and white pulp. The **red pulp** consists of filtering reticular tissue, which lines the venous sinuses that contain blood. **White pulp** consists of aggregates of lymphatic tissue in areas that filter lymph fluid. See Figure 18.3.

Materials

A microscope; prepared slides of lymphatic vessel (l.s.), lymph node, and spleen.

Procedure

1. Using low power, observe the lymph vessel in longitudinal section.
2. Note the valves that are present.
3. Sketch your observations in the space provided.
4. Observe the section of a lymph node and identify the structures labeled in Figure 18.2.
5. Make a sketch of your observations.
6. Observe the section of spleen and identify the structures labeled in Figure 18.3.
7. Sketch your observations.

■ **FIGURE 18.1 The lymphatic system**

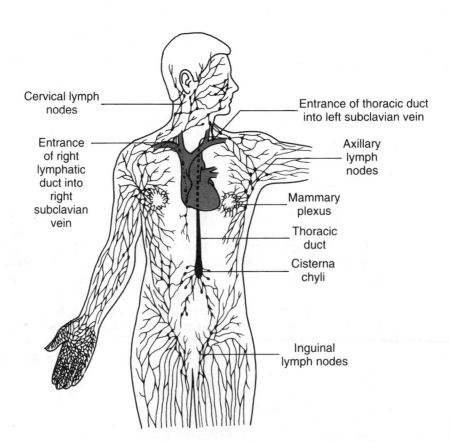

Cervical lymph nodes

Entrance of right lymphatic duct into right subclavian vein

Entrance of thoracic duct into left subclavian vein

Axillary lymph nodes

Mammary plexus

Thoracic duct

Cisterna chyli

Inguinal lymph nodes

■ **FIGURE 18.2 A lymph node**

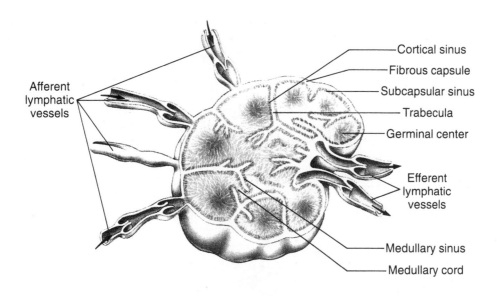

Afferent lymphatic vessels

Cortical sinus
Fibrous capsule
Subcapsular sinus
Trabecula
Germinal center
Efferent lymphatic vessels
Medullary sinus
Medullary cord

Observations

Lymph vessel	Spleen

Lymph node

■ **FIGURE 18.3   The spleen**

(a) Surface view of the spleen. (b) Compartment cross section of the spleen.

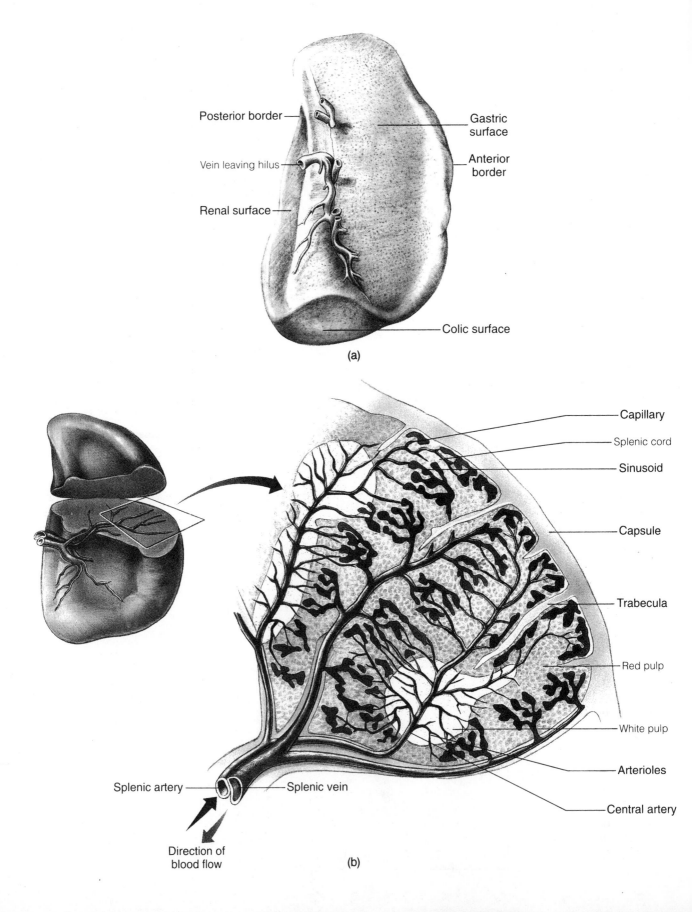

Posterior border

Gastric surface

Vein leaving hilus

Anterior border

Renal surface

Colic surface

(a)

Capillary

Splenic cord

Sinusoid

Capsule

Trabecula

Red pulp

White pulp

Arterioles

Central artery

Splenic artery

Splenic vein

Direction of blood flow

(b)

Conclusions

• Do lymphatic vessels appear to have more valves than veins? Explain.

_____

_____

_____

• How do the white pulp and red pulp of the spleen differ?

_____

_____

_____

 ## Quick Quiz 1

Complete the following statements with the most appropriate word or phrase.

1. The fluid which the lymphatic vessels return to the circulatory system is called _____.

2. The permeability in lymphatic capillaries is _____.

3. Trabeculae are walls which divide the interior of a lymph node into _____.

4. _____ pulp consists of filtering reticular tissue within the spleen.

5. The _____ receives fluid from the legs, abdomen, left side of thorax and head and left arm.

Selected Functions of the Lymphatic System

Background

There are many important functions which are carried out by the lymphatic system. These functions can generally be thought of as being twofold in purpose. The first function is to return interstitial fluid (lymph) that has "escaped" from blood back to the circulatory system. The second major activity is related to the system's participation in the body's defenses. This is accomplished through the action of the lymph nodes, for example, filtering lymph prior to its return to blood circulation. The system also provides for specific immune responses such as the production of antibodies and their circulation.

EXERCISE 1
Lymphatic Circulation

Discussion

As blood circulates through the capillaries, plasma, electrolytes, gases, small organic molecules, and small proteins escape into the tissue forming interstitial fluid. This is because less fluid is returned to blood at the venous end of the capillary network than escapes from blood at the arterial end. Eventually, this fluid is retrieved by the lymphatic system and returned to blood. Before its ultimate return to blood, the fluid is filtered and added to at the lymphatic tissue of the lymph nodes. (Like lymph, blood is also filtered in the spleen by phagocytic cells that line the sinuses.)

Materials

Five white mice; 1-ml syringes; 25-gauge injecting needles; 5% solution of India ink; dissecting instruments; cotton applicator sticks; gooseneck lamp; alcohol swabs; indelible marking pens.

Be sure to review safety precautions before handling fresh or preserved specimens.

Procedure

A. Injection of Experimental Animals

1. Using a code to distinguish the animals, mark four of them with an indelible marking pen. Streak the head of one animal, and the back, abdomen, and leg, respectively, of each of the others. (The fifth mouse will be used to compare tissues after sacrificing and therefore is not coded and injected with the ink solution.) Make a note of your code in Table A.
2. Assign each of the code markings a route of injection: intravenous, intraperitoneal, intramuscular, and subcutaneous.
3. Fill a syringe with 0.05 ml of the ink solution and inject the appropriate animal, according to your code, following the directions in Part B.
4. Return the animal to its cage.
5. Repeat steps 3 and 4 with the remaining three coded animals.

B. Method of Injection (work in pairs)

Intravenous

1. Hold the animal in an injection apparatus so the tail is exposed (see Appendix E).

2. Use a gooseneck lamp to light the area and heat the tail. (The heat will cause the vein in the tail to become readily apparent.)
3. Holding the tail in one hand, vein exposed, cleanse area with alcohol, let dry, and inject 0.05 ml of the ink solution.

Intraperitoneal

1. Using your index finger and thumb, grasp the animal ventral side up at the back of the neck and use your little finger to hold the tail down and away from the abdomen.
2. Moisten the hair on the abdomen with alcohol to facilitate injection.
3. Have your lab partner gently pull the skin of the abdomen up away from the viscera.
4. Insert the needle of the syringe straight into the midsection of the abdominal cavity and inject 0.05 ml of the ink solution.

Intramuscular

1. Hold the animal in the manner described for intraperitoneal injection.
2. Moisten the hair on the dorsal thigh with alcohol.
3. Insert the needle of the syringe into the thigh muscle and inject 0.05 ml of the ink solution.

Subcutaneous

1. Holding the animal as previously described in the intraperitoneal method, moisten the hair on the abdomen with alcohol.
2. Insert the needle of the syringe just under the skin of the abdomen and angle the needle so that it runs parallel, but beneath the skin. Inject 0.05 ml of ink solution.

C. Examination of Animals

1. Sacrifice the experimental animals according to the instructor's directions.
2. Dissect the animal and examine the superficial area surrounding the injection site for traces of the ink solution. Wetting the fur around areas of incisions will allow for neater dissection. (*Note:* The presence of ink can be ascertained by noting the presence of black spots in the tissue.)
3. Examine the lymph nodes in the cervical, axillary, and inguinal regions for the presence of ink particles.
4. Record the results by completing Table A.

Observations

■ TABLE A   Circulation Pathways[a]

INJECTION ROUTE	CODE	INJECTION SITE	REGIONAL LYMPH NODES	SPLEEN	LIVER	KIDNEY	HEART	LUNG
Subcutaneous								
Intravenous								
Intramuscular								
Intraperitoneal								

[a] + = ink particles present; – = no ink particles observed.

Conclusions

- Considering the circulation of blood, tissue fluid, and lymph, explain the results observed.

  _____

  _____

  _____

  _____

  _____

  _____

- Is there a difference in the distribution of ink particles in relation to the route of injection? Explain.

  _____

  _____

  _____

  _____

  _____

  _____

The Respiratory System

OVERVIEW

In this unit you will have the chance to identify the respiratory structures in the fetal pig and human, and then take measurements concerning pulmonary function.

OUTLINE

Anatomy of the Respiratory System

Background

The respiratory system functions in the conduction and exchange of gases in the body. Entrance into the upper respiratory tract can be achieved by two routes: nasal and oral. The **nasal route** carries only air and is entered at the nostrils. The **oral route** carries food and air and is entered at the mouth. These two routes join at the back of the mouth. They divide once again just above the larynx into separate respiratory and digestive systems.

As air passes through the respiratory passages, it is warmed, humidifed and cleansed. These passages divide, in the lung, like the branches of a tree. Ultimately this "tree" of airways ends in the **alveolus**, a microscopic (55–65 μm diameter) sac where oxygen and carbon dioxide exchange occurs with the pulmonary capillary blood

supply. There are an estimated 3–4 billion alveoli in the human lungs, occupying an area estimated at 70–100 m². See Figure 19.1a, b, and c.

EXERCISE 1
Microscopic Anatomy: Selected Tissues of the Respiratory System

Discussion

Histologically the respiratory system passes from a pseudostratified ciliated columnar epithelium in the nasal cavity to a simple epithelium with no goblet cells in the alveoli. See Table 19.1 and accompanying diagrams.

The system is readily understandable if it is separated into a conducting division and a respiratory division. The **conducting division** includes all airways of the system down through the **terminal bronchioles**. No air exchange is allowed in the conducting division because the

■ FIGURE 19.1 Respiratory tract

(a) Sagittal view of the head and neck. (b) Lower respiratory tract: the lungs. (c) Lower respiratory tract: the termination of bronchioles into alveoli.

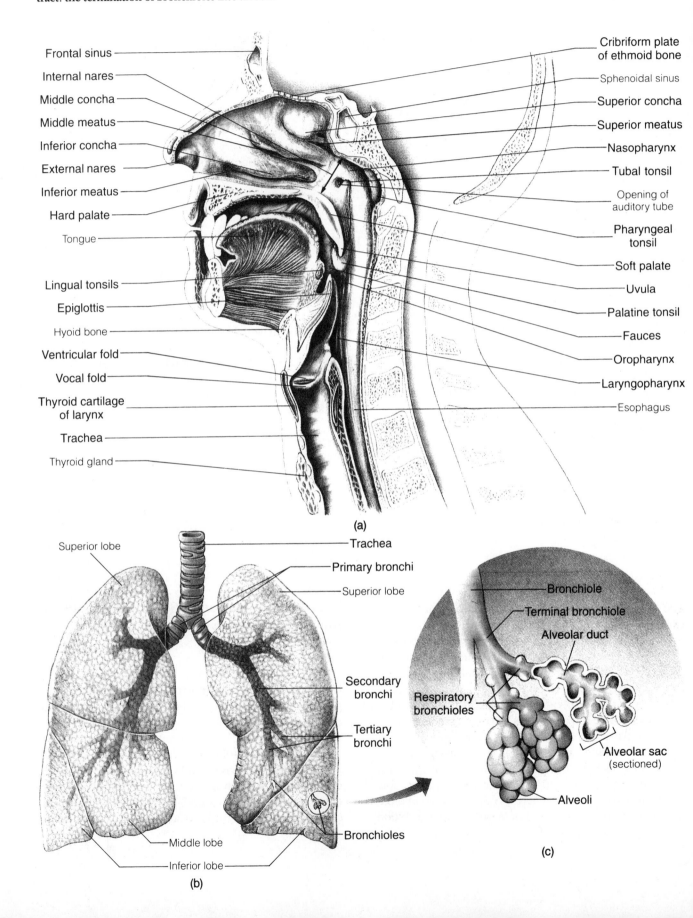

Frontal sinus
Internal nares
Middle concha
Middle meatus
Inferior concha
External nares
Inferior meatus
Hard palate
Tongue
Lingual tonsils
Epiglottis
Hyoid bone
Ventricular fold
Vocal fold
Thyroid cartilage of larynx
Trachea
Thyroid gland

Cribriform plate of ethmoid bone
Sphenoidal sinus
Superior concha
Superior meatus
Nasopharynx
Tubal tonsil
Opening of auditory tube
Pharyngeal tonsil
Soft palate
Uvula
Palatine tonsil
Fauces
Oropharynx
Laryngopharynx
Esophagus

(a)

Superior lobe
Trachea
Primary bronchi
Superior lobe
Secondary bronchi
Tertiary bronchi
Middle lobe
Inferior lobe
Bronchioles

(b)

Bronchiole
Terminal bronchiole
Alveolar duct
Respiratory bronchioles
Alveolar sac (sectioned)
Alveoli

(c)

walls are too thick. This is known as **anatomic dead space**. The **respiratory division** allows air exchange across its thin walls and includes the structures distal to the terminal bronchiole.

Materials

A microscope; slides of nasal epithelium, trachea (c.s.), bronchiole (c.s.), respiratory bronchiole (l.s.), and alveoli.

■ **TABLE 19.1 Microscopic Features of the Respiratory System**

STRUCTURE	DESCRIPTION
*Nasal cavity	Lined with *pseudostratified ciliated columnar epithelium; goblet cells, submucosal glands*; surface area increased by nasal conchae
Larynx	"Box" of cartilages; contains vocal cords; mucosa *stratified squamous* or *columnar*
*Trachea	Tube in thoracic cavity 10–12 cm long; 2–2.5 cm diameter; *pseudostratified ciliated columnar epithelium; goblet cells; submucosal glands; cartilage C-rings; smooth muscle and elastin*
Bronchi	Extra- and intrapulmonary sections; *simple ciliated columnar epithelium; goblet cells; submucosal glands; cartilage plates; smooth muscle and elastin*
Bronchioles	Size down to about 0.3 mm diameter; *ciliated cuboidal epithelium;* no glands; no cartilage; loses many goblet cells; *smooth muscle and elastin*
Respiratory bronchioles	Less than 0.5 mm diameter; patchy *cilia; cuboidal to squamous epithelium; smooth muscle* and *elastin; some alveolar outpocketings*
*Alveolar ducts	Extremely thin, *fibroelastic* walls; *smooth muscle* fibers at entrance to *alveoli*
*Alveolar sacs	Clusters of 15–20 *alveoli;* type I epithelial cells in wall for gas exchange; type II cells secrete *surfactant;* each alveolus surrounded by *pulmonary capillaries* and separated from capillary *basal lamina* by interstitium of 150–200 Å; diameter about 55–65 μm; may contain phagocytes in wall (septum)

*(See photos below)

Nasal cavity

Alveolar ducts and sacs

Trachea

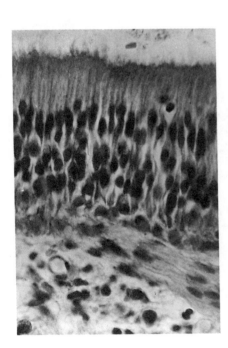

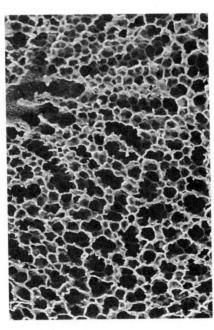

Procedure

1. Use high power to examine the prepared slides of various sections.
2. Identify the structures set in italics in Table 19.1.

Conclusions

● Why do cilia continue more distally in the respiratory system than goblet cells?

_____

_____

_____

_____

● Looking at the composition of the airway wall in the conducting division, why do you suppose air exchange does not occur?

_____

_____

_____

_____

EXERCISE 2
Structure of the Larynx

Discussion

The larynx can be thought of as a box whose walls are made of nine cartilages. The major single cartilage of the ventral side of the larynx is the "Adam's apple," or promi-

---

■ **FIGURE 19.2 External view of the larynx**
(a) Anterior. (b) Posterior.

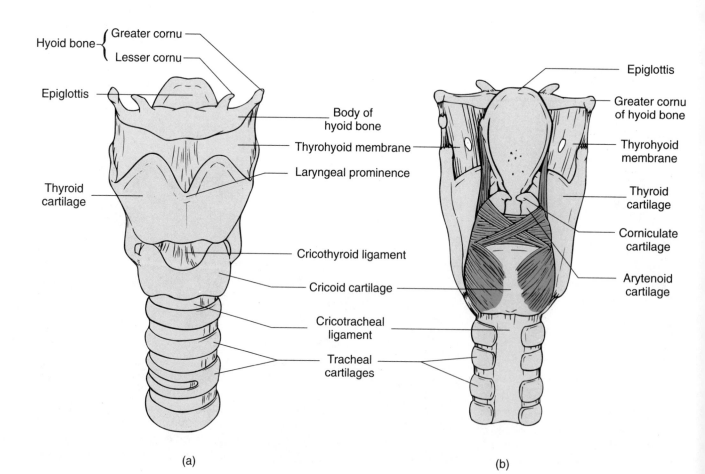

(a)   (b)

■ **FIGURE 19.3 Internal views of the larynx**

(a) Sagittal. (b) Superior – folds closed. (c) Superior – folds opened.

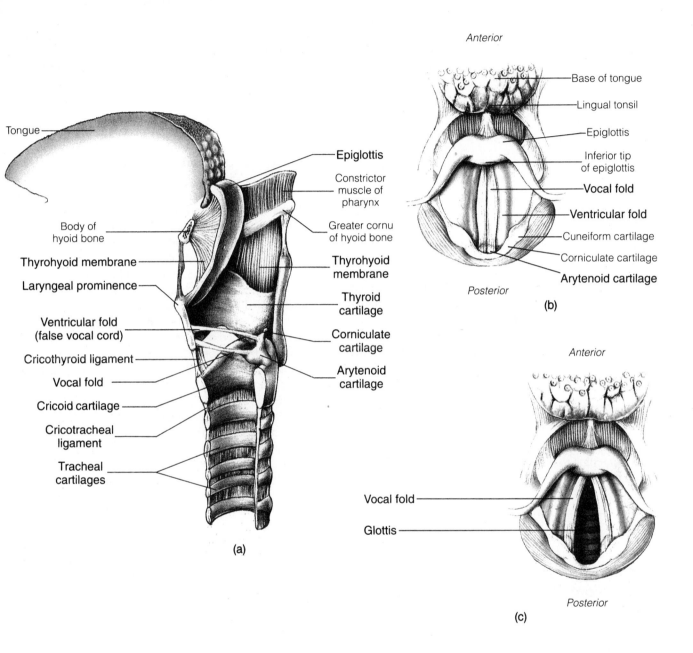

nent shield-shaped **thyroid cartilage**. The **epiglottis**, behind the base of the tongue, is attached to the inside of the thyroid cartilage and forms the "top lid" of the larynx. It is a single, leaf-shaped structure composed of hyaline and elastic cartilage. The epiglottis is drawn over the entrance to the larynx as it is elevated in swallowing, to prevent food from entering the airway. A third unpaired cartilage is the **cricoid**, which makes a complete ring around the inferior end of the larynx, and is wider in back. Paired cartilages of the larynx include the pyramid-shaped **arytenoid** (resting on top of the posterior border of the cricoid), and the **cuneiform** and the **corniculate** (both closely related to the arytenoid cartilages).

The **vocal cords** extend in an anterior-posterior direction from the thyroid cartilage to the arytenoid cartilages. Two pairs of folds on the mucous membrane extend inward from the lateral walls of the larynx: the upper ventricular folds (**false vocal cords**) and lower **true vocal cords**. The space between the lateral portions of the epiglottis and the vocal folds is the **vestibule**. The pocket of the wall between the upper and lower cords is known as the **ventricle**, while the space between the true vocal cords is called the **glottis** (or **rima glottidis**). The size of this space is regulated by tension on the true vocal cords exerted by the arytenoid cartilages as they rotate. This is controlled in turn by various intrinsic muscles of the larynx, which are attached to the cartilages. Air is forced through the glottis during expiration, and as the tension of the cords changes, the pitch of the sound that emanates can be varied voluntarily. Normally the glottis closes reflexively during swallowing, as a means of protection against aspiration of foreign material. In addition, the larynx closes in spasms, creating a cough, in response to irritants on the respiratory mucosa. Glottic closure is also necessary to develop pressure within the abdomen, as in defecation. See Figures 19.2 and 19.3.

Materials

A model of the larynx.

Procedure

1. Examine the model of the larynx externally and identify the following cartilages: thyroid, cricoid, epiglottis, arytenoid, cuneiform, and corniculate.
2. Identify the following internal structures: true and false vocal cords, vestibule, ventricle, and rima glottidis.

Conclusion

- As the voice deepens or disappears in laryngitis, what is happening to the anatomy you observe?

_____

_____

_____

EXERCISE 3
Macroscopic Anatomy: The Fetal Pig

Discussion

The respiratory system begins at the nostrils or **external nares**, and passes through the **nasal cavity** and **nasopharynx** to the **oropharynx**. Follow Figure 19.4. At this level, the nasal route joins the oral route, which begins at the mouth and passes through the oral cavity to the oropharynx. The roof of the oral cavity is formed by the **palate**, the hard portion of which forms the floor of the nasal cavity. Posterior to the hard palate is the soft palate, which forms the floor of the nasopharynx into which the eustachian tubes open. Both routes share a common passage through the **laryngopharynx**, and then divide into separate paths. The route that carries only air is the remainder of the respiratory tract, while that carrying food is the digestive tract. The **epiglottis**

lies at the entrance to the lower respiratory tract and covers the **larynx** below it during swallowing. Posterior to the larynx is the **trachea**, characterized by C-shaped rings of cartilage that maintain the open structure of the trachea at all times. The trachea divides at an area called the **carina**, into two main **bronchi**, each of which further divides in the tissue of the lungs. The lungs lie in the thoracic cavity and are covered by a serosal membrane, the **visceral pleura**, which reflects on itself and lines the inner thoracic wall at the **parietal pleura** (see Figure 19.5). As the trachea divides, so does the blood supply of the lungs. At the root or hilum of the lungs the bronchi, pulmonary arteries and pulmonary veins can be seen. The bronchi divide ultimately into bronchioles, which have no cartilage supports. Bronchioles continue to divide many times, passing into very small structures.

The lung of the pig is divided into four lobes on the right side and three on the left side. The major lobes are the apical, cardiac and diaphragmatic. In addition the right lung has an intermediate lobe located under the apex of the heart. Each lung is served by a main bronchus which then divides as it enters the lobe of the lung.

Two sets of muscles operate in normal ventilations. The dome-shaped **diaphragm** is a sheet of muscle across the lower border of the thorax. It is innervated by the **phrenic nerve**. The esophagus, vena cava, and aorta pass through the diaphragm. When this muscle contracts it moves downward, increasing the volume of the thoracic cavity. The **intercostal muscles** between the ribs contract and lift the rib cage forward and upward, thereby expanding the thoracic cavity, causing inspiration. These muscles relax in normal expiration as the volume of the thoracic cavity decreases.

Materials

A preserved and injected fetal pig; dissecting tray and instruments.
Be sure to review safety precautions before handling fresh or preserved specimens.

Procedure

1. Locate the paired openings, the external nares which lead into the nasal cavity.
2. Extend the mouth opening by cutting the mandible at the jaw angle with a scalpel or bone shears. Observe the anterior hard palate and posterior soft palate.
3. Make a longitudinal medial slit in the soft palate which exposes the nasopharynx. The Eustachian tube extends from the wall of the nasopharynx to the middle ear.
4. Observe the epiglottis, a flap of cartilage near the base of the tongue and on top of the larynx.

■ **FIGURE 19.4 (a) The upper respiratory system of the fetal pig   (b) The respiratory system of the fetal pig**
Refer to the color photo gallery.

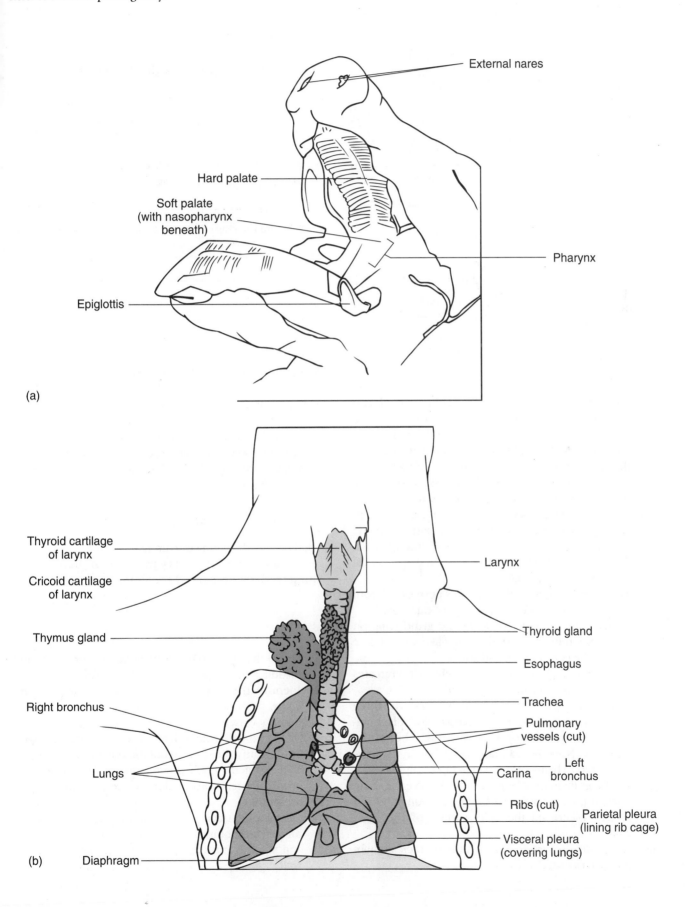

(a)

External nares

Hard palate

Soft palate
(with nasopharynx
beneath)

Pharynx

Epiglottis

(b)

Thyroid cartilage
of larynx

Cricoid cartilage
of larynx

Thymus gland

Right bronchus

Lungs

Diaphragm

Larynx

Thyroid gland

Esophagus

Trachea

Pulmonary
vessels (cut)

Left
bronchus

Carina

Ribs (cut)

Parietal pleura
(lining rib cage)

Visceral pleura
(covering lungs)

■ **FIGURE 19.5 The pleura**

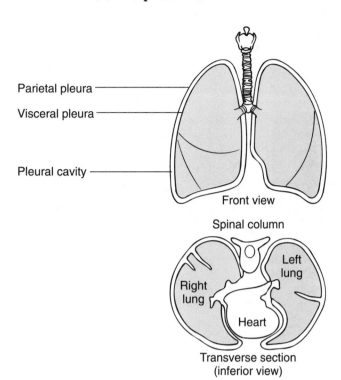

Parietal pleura

Visceral pleura

Pleural cavity

Front view

Spinal column

Left lung

Right lung

Heart

Transverse section
(inferior view)

5. Locate the oropharynx at the back of the mouth and the laryngopharynx below the oropharynx.
6. Pass a dull probe into the larynx and one into the esophagus just dorsal to the laryngopharynx. Note the respiratory tract is ventral to the esophagus.
7. Examine the larynx. It may be necessary to remove some tissue such as thymus gland and muscle to expose its surface. Identify the thyroid and cricoid cartilages. Make a longitudinal incision along the midventral surface of the larynx wall and pull back the two halves. Observe the false and true vocal cords on the inner wall.
8. Identify the trachea which extends posteriorly from the larynx. Note its C-shaped rings of cartilage.
9. Observe the reddish-brown thyroid gland lying on top of the trachea. Note the string-like vagus nerve next to the common carotid artery.
10. Follow the trachea to the carina where it bifurcates into the main right and left bronchi.
11. The bronchi lead into the lungs. Observe the three lobes on both sides: apical, cardiac and diaphragmatic and the additional intermediate lobe on the right side. Note the lungs are rather firm in the fetus compared to being more spongy in the adult.
12. The lungs lie in the pleural cavity. Observe the shiny visceral pleura on the surface of the lung and the parietal pleura on the inner surface of the chest wall.
13. Locate the diaphragm between the thoracic and abdominal cavities.

14. Identify the white thread-like phrenic nerve that extends along the pericardium to the diaphragm.

Conclusions

- What structures pass through the diaphragm?

  _____

  _____

  _____

- What structure that you observed closes off the naso-pharynx during swallowing?

  _____

- Why are the rings of cartilage in the trachea C-shaped?

  _____

  _____

  _____

EXERCISE 4
Macroscopic Anatomy: The Sheep Pluck

Discussion

A "pluck," consisting of the trachea, lungs, larynx, and heart (and sometimes pieces of the diaphragm), is removed from an animal after slaughter.

Materials

Sheep pluck; dissecting instruments and tray.
   Be sure to review safety precautions before handling fresh or preserved specimens.

Procedure

1. Orient the pluck so that the trachea is nearest you and the lungs and diaphragm are farthest from you.
2. Identify the larynx, trachea, main stem bronchi, lungs, heart, aorta, vena cavae, diaphragm, and pulmonary vessels.
3. Identify the thyroid, cricoid, and epiglottal cartilages; also identify the thyrohyoid membrane and cricothyroid ligament. Look down into the larynx and observe the vocal cords and the space in between them, the glottis.
4. Observe the trachea; slice the trachea longitudinally for several centimeters and observe the cartilage rings and smooth muscle; feel the inside of the tracheal wall.

5. Observe the lungs; count the right and left lobes; observe the visceral pleura on the lung surface; follow the pulmonary vessels between the lungs and heart, noting exit from and entry into the heart.

6. Note the aorta and locate the main branches as well as the ligamentum arteriosum.

7. Observe the bronchi as they angle toward the lungs; note the branches to the right and left lungs; cut along the length of a bronchus as far into the lung as possible and note the branching and the presence (or lack of) cartilage.

■ **FIGURE 19.6 Lobes of the lung**

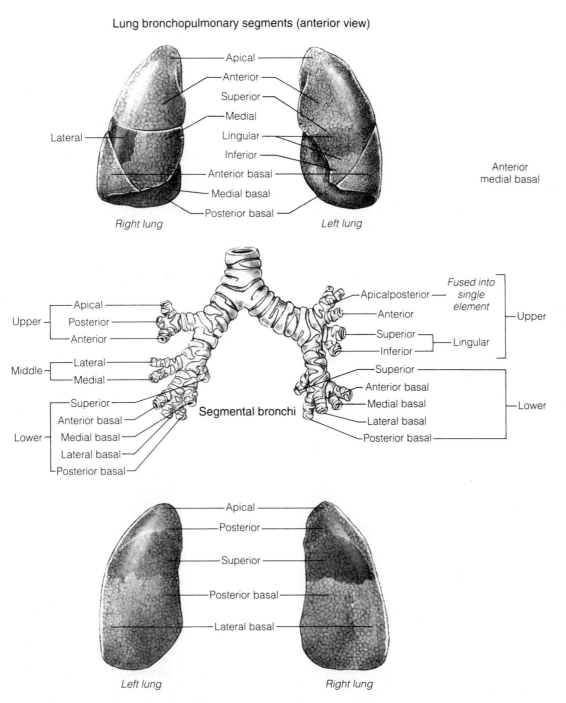

Lung bronchopulmonary segments (anterior view)

Right lung   Left lung

Apical
Anterior
Superior
Medial
Lingular
Inferior
Anterior basal
Medial basal
Posterior basal

Lateral

Anterior medial basal

Segmental bronchi

Upper — Apical, Posterior, Anterior
Middle — Lateral, Medial
Lower — Superior, Anterior basal, Medial basal, Lateral basal, Posterior basal

Apicalposterior — Fused into single element
Anterior
Superior — Lingular
Inferior

Upper

Superior
Anterior basal
Medial basal
Lateral basal
Posterior basal

Lower

Lung bronchopulmonary segments (posterior view)

Apical
Posterior
Superior
Posterior basal
Lateral basal

Left lung   Right lung

Conclusions

- How does the number of lobes of the sheep lungs compare to that of the human being?

_____

_____

_____

- How does the division of the main stem bronchi compare between sheep and human?

_____

_____

_____

- Why does the inside of the trachea feel as it does? Explain.

_____

_____

_____

_____

EXERCISE 5
Macroscopic Anatomy: The Human

Discussion

The structures of the respiratory system are essentially the same in human, sheep, and fetal pig. Variations occur in the orientation of features in the head, and in the number of lobes of the lungs.

In humans the right primary bronchus branches into the three right lobes of the lung via three secondary bronchi. The left primary bronchus comes off the trachea at a greater angle than the right, and branches into the two left lobes of the lung via two secondary bronchi. Secondary bronchi further divide into tertiary bronchi, which serve the various bronchopulmonary segments of the lung. See Figure 19.6. The relationship of the pleura to the pericardium is illustrated in Figure 19.5. Notice that as the visceral pleura folds on itself in the area of the mediastinum it becomes the pericardium surrounding the heart. See Appendix B for landmarks relating to the respiratory system.

Occasionally the term **"lobule"** is used to refer to the unit of structure of the lung: it is pyramidal, with the base of the lung surface and the apex pointing toward the hi-

lum. A terminal bronchiole and the pulmonary artery branch enter the apex of the lobule. Incomplete separation of lobules occurs in the human by connective tissue **septa**, which are continuous both with the connective tissue around the bronchi and major vessels (and ultimately to the hilum) as well as the deep connective tissue layers of the pleura. Lymphatic capillaries and branches of the pulmonary vein run in the septa.

Materials

Charts of the human respiratory system; human torso; sagittal section of human skull.

Procedure

1. Using Figure 19.1, identify structures of the human respiratory system from external nares to alveolus.
2. Using the skull, identify the sinuses and nasal cavity structures outlined in Table 5.6 and Figure 19.1a.

Conclusions

- What bones of the skull form the nasal cavity and nasal septum?

_____

_____

_____

- Why do particles more often lodge and obstruct airflow in the right bronchus than the left bronchus?

_____

_____

_____

_____

❖ Quick Quiz 1

1. Which structure did you observe passing through the diaphragm?
    (a) aorta
    (b) esophagus
    (c) inferior vena cava
    (d) all of these

2. Where did you find the eustachian tubes (from the middle ear) opening up?
   (a) behind the hard palate
   (b) behind the soft palate
   (c) in the nasopharynx
   (d) two of these
3. Which bone makes up the floor of the nasal cavity?
   (a) palatine bone
   (b) palatine process
   (c) a portion of the maxilla
   (d) all of these
4. Which of these is an appropriate description of the diaphragm?
   (a) It consists of two hemidiaphragms.
   (b) It is served by the phrenic nerve.
   (c) It separates the thoracic cavity from the abdominal cavity.
   (d) All of these are correct.
5. Which of these appropriately describes the true vocal cords?
   (a) They are separated from the false vocal cords by the vestibule.
   (b) The space between the cords is known as the glottis.
   (c) The cords run from the thyroid cartilage anteriorly to the arytenoid cartilages.
   (d) All of these.

❖ Functions of the Respiratory System

Background

The exchange of gases at the cellular level mainly depends on the integrated functioning of the cardiovascular and respiratory systems. Proper delivery of fresh air to the circulating blood and control of this delivery is examined in assessing respiratory functions.

The tests that examine the volume of air in the lung during various respiratory maneuvers and rates of airflow are known as **pulmonary function tests**. Examining a patient and comparing his values to known standards allows certain patterns of impaired lung function to be seen. These coincide with the relative health or disease of the respiratory system. These tests are diagnostic tools and assist in evaluations regarding surgery, use of anesthesia, surveys of disease in the community, and evaluation for workmen's compensation.

Two major forces must be considered in the movement of air through the tracheobronchial tree.

1. The ease with which the lung distends is called **compliance**, and it is inversely related to the elasticity of the lung-thorax system. Compliance is a measure of how easy it is to fill the lungs.
2. The **resistance** to air as it flows is affected by several factors (e.g., rate of airflow and size of passageways), and it is important to remember that it is greater during the air *outflow*, when airways normally are somewhat reduced in diameter.

Pulmonary function tests can examine either the **static** (**volumes** and **capacities**) or the **dynamic** (**rates** of airflow) characteristics of the lung-thorax system. *Diminished lung volume* with normal expiratory flow rates may indicate a reduced distensibility of the lung-thorax system (as in pulmonary fibrosis). Disorders of this type are termed **restrictive** disorders. *Diminished forced expiratory flow rates* may indicate either reduced elasticity of the lung or obstruction to the airflow. These are termed **obstructive** disorders.

Certain individuals may have static values in the normal range, but may not be able to *use* these capacities effectively. Their respiratory efforts may be impaired because the velocity of air outflow is compromised. These changes would be seen as reduced values in tests of dynamic lung volumes.

EXERCISE 1

Static Pulmonary Volumes and Capacities

Discussion

Figure 19.7 is a graphic representation (a spirogram) of the different volumes of air contained in the lungs under various breathing conditions. The term **capacity** is used to refer to two or more volumes. Each of these volumes and capacities is described in Tables 19.2 and 19.3. It is important to remember that these volumes vary even in normal people, and depend on age, size, and sex.

Most experimental and clinical laboratories are equipped with a **spirometer**, an instrument used to measure pulmonary lung volumes and capacities as well as dynamic pulmonary functions and oxygen consumption. Most spirometers consist of a closed system of one-way tubes connected to a canister that contains an inverted bell. The bell rides up and down in a water sleeve (see Figure 19.8). Through a pulley system, the inverted bell is connected to a pen that records all respiratory movements. Inspiration draws air from the spirometer, lowering the bell and raising the pen. Expiration has the opposite effect. The recording paper is graduated horizontally for time (depending on paper speed) and vertically in milliliters, enabling the observer to read lung volumes and capacities with each ventilatory maneuver.

■ **FIGURE 19.7 Pulmonary volumes and capacities**

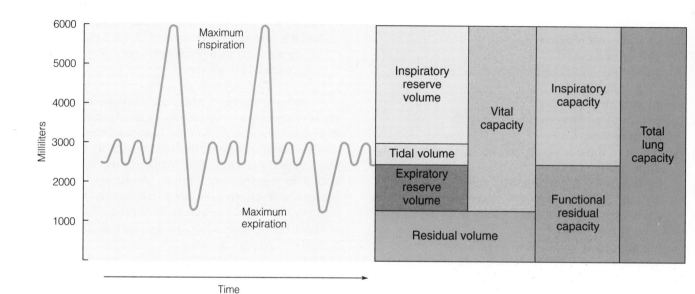

■ **TABLE 19.2 Pulmonary Volumes**

CAPACITY	DESCRIPTION	APPROXIMATE AMOUNT (ml)
Tidal volume	Represents the volume of air that is normally inspired and expired during relaxed breathing	500
Inspiratory reserve volume	Represents the volume of air that can be inspired in addition to the normal tidal volume; this volume can be diminished in a **restrictive** disease (e.g., fibrotic lung disease)	3000
Expiratory reserve volume	Represents the volume of air that can be expired at the end of a normal tidal volume; it is limited by a number of factors: how high the diaphragm moves on expiration, tendency for small airway closure upon a forced expiration effort, chest wall resistance to further volume decrease, and strength of expiratory muscles	1200
Residual volume	Represents the volume of air left in the lungs after a maximal expiration; this air cannot be exhaled; this volume, which normally increases with age, is usually about 25–30% of total lung capacity; when elastic recoil diminishes or there is an obstruction to air outflow, this volume rises	1200
Minimal air volume	Represents the air that remains in the distal parts of the lung when the lungs are deflated	200

■ **TABLE 19.3 Pulmonary Capacities**

CAPACITY	DESCRIPTION	AMOUNT (ml)
Inspiratory capacity	Represents a summation of the tidal volume and the inspiratory reserve volume; it is the maximum amount of air that can be inspired after a normal expiration; it normally measures about 60% of total lung capacity	3500
Expiratory capacity	Represents the summation of tidal volume and expiratory reserve volume; it is the maximum amount of air that can be expired after the end of a normal inspiration	1700
Functional residual capacity	Represents a summation of the expiratory reserve volume and the residual volume; it is the amount of air that remains in the lungs after a normal expiration; this normally measures about 40% of total lung capacity and is smaller when lying down or if airflow is diminished (by obstruction or loss of elasticity)	2400
Vital capacity	Represents the sum of the expiratory reserve volume, tidal volume, and inspiratory reserve volume; it is the maximum amount of air that can be expired after a maximum inspiration; normally measures about 70–75% of total lung capacity and offers a way of viewing distensibility of the lung-thorax system	4700
Total lung capacity	Represents the sum of the tidal volume, expiratory reserve volume, inspiratory reserve volume, and residual volume; it is the total amount of air contained in the lungs after a maximum inspiration; limited by the strength of the respiratory muscles and the elastic resistance of the lungs and chest wall	About 6000

■ **FIGURE 19.8 The spirometer**

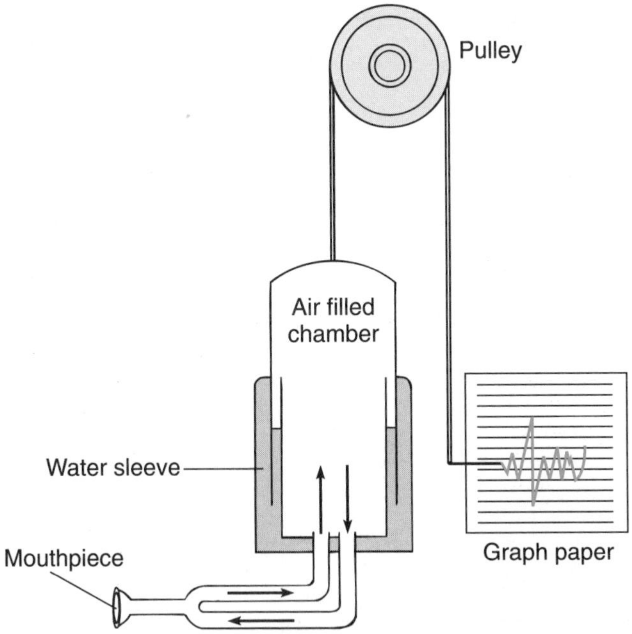

Materials

A respirometer; mouthpieces; hand-held spirometer. *Note:* [Have your instructor demonstrate the use of a respirometer and then proceed with the following exercises. Some of these values can also be obtained by using the student (hand-held) spirometer.] Work with a lab partner for all tests.

Procedure

Set drum speed at 32 mm/min and open connector top to the respirometer.

1. *Tidal Volume and Minute Ventilation* Have the subject attach a nose clip to maximize airflow through the mouth. Breathe through the mouthpiece with the metal tubing connector tap open to the atmosphere. Record for a few minutes and note the tidal volume in Table A.

2. *Respiratory Rate and Minute Volume*   From your recording determine the length of time elapsed during one normal breath. Record in Table A. Calculate the minute volume using these data and record the results in Table A.

3. *Expiratory Reserve Volume*   After a series of normal tidal volumes, have the subject expire maximally after a normal expiration. Observe and record the expiratory reserve volume.

4. *Inspiratory Reserve Volume*   After a series of normal tidal volumes, have the subject inspire maximally after a normal inspiration. Observe and record the inspiratory reserve volume.

5. *Inspiratory Capacity*   After a series of normal tidal volumes, have the subject inspire maximally after a normal inspiration. Record the volume.

6. *Vital Capacity*   Instruct the subject to inspire maximally, and then to expire maximally. Perform this maneuver several times and record the highest volume.

7. Dispose of the mouthpiece according to the instructor's directions.

Conclusions

• Compare experimental values with those read from the nomograms (Figures 19.9 and 19.10). Which capacities in Table A could have been estimated by mathematical addition?

_____

_____

_____

• Which volumes described in the Discussion section cannot be measured by mathematical addition?

_____

_____

_____

_____

• Is all the air computed for minute ventilation involved in gas exchange? Explain.

_____

_____

_____

_____

_____

Observations

■ **TABLE A Pulmonary Volumes and Capacities**

VOLUME OR CAPACITY	EXPERIMENTAL	STANDARD VALUE
Tidal volume		
Respiratory rate		
Minute volume		
Expiratory reserve volume		
Inspiratory reserve volume		
Inspiratory capacity		
Vital capacity		

- Would vital capacity be increased or decreased in a "side-to-side" abnormal curvature of the spine? Explain.

_____

_____

_____

_____

- How does the residual volume change in a disease which obstructs air outflow (like emphysema)? Explain.

_____

_____

- Why is a clip placed on the nose for these procedures?

_____

_____

_____

■ **FIGURE 19.9 Respiratory nomogram: Male**

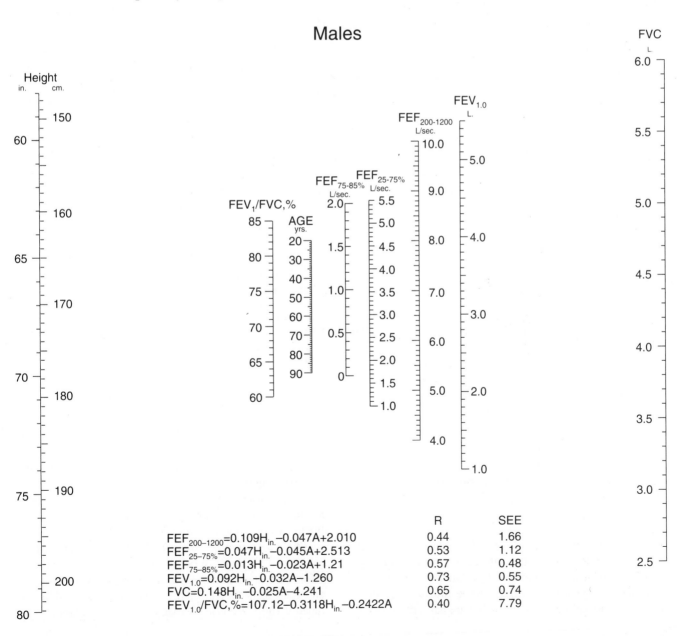

	R	SEE
$FEF_{200-1200} = 0.109H_{in.} - 0.047A + 2.010$	0.44	1.66
$FEF_{25-75\%} = 0.047H_{in.} - 0.045A + 2.513$	0.53	1.12
$FEF_{75-85\%} = 0.013H_{in.} - 0.023A + 1.21$	0.57	0.48
$FEV_{1.0} = 0.092H_{in.} - 0.032A - 1.260$	0.73	0.55
$FVC = 0.148H_{in.} - 0.025A - 4.241$	0.65	0.74
$FEV_{1.0}/FVC,\% = 107.12 - 0.3118H_{in.} - 0.2422A$	0.40	7.79

■ **FIGURE 19.10 Respiratory nomogram: Female**

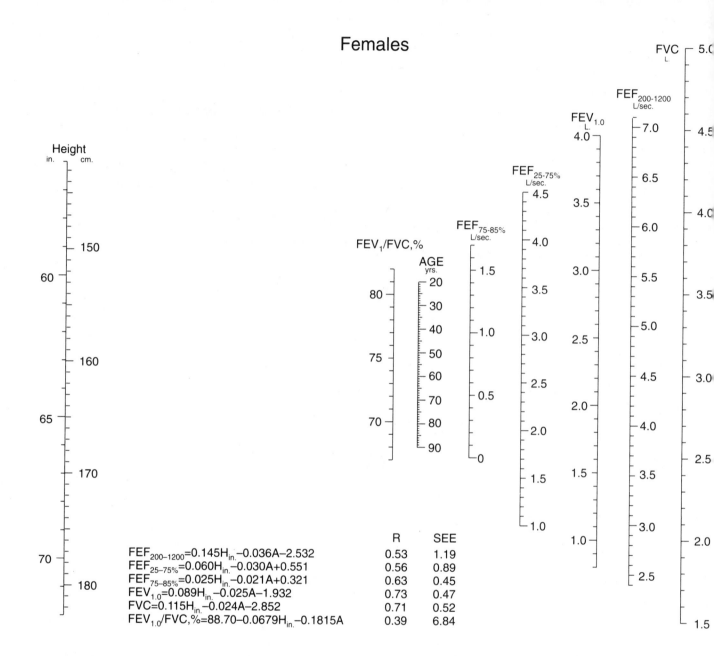

Females

$$FEF_{200-1200}=0.145H_{in.}-0.036A-2.532$$
$$FEF_{25-75\%}=0.060H_{in.}-0.030A+0.551$$
$$FEF_{75-85\%}=0.025H_{in.}-0.021A+0.321$$
$$FEV_{1.0}=0.089H_{in.}-0.025A-1.932$$
$$FVC=0.115H_{in.}-0.024A-2.852$$
$$FEV_{1.0}/FVC,\%=88.70-0.0679H_{in.}-0.1815A$$

	R	SEE
$FEF_{200-1200}$	0.53	1.19
$FEF_{25-75\%}$	0.56	0.89
$FEF_{75-85\%}$	0.63	0.45
$FEV_{1.0}$	0.73	0.47
FVC	0.71	0.52
$FEV_{1.0}/FVC,\%$	0.39	6.84

EXERCISE 2
Dynamic Pulmonary Volumes and Rates

Discussion

Information about resistance to airflow may be obtained from tests that monitor the rate of air outflow during forced ventillatory movements. Resistance to air outflow is generally increased in **obstructive** diseases like emphysema and asthma. The following are basic definitions of some of these measurements. Figure 19.11 illustrates these spirograms.

MVV (maximum voluntary ventilation) The amount of air that can be moved in 1 min when a subject breathes as deeply and rapidly as possible.

FEV (forced expiratory volume) A measure of the amount of air that can be moved by a forced expiratory effort in a given time. A subject inspires to full volume and then forces a complete expiration. The total air exhaled is the forced tidal capacity (FVC). By graphing this volume against time, a rate of forced exhalation can be obtained in liters per second.

FEV 1.0%   The amount of air moved in the first second of a forced expiratory effort. A healthy person between the ages of 20 and 30 can usually expire 83% of his total FVC in the first second. This is the $FEV_{1.0\%}$. Values can likewise be obtained for the $FEV_{2.0\%}$ and $FEV_{3.0\%}$, which are 93 and 97%, respectively.

MEFR (maximum expiratory flow rate)   The MEFR is calculated by measuring the average rate of airflow in the midportion of the expiratory spirogram. It is thought to be a very accurate way to determine resistance to airflow. The rate of airflow in the upper airways (large airways) is measured by using the slope of the forced vital capacity between 200 and 1200 ml of expired air. This is known as the $FEV_{200-1200}$. The rate of airflow in the middle airway (medium size) is measured by using the slope of the forced vital capacity along the portion of expired air. This is known as the $FEV_{25-75\%}$. Both values vary in the same manner.

Materials

A spirometer; mouthpieces.

Procedure

1. Have the subject breathe as deeply and rapidly as possible into the spirometer for 20 sec. Record the volume in Table B.
2. Have the subject take a maximum inspiration. Turn the paper speed to 1600 mm/min (1920 mm/min if possible), and have subject exhale as forcefully and rapidly as possible. Record the data in Table B.
3. Compare these values with those on the nomogram (both graphic and calculated) of Figures 19.9 and 19.10.
4. Dispose of the mouthpiece according to the instructor's directions.

■ **FIGURE 19.11 A spirogram**

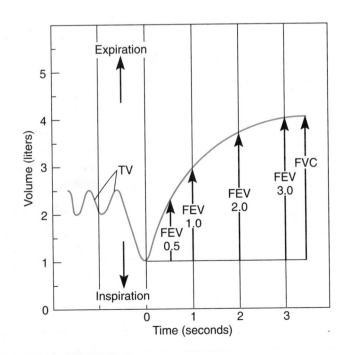

Observations

■ TABLE B Respiration Data

	EXPERIMENTAL VALUES	ACCEPTED VALUES	CALCULATED VALUES
MVV Volume of air moved in 20 sec	_____	_____	_____
Volume of air moved in 1 min	_____	_____	_____
FVC Volume of air exhaled in forced effort	_____	_____	_____
$FEV_{1.0}$ Volume of air exhaled in first second	_____	_____	_____
$FEV_{1.0\%}$ Volume of air exhaled in first second/FVC	_____	_____	_____
$FEV_{2.0}$ Volume of air exhaled in second second	_____	_____	_____
$FEV_{2.0\%}$ Volume of air exhaled in second second/FVC	_____	_____	_____
$FEV_{3.0}$ Volume of air exhaled in third second	_____	_____	_____
$FEV_{3.0\%}$ Volume of air exhaled in third second/FVC	_____	_____	_____
$FEV_{200-1200}$ Rate of airflow between 200–1200 ml	_____	_____	_____
$FEV_{25-75\%}$ Rate of airflow between 25 and 75% expiration	_____	_____	_____

Conclusions

- How many seconds elapse before all air is exhaled?

  _____

- What is the normal time?

  _____

- How would this value be affected in obstructive disease and why?

  _____

  _____

  _____

  _____

- How would this value be affected in restrictive disease and why?

  _____

  _____

  _____

  _____

- How does your $FEV_{1.0\%}$ compare to the normal value? How would this value be altered in obstructive disease? Explain.

  _____

  _____

  _____

  _____

  _____

  _____

- How does your $FEV_{25-75\%}$ compare to the normal value? How would this value be altered in restrictive disease? Explain.

  _____

  _____

  _____

  _____

  _____

EXERCISE 3
The Effect of Carbon Dioxide on the Respiratory Center

Discussion

Hyperventilation washes much of the carbon dioxide out of the blood. The increases in the oxygen content of the arterial blood after hyperventilation are negligible, but the carbon dioxide level may fall from a normal value of 44 mm Hg to as low as 15 mm Hg.

Hyperventilation may result in a feeling of dizziness because of cerebral anoxia (due to either decreased blood pressure, to an increase in pH, or to the constriction of cerebral vessels caused by the diminished carbon dioxide content of the blood). In this experiment we demonstrate the effects of varying levels of carbon dioxide on the control of respiration.

Materials

A plastic (or paper) bag; straws; clock.

Procedure

Tests are performed by each student individually. The nose should be pinched closed.

1. Breathe quietly for 3 min and then note how long the breath can be held after a quiet inspiration. Record the time in the Observations section.
2. Hyperventilate (i.e., breathe deeply and rapidly) 20 times and then determine how long the breath can be held. Do not continue the experiment if a pronounced feeling of dizziness results. Record the time.
3. Place a plastic bag over the mouth and nose. Hyperventilate into the bag 20 times. Note how long the breath can be held after hyperventilation into the bag. Record.
4. Run around the area rapidly for 1 to 2 min. Determine how long the breath can be held immediately after the exercse. Record.

Observations

Time breath can be held:

a. Normally   _____

b. After hyperventilation   _____

c. After breathing into bag   _____

d. After exercise   _____

Conclusions

• Explain the control over normal (non-disease-state) respiration. In discussing the results obtained in the experiment, make use of the fact that the respiratory center is very sensitive to slight changes in the carbon dioxide content of the blood.

_____

_____

_____

_____

_____

_____

_____

_____

_____

• How do you explain the change in time comparing:

a. Steps 1 and 2 _____

b. Steps 1 and 3 _____

c. Steps 1 and 4 _____

 Quick Quiz 2

Match column A to column B.

	A		B
_____	1. normal value for $FEV_{1.0\%}$	(a)	FRC
		(b)	80-85%
_____	2. normal value for VC for 30 year-old male six feet tall	(c)	1200 ml
		(d)	5060 ml
		(e)	93%
		(f)	500 ml
_____	3. normal expiratory reserve volume		
_____	4. increases in emphysema		
_____	5. normal tidal volume		

The Digestive System

OVERVIEW

In this unit you will become familiar with the overall macroscopic anatomy of the organs which comprise the digestive tube and the accessory structures associated with it. You will also study the microscopic structure of several of these organs. Additionally, you will conduct experiments that duplicate the biochemical processes that occur in the human digestive tract.

OUTLINE

Anatomy of the Digestive System

Background

A sandwich and a slice of blackberry pie sound good for lunch. Just the thought may cause a little saliva to flow in your mouth. The values of these foods, however, go far beyond the satisfying good taste: they release energy for your body and provide building materials, once they have been digested.

The digestive system carries out four functions: the ingestion and digestion of food, absorption of food breakdown products, and egestion of undigested foods and wastes. These functions are carried out in a tubular tract that is adapted for the functions it performs. The tract opens to the exterior at each end and has a muscular wall throughout to aid in the propulsion and mixing motions of food. In addition, the innermost epithelial cells of the tube, as well as some accessory organs, secrete substances that aid the chemical breakdown of food. Last, the digestive tube is adapted for absorption. The many folds of the inner wall increase the surface area across which digested materials pass.

The tubular structure is modified into the following organs: mouth, esophagus, stomach, small and large intestines, rectum, and anus. Three accessory organs secrete into this tract: the salivary glands; the pancreas, and the liver. The teeth and the tongue are additional accessory organs of the mouth. In this section both the microscopic and macroscopic details are examined. See Table 20.1 and Figures 20.1 and 20.2.

■ **TABLE 20.1 Histology of the Digestive Tract (Esophagus–Anus)**

TISSUE LAYER	DESCRIPTION
Mucosa	Innermost protective or secretory epithelium on loose connective tissue, the **lamina propria**, which contains lymph tissue and blood vessels for absorption of digested foods; surrounded by the **muscularis mucosa**, a thin layer of smooth muscle that acts to mix food
Submucosa	Dense or loose connective tissue, binding mucosa muscularis; highly vascular, aids in absorption of digested food; contains nerve **plexus of Meissner** served by automatic nerves (vagus and a few sympathetic fibers control as in muscularis; see below); may contain bases of glands that extend to and open on free border of mucosa
Muscularis	Mostly a smooth muscle layer (skeletal in upper esophagus and anus); consists of two sheets: a tightly wound "circular" layer and a loosely encircling longitudinal" layer; circular layer thickened into sphincters; engages in mixing and propulsion of food; **Auerbach's plexus (myenteric)** is the major intrinsic nerve network, which exhibits both intrinsic coordination (in the absence of external nervous stimulation) and extrinsic control (in response to autonomic nervous control, as for submucosa above)
Serosa or adventitia	Relatively dense connective tissue wrapping around the tract; above the diaphragm the adventitia acts merely to connect structures to other structures; below diaphragm the serosa (or serous membrane) is combined with a layer of fluid-secreting mesothelial cells

■ **FIGURE 20.1 The human digestive system**

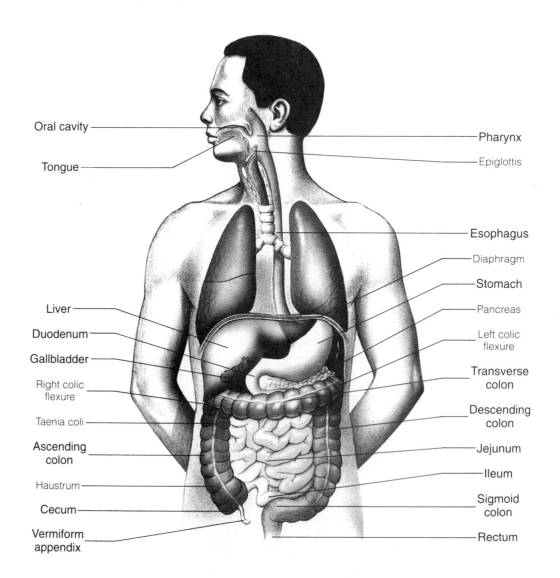

■ **FIGURE 20.2   Generalized cross section of the digestive tract**

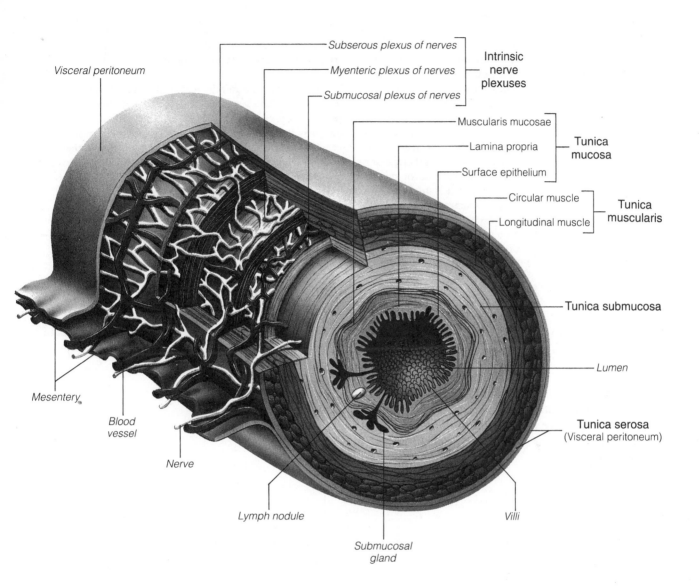

EXERCISE 1
Microscopic Anatomy

Discussion

The tubular tract of the digestive system is specialized in the following manner: cells and glands secrete appropriate digestive enzymes into the tract; nerves control secretions, and the contraction of muscles in the wall; infoldings of the lining promote absorption; membranes serve to attach the tract to other structures. The wall of the tract is divided into four layers. Working from the inner layer outward they are: the mucosa, the submucosa, the muscularis, and the serosa or adventitia. Some variations in the histology are listed in Table 20.2 and illustrated in Figure 20.3.

Materials

A microscope; slides of tongue (with taste buds), esophageal–stomach junction, small intestine (duodenum, ileum, jejunum), colon.

Procedure

1. Examine each of the slides for the structures listed in Table 20.1. Identify all four layers in the slides: mucosa, submucosa, muscularis, and serosa or adventitia.
2. Examine each of the slides for the structures specific to that area, as listed in Table 20.2 and illustrated in Figure 20.3.

■ **TABLE 20.2 Variations of Tissue in the Digestive Tract**

ORGAN	DESCRIPTION
Tongue	Stratified muscle covered with a mucosa, containing mucous and serous (**Ebner's**) glands, **papillae** (raised areas containing taste buds of barrel-shaped **sustentacular cells**), and 4–16 **neuroepithelial** taste cells with short taste hairs on the cell's free surface projecting into the cavity of the outer taste pore
Esophageal-stomach junction	Mucosa of the esophagus is stratified squamous epithelium and the wall is muscular with thick folds; the circular layer is thickened into the **cardiac sphincter** at the entrance to the stomach; upon entering stomach there is a sharp transition of the epithelium of the mucosa to "pits," which fold deep into the muscularis mucosal layer (there is a third, inner oblique layer of muscle in the stomach); the epithelial cells are of several types: **mucus-secreting cells**[a] in the neck of the "pit" or gland (columnar); spherical or pyramidal parietal cells at the base of the "pit" secreting hydrochloric acid; pyramidal chief or zymogenic cells secreting digestive enzymes (darker granules); the circular layer of muscle is thickened into the **pyloric sphincter** at the exit of the stomach
Duodenum	Mucosa is characterized by fingerlike projections of villi and depressions, the latter extending through the muscularis mucosa with their bases in the submucosa; the cells in the depressions are mucus-secreting cells that form **Brunner's glands**; the epithelial cells are columnar with a brush border of **microvilli**; villi have a core of lamina propria
Small intestine	Villi continue; the glands called **crypts of Lieberkühn**, secrete digestive enzymes at their bases; muscle layers of muscularis are clearly differentiated into an inner circular layer and outer longitudinal layer; components of **Auerbach's plexus** may be seen between muscle layers; submucosa shows masses of purple lymphoid tissue or **Peyer's patches**
Large intestine	No villi; crypts lined almost entirely by goblet cells

[a]Many are goblet cells.

3. Draw and label diagrams of the tissue in the Observations section. Note specialized cells and/or glands of the area. Observe and draw any other appropriate histological sections (appendix, rectum, etc.) that are available.

Small intestine (note area)

Observations

Tongue

Large intestine

Esophagus–stomach

Other _____

■ **FIGURE 20.3 Some tissues of the digestive tract**
(a) Small intestine. (b) Gastric glands.

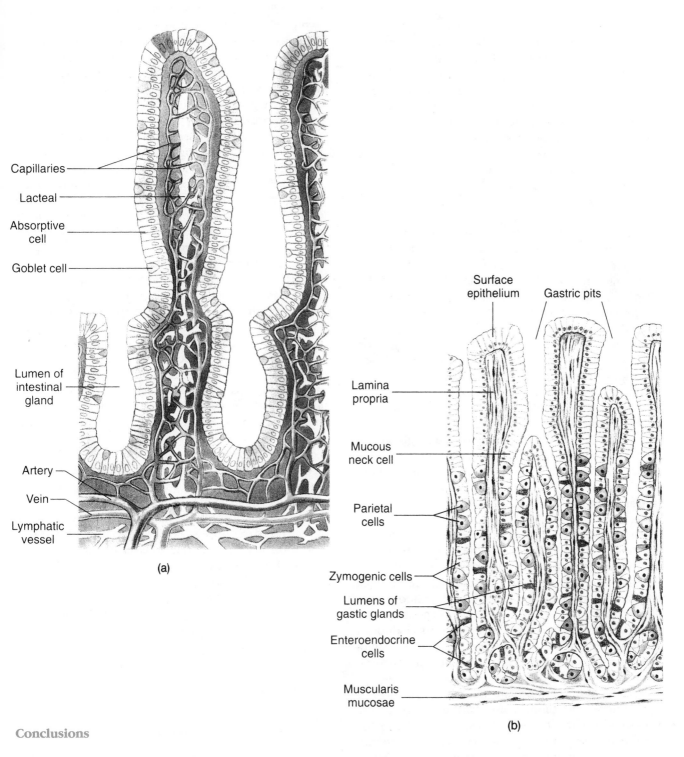

(a)

(b)

Conclusions

- **What type of epithelium is present in the esophagus? Why?**

- Does the histology of the organ relate to its activities? Explain.

EXERCISE 2

Macroscopic Anatomy: The Fetal Pig

Discussion

The upper alimentary canal begins at the mouth, bounded by the **lips** or **labia**. This leads into the **oral** or **buccal cavity**, which ends posterior to the **teeth** as it narrows at muscular infoldings of the lateral wall, called the **fauces**. The oral cavity is bounded by the hard and soft palates above, the cheeks on the sides, and the tongue and tissue below. The cavity itself is divided into the **vestibule** (the space between teeth and cheeks) and the **oral cavity proper**. The anterior **hard palate** separates the oral and nasal cavities whereas the posterior **soft palate** separates the oral cavity from the nasopharynx.

The **pattern of dentition** (incisors, canines, premolars, molars) in the fetal pig is 3-1-4-0 while that for the adult pig is 3-1-4-3. The muscular **tongue** is attached to the floor of the mouth by a membrane, the **lingual frenulum**. The surface of the tongue is covered by a variety of projections, the **papillae** in which taste buds are embedded.

The **salivary glands** (Figure 20.4) are accessory organs of digestion, the secretions of which are released into the oral cavity (Figure 20.5). There are three pairs of salivary glands which connect to the oral cavity by ducts. The largest is the triangular **parotid gland** located near the base of the ear. **Stensen's duct** connects the parotid gland with the mouth. The **submandibular gland** and the associated **Wharton's duct** lies beneath the parotid gland. The third salivary gland is the **sublingual gland** located anterior to the lower border of the submandibular gland. The duct from this gland passes into the mouth under the tongue.

The mouth connects with the **esophagus** after passing through the **oropharynx** at the back of the mouth and the **laryngopharynx**, just below the oropharynx. The oro- and laryngopharynx are common to both the respiratory and digestive systems. They lead into the trachea (respiratory) and the esophagus (digestive). The esophagus lies dorsal to the trachea and passes through the diaphragm to enter the cardiac region of the stomach. A modified inner ring of muscle, the **cardiac sphincter** controls entry into the stomach. The four major sections of the **stomach** are: **cardiac portion** (at the entrance); the **fundus** (an upward extension on the pig's left side of the stomach); the **body** (bounded by the lesser curvature cranially and the greater curvature caudally); and the **pyloric portion** (the narrowed end of the stomach). The pyloris has a ring of smooth muscle internally, the **pyloric sphincter**, which controls the entrance of chyme (food that has been partly digested in the stomach) into the small intestine. The stomach lining is thrown into folds called **rugae**.

The **small intestine** is a long, coiled tube that fills a good portion of the abdominal cavity. It is divided into three sections: the short, anterior curved **duodenum**; the middle **jejunum**; and the terminal **ileum**. The latter two portions cannot be distinguished externally but are the longest portions. The small intestine is lined with numerous fingerlike projections called **villi**. The duodenum receives secretions from both the pancreas and liver. The **pancreas** is an elongated lobular structure that drains its digestive secretions into the duodenum via a small pancreatic duct.

The **liver** is the largest gland in the body. It consists of five lobes: right lateral, right central, left central, left lateral, and a small caudate lobe. The liver is attached to the diaphragm and ventral body wall by the **falciform ligament**. The **coronary ligament** joins the dorsal section of the liver to the diaphragm. The umbilical vein enters to the left of the **gall bladder**, a saclike organ lying on the undersurface of the right lateral lobe of the liver. The gall bladder stores bile which it receives from the liver via **hepatic ducts**. The **cystic duct** leads out of the gall bladder and joins with the hepatic duct from the liver to continue as the **common bile duct** to the duodenum.

The small intestine joins the **large intestine (colon)**, forming the **cecum**, a blind sac. Internally a ring of smooth muscle, the **ileocecal valve** can be seen. The first part of the large intestine is the **spiral colon**, a compact coiled mass on the left side of the abdomen. This leads into the posterior **rectum** which descends as a straight tube through the pelvic cavity to terminate as the **anus**.

Within the abdominal cavity (Figure 20.6) there are several membranes that connect organs to each other and anchor them to the body wall. The **peritoneum** is the double membrane lining the abdominal cavity. The outer **parietal** layer lines the body wall and the inner **visceral** layer covers the organs. Between the dorsal body wall and small intestine is a double layer of the peritoneum called the **mesentery**. Lymph nodes can be found where the mesentery connects with the body wall. Running between the stomach and spleen is the **greater omentum** and extending from the stomach and duodenum to the liver is the **lesser omentum**.

Materials

A preserved and injected fetal pig; dissecting tray and instruments.

Be sure to review safety precautions before handling preserved or fresh specimens.

■ **FIGURE 20.4 The salivary glands of the fetal pig**
Refer to the color photo gallery.

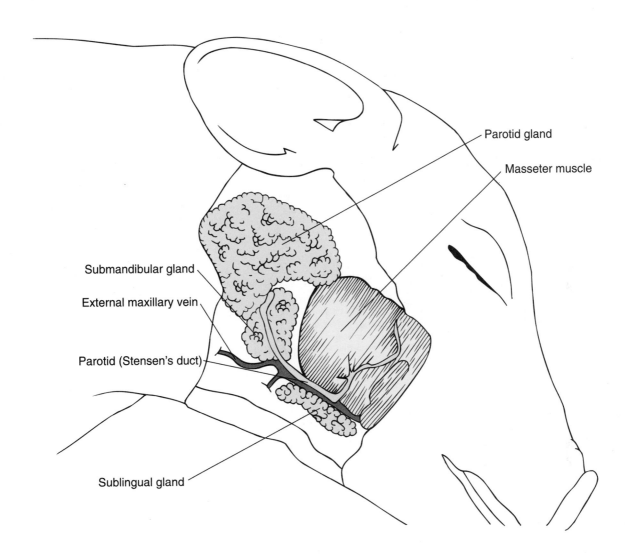

Procedure

1. Observe the entrance to the digestive tract, the mouth which leads into the oral cavity.
2. Entering into the mouth are the secretions of the salivary glands. To locate these, cut a triangular piece of tissue on the side of the jaw. See Figure

20.4. Posterior to the masseter muscle and below the ear, locate the large triangular shaped parotid gland. Leading from this gland locate Stenson's duct which crosses the masseter muscle and enters the oral cavity. Observe the more compact submandibular gland beneath the parotid gland. The small sublingual gland lies anterior to the submandibular gland at the base of the tongue.

■ **FIGURE 20.5   The oral cavity of the fetal pig**
Refer to the color photo gallery.

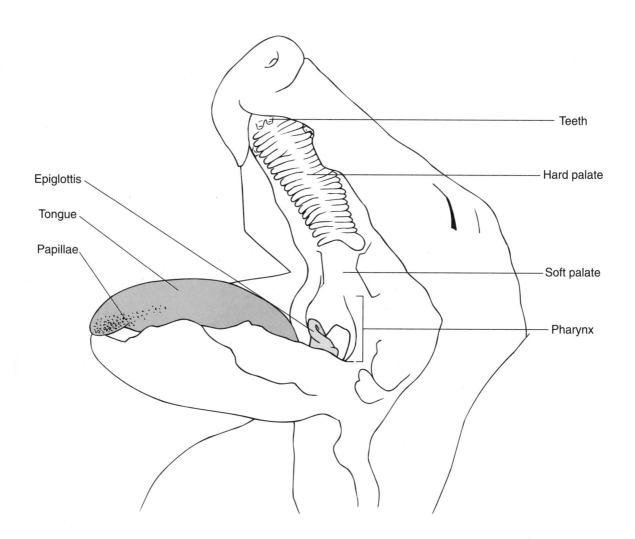

Epiglottis

Tongue

Papillae

Teeth

Hard palate

Soft palate

Pharynx

3. With a pair of scissors or bone shears, extend the angle of the jaw to expose the oral cavity. Observe the tongue and its associated papillae, and lingual frenulum. Small teeth may be present.

4. Observe the anterior hard palate and posterior soft palate. In the pig the uvula is absent which is the posterior continuation of the soft palate in the human. See Figure 20.5.

5. The fauces marks the entrance of the oral cavity into the oropharynx which is continuous with the laryngopharynx.

6. Note the epiglottis covering the entrance into the larynx.

7. If you have not done so already, open the thoracic and abdominal cavities by making a midventral inci-

sion from the neck to the umbilical cord. Make an incision on both sides of the umbilical cord and continue the incisions to the legs. To expose the abdominal cavity gently pull the umbilical cord towards you and carefully cut the umbilical vein close to the cord. Make a pair of lateral incisions on either side of the diaphragm. See Figure 20.6.

8. Locate and observe the esophagus dorsal to the trachea. Follow it passing through the diaphragm to the stomach.

9. View the abdominal cavity below the diaphragm. Observe the shiny parietal peritoneum lining the abdominal wall and the visceral peritoneum on the surface of the organs.

■ FIGURE 20.6 Abdominal organs of the fetal pig
Refer to the color photo gallery.

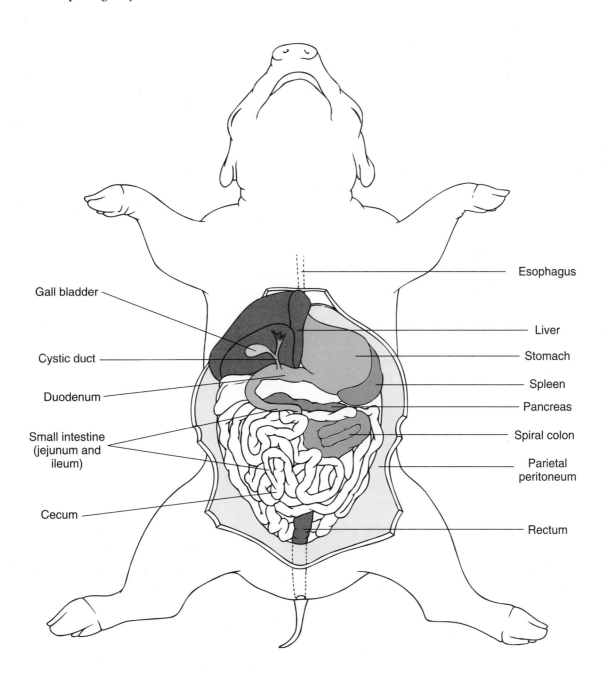

Gall bladder

Cystic duct

Duodenum

Small intestine
(jejunum and
ileum)

Cecum

Esophagus

Liver

Stomach

Spleen

Pancreas

Spiral colon

Parietal
peritoneum

Rectum

10. Lift the liver and observe the stomach on the left side of the body. Locate the entrance of the esophagus into the stomach at the cardiac region. Identify the fundus, body, greater curvature, lesser curvature, and lower pyloric region.

11. Slice open the stomach along the greater curvature. Note the green material called meconium which is bile-stained mucus, and epithelial cells sloughed off

from the digestive tract and amniotic fluid swallowed by the fetus.

12. Cut through the stomach to observe the cardiac and pyloric sphincters.

13. Identify the duodenum of the small intestine as it leaves the pylorus. Observe the remaining portions of the small intestine, the jejunum and ileum. Lift up a portion of the small intestine and note the mesen-

tery. You may observe some small lymph nodes in the mesentery along with blood vessels and lymphatic vessels.

14. Trace the ileum to where it joins with the large intestine or colon. Note the blind sac, the cecum.

15. Identify the compact coiled spiral colon, followed by the rectum which terminates in the external opening, the anus.

16. Observe the liver and locate the five lobes: right lateral, right central, left central, left lateral and caudate lobes.

17. Lift up the right lobe of the liver and observe the saclike gall bladder. Follow the cystic duct from the gall bladder to where it joins with the hepatic duct from the liver to form the common bile duct.

18. Locate the pancreas lying in the curve between the stomach and duodenum and then extending to the left. The pancreatic duct is small and difficult to locate.

19. Observe the spleen, the dark elongated organ on the left side of the abdominal cavity.

Conclusions

• What seems to be the function of the rugae in the stomach?

_____

_____

_____

• What is the anatomical relationship of the liver and gall bladder and stomach and pancreas?

_____

_____

_____

_____

• Describe how bile gets from the liver and gall bladder into the duodenum.

_____

_____

_____

_____

• What non-digestive structures were observed in the mesentery?

_____

_____

EXERCISE 3

Comparison of Human and Fetal Pig Digestive Systems

Discussion

The digestive tract in the fetal pig is essentially a good model for the tract in humans. There are, however, a few differences worth mentioning. These are summarized in Table 20.3.

Materials

Models and charts of the human digestive tract.

Procedure

Identify the structures labeled on Figures 20.4–20.6 on the fetal pig and locate the comparable human structures on the models and/or charts available.

■ **TABLE 20.3 Some Differences between Fetal Pig and Human Digestive Systems**

AREA	HUMAN	PIG
Tongue	Smooth with posterior projections called uvula	Not as smooth as human and uvula is absent
Permanent teeth (dentition pattern)	2-1-2-3	3-1-4-3 (adult)
Liver	Four lobes: right, left, caudate, and quadrate	Five lobes: right and left lateral, right and left central and caudate
Large intestine (colon)	Consists of ascending, transverse, descending and sigmoid portions, along with existence of taenia coli muscle strips and haustra pouches	Consists of spiral colon and dorsal rectum; longer than human colon
Appendix	Present	Absent

Conclusions

- What structural variations between the fetal pig and human did you observe?

_____

_____

_____

_____

❖ Quick Quiz 1

1. The submucosa has:
   (a) Meissner's plexus
   (b) dense connective tissue
   (c) both a and b
   (d) neither a nor b
2. Which of the pairs below is/are mismatched?
   (a) rugae — small intestine
   (b) villi — stomach
   (c) papillae — duodenum
   (d) all of these
3. The fetal pig has how many pairs of salivary glands
   (a) 2
   (b) 3
   (c) 5
   (d) 6
4. Internally, the passage between the small and large intestine is marked by which of these?
   (a) ileocecal valve
   (b) pyloric valve
   (c) ampulla of Vater
   (d) fauces
5. Which of these is present in the fetal pig but not in the human?
   (a) gall bladder
   (a) parotid gland
   (c) appendix
   (d) spiral colon

❖ Physiology of the Digestive System

Background

To maintain homeostasis, the cells of the body must be provided with nutrients and wastes must be removed. The major nutrients required are proteins, carbohydrates, lipids, water, minerals, and vitamins. The sources of these nutrients are the fluids and foodstuffs that are ingested in the diet. Since many nutrients are ingested in nonabsorbable form, the complex foodstuffs must first undergo digestion to simpler forms, which can be absorbed and utilized. Of the six major nutrients only the complex proteins, carbohydrates, and lipids must be digested. Figure 20.7 outlines the process of chemical digestion. Water, and many minerals and vitamins, can be absorbed and utilized in the form in which they are ingested.

There are two types of digestion: **mechanical digestion**, which involves physical breakdown processes, and **chemical digestion**, which involves the **hydrolytic** breakdown of complex molecules. Mechanical digestion includes chewing (**mastication**), swallowing (**deglutition**), mixing, and movement of food along the digestive

---

■ **FIGURE 20.7 The chemical digestive process**

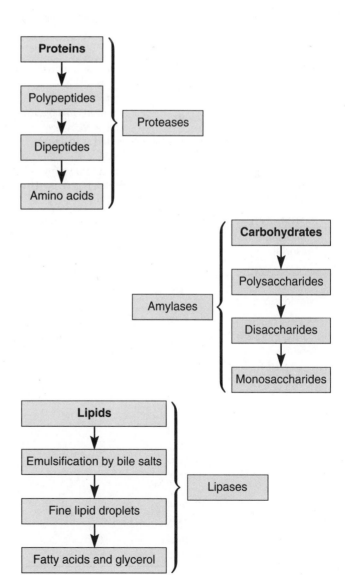

tract (**peristalsis**). Mechanical digestion facilitates chemical digestion by breaking the foods into smaller particles and thereby providing a greater surface area for chemical activity, as well as aiding in the mixing of food with digestive juices. Chemical digestion involves the progressive breakdown of complex compounds by the addition

■ **TABLE 20.5   The Major Digestive Enzymes**

ENZYME	SOURCE	UNIQUE FEATURES OF ENZYME	SITE OF ACTION OF ENZYME	SUBSTRATE	END PRODUCT
SALIVARY AMYLASE (ptyalin)	Salivary glands	Functions maximally in pH range 6.0–7.5	Mouth	Starch	Dextrins and maltose
GASTRIC PROTEASES Pepsin	Gastric glands	Secreted as inactive pepsinogen; activated by the low pH of hydrochloric acid	Stomach	Protein	Polypeptides
GASTRIC LIPASE	Gastric glands	Requires low pH; has limited role in adult stomach	Stomach	Emulsified fats	Fatty acids and glycerol
Rennin	Gastric glands	Functional in infants	Stomach	Casein in milk	Paracasein, Ca^{2+}, curd
PANCREATIC PROTEASES Trypsin	Pancreas	Secreted as inactive trypsinogen	Small intestine	Polypeptides and proteins	Small polypeptides and dipeptides
Chymotrypsin	Pancreas	Secreted as inactive chymotrypsinogen; activated by trypsin	Small intestine	Polypeptides and proteins	Small polypeptides
Carboxypeptidase	Pancreas	Secreted as inactive small procarboxypeptidase, which is activated by trypsin	Small intestine	Polypeptides and proteins	Small polypeptides
PANCREATIC LIPASE (steapsin)	Pancreas	Active at a pH of about 8.0	Small intestine	Emulsified fats	Fatty acids and glycerol
PANCREATIC AMYLASE (amylopsin)	Pancreas	Active at a pH of about 8.0	Small intestine	Starch, glycogen, and dextrins	Maltose
INTESTINAL PROTEASES Peptidases (amino, tetra, tri & di)	Intestinal glands	Group of enzymes that complete the final breakdown of peptides	Small intestine	Small polypeptides and dipeptides	Peptides and free amino acids
INTESTINAL AMYLASES Disaccharidases Maltase	Intestinal glands		Small intestine	Maltose (disaccharide)	Glucose (monosaccharide)
Lactase	Intestinal glands		Small intestine	Lactose (disaccharide)	Glucose and galactose (monosaccharides)
Sucrase	Intestinal glands		Small intestine	Sucrose (disaccharide)	Glucose and fructose (monosaccharides)
INTESTINAL LIPASE	Intestinal glands		Small intestine	Emulsified fat	Fatty acid and glycerol

of water. This activity is facilitated by digestive enzymes that act as catalysts to speed up the hydrolysis of food molecules.

The enzymes that catalyze the chemical digestive process bind to the substrate (e.g., carbohydrates) to form an **enzyme-substrate complex**. It is thought that this union alters the **energy of activation** needed to initiate the reaction and thus affects the speed at which the reaction occurs. The enzymes do not *cause* a reaction, but rather change the energy requirements for the reaction. They can speed up or slow down a reaction. Enzymes are also *specific;* that is, certain enzymes interact only with specific substrates. Enzymes are *not used up* in catalyzing a reaction, but rather are liberated after the reaction to catalyze additional similar reactions. In the digestive process there are three major kinds of enzymes: **proteases, amylases**, and **lipases**, which act on proteins, carbohydrates, and lipids, respectively.

Digestion of foods is a progressive process in which large complex molecules are gradually broken down to their smaller components, which can then be absorbed. The end products of these processes—amino acids, monosaccharides, fatty acids, and glycerol—can now be absorbed by the body. They are transported to areas where they are needed and utilized in forming new cellular components, or for energy storage or release.

EXERCISE 1
Digestion of Protein

Discussion

As indicated in Table 20.5, digestion of proteins occurs in the stomach and small intestine. Large protein molecules consisting of long chains of amino acids are gradually broken down into smaller polypeptides until the process is completed and yields individual amino acids. These amino acids are absorbed by the blood capillaries and transported to areas of the body where protein synthesis is taking place. They also can be "converted" into other metabolites and used for energy.

A. Pepsin

Materials

Clean test tubes and rack; solutions of 1% pepsin; 1% boiled pepsin; 0.5% sodium hydroxide; 0.8% hydrochloric acid; Hydrion paper; hard-boiled egg; 10-ml graduate cylinders; 1% copper sulfate solution; 37°C water

bath; Pasteur pipettes; 10% sodium hydroxide; wax marking pencil.

Note: It is imperative that all glassware be clean to avoid contamination of the chemical reactions.

Procedure

1. Using the wide end of a Pasteur pipette as a borer, remove from the hard-boiled egg a cylinder of egg white about 1.0 cm long. Cut the cylinder into five equal discs.
2. Label five test tubes 1 through 5. Place these in a rack and add the reagents to each of the tubes according to the chart in step 3.
3. Reagents

TUBE	1% PEPSIN	1% BOILED PEPSIN	0.8% HCl	0.5% NaOH	WATER
1	5.0 ml		5.0 ml		
2	5.0 ml				5.0 ml
3		5.0 ml	5.0 ml		
4	5.0 ml			5.0 ml	
5			5.0 ml		5.0 ml

4. Use the Hydrion paper to determine the pH of the material in each tube. Record in Table A.
5. Add a disc of egg white to each of the tubes and incubate in a 37°C water bath for 2 hr. Gently shake the tubes every 30 min.
6. After the incubation period, test the solution in each tube to see if the protein has been digested. Test with the Biuret reagent as indicated in steps 7–9.
7. Remove the disc of egg white from each tube. Determine the amount of digestion by observing the amount of disintegration of the disc. Make a qualitative estimation of the degree of digestion on a scale of 1 to 5, with 5 being complete digestion. Record the results in Table A and discard the disc.
8. Add 3 ml of 10% solution hydroxide to each tube and mix gently.
9. Add 1% copper sulfate to each of the tubes, drop by drop. Notice any color changes and record the results in Table A.

Note: The intensity of the color reaction is proportional to the number of peptide linkages: a blue indicates complete digestion, purple indicates the presence of undigested protein, and pink indicates large number of undigested tripeptides and dipeptides.

Observations

■ TABLE A Protein Digestion: Pepsin

OBSERVATION	TUBE				
	1	2	3	4	5
pH					
Biuret reaction					
Degree of digestion					

Conclusions

● In which tube(s) did digestion occur?

_____

● What is the evidence to support this conclusion?

_____

_____

_____

● Write a word description/equation for the reaction in the tube(s) where digestion occurred.

_____

_____

_____

● Which tube in this experiment most closely simulates the region of the body where pepsin is active? Explain.

_____

_____

_____

● How does the experimental evidence support your conclusion?

_____

_____

_____

● What was the significance of tube 5 in this experiment? Explain.

_____

_____

_____

B. Pancreatin

Materials

Clean test tubes and rack; wax marking pencil; hard-boiled egg; Pasteur pipette; graduate cylinder; 37°C water bath; solutions of 2.0% pancreatin, 2% boiled pancreatin, 0.8% hydrochloric acid, 0.2% sodium hydroxide, 10% sodium hydroxide, 1% copper sulfate; Hydrion paper.

Procedure

1. Prepare discs of egg white as described in step 1 of Part A.
2. Label four clean test tubes 1 through 4. Place them in a rack and add the reagents to each of the tubes according to the chart in step 3.
3. Reagents

TUBE	2% PAN-CREATIN	2% BOILED PAN-CREATIN	WATER	0.8% HCl	0.2% NaOH
1	3.0 ml				3.0 ml
2	3.0 ml			3.0 ml	
3		3.0 ml			3.0 ml
4			3.0 ml		3.0 ml

4. Use the Hydrion paper to determine the pH of the materials in each tube. Record in Table B.
5. Add a disc of egg white to each of the tubes and incubate in a 37°C water bath for 2 hr. Gently shake the tubes every 30 min.
6. After the incubation period, test the solution in each tube to see if digestion has occurred. Use the same procedure as in steps 7 through 9 in Part A.

Observations

■ TABLE B Protein Digestion: Pancreatin

OBSERVATION	TUBE			
	1	2	3	4
pH				
Biuret reaction				
Degree of digestion				

Conclusions

* Is there a difference in the optimal pH for pepsin and pancreatin? Relate this to the pH found in different sections of the digestive tract.

_____

_____

_____

_____

* Which tube in Part B most closely simulates the region of the body where pancreatic enzymes are functioning? How does the experimental evidence support your conclusions?

_____

_____

_____

_____

C. Rennin

Materials

Test tubes and rack; wax marking pencil; milk; 2% albumin solution; graduate cylinders; 37°C water bath; rennin powder; beakers.

Procedure

1. Label three clean test tubes 1 through 3. Place them in a rack and add the reagents to each tube according to the chart in step 2.
2. Reagents

TUBE	MILK	ALBUMIN SOLUTION	RENNIN POWDER
1	5.0 ml		
2	5.0 ml		"Pinch"
3		5.0 ml	"Pinch"

3. Incubate all tubes in a 37°C water bath for 15 to 20 min. Check the tube throughout the incubation period for any changes.
4. After 20 min, pour the contents of each tube into a separate beaker and record the results in Table C.

Observations

■ **TABLE C Protein Digestion: Rennin**

OBSERVATION	TUBE		
	1	2	3
Consistency			

Conclusions

* Write an equation to describe the digestive process that rennin catalyzes.

_____

_____

* What is the substrate for rennin?

_____

* In which tube(s) did digestion not occur? Explain.

_____

_____

_____

* What is the major constituent of the curd of coagulated milk?

_____

_____

_____

* What completes the job that rennin starts in the coagulation of milk?

_____

_____

_____

* Since rennin is not found in the adult, what do you think helps adults begin to "digest" milk?

_____

_____

_____

* What function does rennin's digestion of paracasein serve in an infant?

_____

_____

_____

EXERCISE 2
Digestion of Carbohydrates

Discussion

Carbohydrates serve as a major energy source that fuels the myriad reactions body cells carry out. When complex carbohydrates are ingested, they undergo digestion to produce **monosaccharides**, which are simple sugars, like glucose. The monosaccharides are absorbed by the blood capillaries in the small intestine and transported to cells for direct production of ATP or storage as glycogen.

A common test for the presence of monosaccharides is Benedict's test. The copper sulfate reagent turns green, yellow or orange in the presence of progressive amounts of a **reducing sugar** (a sugar such as glucose). In the presence of starch, iodine turns blue-black.

Materials

Test tubes and rack; 37°C water bath; boiling water bath; solutions of: 10% pancreatin, 0.2% sodium hydroxide, 1% starch; Benedict's reagent, Lugol's reagent or iodine solution; wax marking pencil.

Procedure

1. Label two test tubes 1 and 2. Place these in a rack and add the reagents to each according to the chart in step 2.
2. Reagents

TUBE	PAN-CREATIN	0.2% NaOH	1% STARCH	WATER
1	3 ml	2 ml	3 ml	
2		2 ml	3 ml	3 ml

3. Mix the contents of each tube and incubate in a 37°C water bath for 20 min.
4. While these tubes are incubating, label two other test tubes 1a and 2a.
5. At the end of the incubation period, place 3 ml of the contents from tube 1 into the tube marked 1a and 3 ml from tube 2 into tube 2a.
6. Test the second set of tubes (1a and 2a) for the presence of starch by adding 2 drops of iodine solution. Record the results in Table D.
7. Test the first set of tubes (1 and 2) for the presence of monosaccharides using the Benedict's reagent as follows:
   a. To each tube add 2 ml of Benedict's reagent.

b. Heat the tubes in a boiling water bath for a few minutes.
   c. Observe any color change.
8. Record the results in Table D.

Observations

■ **TABLE D   Carbohydrate Digestion**

OBSERVATION	TUBE	
	1a	2a
Reaction with iodine		
	1	2
Reaction with Benedict's reagent		

Conclusions

- In which tube(s) did digestion occur?

  _____

- Explain using a word equation.

  _____

  _____

- Assuming the intensity of a positive starch-iodine interaction varies with the amount of starch present, describe the results that would be obtained at 42°C. Why?

  _____

  _____

  _____

- What constituent of pancreatin is active on the starch in this experiment?

  _____

  _____

  _____

- Is this type of constituent functional in more than one area of the digestive tract? Explain.

  _____

  _____

  _____

EXERCISE 3
Digestion of Lipids

Discussion

Lipids are relatively insoluble in water. This propery presents a unique problem for their digestion because the digestive process involves the hydrolysis of soluble food-stuffs through the intermediate action of enzymes. For fats to be effectively digested by lipases, they must be **emulsified** into fine fat droplets.

The emulsification of fats is accomplished through the action of **bile salts**. These salts are secreted into the duodenum in bile, which is produced by the liver and stored in the gall bladder. The presence of fats in the digestive tract results in the release of bile from the gall bladder. Fat digestion begins with the action of lipase resulting in the formation of smaller fat globules. The bile salts act like a detergent on these globules, decreasing the surface tension of the particles and preventing them from coalescing and reforming large fat globules. The smaller fat particles present a greater surface area over which the lipases can catalyze the hydrolysis of fats to **fatty acids** and **glycerol**.

A. Surface Activation: The Reduction of Surface Tension

Materials

Test tubes and rack; sulfur powder; 2% solution of bile salts.

Procedure

1. Place two clean test tubes in a rack and label them 1 and 2.
2. Add 5 ml of water to tube 1 and gently sprinkle sulfur powder into the tube. Observe and record the results in Table E.
3. To the second tube add 5 ml of water and 1 ml of bile salts. Gently sprinkle sulfur powder into the tube and observe what happens. Record the results in Table E.

Observations

■ **TABLE E Surface Tension**

OBSERVATION	TUBE	
	1	2
Location of sulfur powder		

Conclusions

• Explain the results observed in the procedure above.

_____

_____

_____

B. Comparison of Emulsifying Agents

Materials

Test tubes and rack; graduate cylinders; 5-ml pipettes; Pasteur pipettes; 2% bile salts solution; 2% soap solution; mineral oil; distilled water; wax marking pencil.

Procedure

1. Label three test tubes 1 through 3 and place them in a rack.
2. Add reagents to each tube according to the following chart.
3. Reagents

TUBE	DISTILLED WATER	BILE SALTS	SOAP SOLUTION	MINERAL OIL
1	5 ml			15 drops
2			5 ml	15 drops
3		5 ml		15 drops

4. Note the appearance of the oil in relation to the other fluid in the tube. Record your observations in Table F.
5. Shake each tube and observe what happens to the oil. Note the way the oil is dispersed and the size of the particle.
6. Place the tube back in the rack and observe again after 3 min. Record the results in Table F.

Observations

■ **TABLE F Emulsifying Agents**

OBSERVATION	TUBE		
	1	2	3
Immediately after shaking			
3 min after shaking			

Conclusions

- After shaking, which tube had the finest oil particles? Explain why.

  _____

  _____

- After shaking, which tube had the coarsest oil particles? Explain why.

  _____

- How do bile salts work?

  _____

  _____

  _____

- Which tube simulates the area of the digestive tract where most lipid digestion occurs? Explain.

  _____

  _____

  _____

- Where is bile produced and what are its chief constituents?

  _____

  _____

  _____

C. Lipase

Materials

Test tubes and rack; wax marking pencil; milk; litmus powder; 1% lipase solution; 1% boiled lipase solution; 2% bile salt solution; distilled water; 37°C water bath.

Procedure

1. Label three test tubes 1 through 3 and place them in a rack.
2. Add reagents to each tube according to the following chart in step 3.
3. Reagents

TUBE	MILK	LITMUS POWDER	1% LIPASE	1% BOILED LIPASE	DISTILLED WATER	BILE SALTS
1	3 ml	Enough to produce blue color	3 ml		3 ml	
2	3 ml	Enough to produce blue color		3 ml	3 ml	
3	3 ml	Enough to produce blue color		3 ml		3 ml

4. Incubate the tubes in the 37°C water bath and observe at 10-min intervals for color changes.

Note: Litmus is blue in alkaline environment and pink in acid environment.

5. Record your results in Table G.

Observations

■ TABLE G Lipid Digestion

OBSERVATION	TUBE 1	2	3
Initial color			
Color after 10 min			
Color after 20 min			
Color after 30 min			

Conclusions

- Which tube(s) produced a color change indicating that an acid environment had been produced? Explain.

_____

_____

_____

_____

_____

 ## Quick Quiz 2

1. Mechanical digestion includes mastication, deglutition and peristalsis. True or False?
2. Chemical digestion is facilitated by mechanical digestion. True or False?
3. Except for gastric juices, most digestive juices work effectively in an alkaline pH range. True or False?
4. Boiling digestive enzymes increases their ability to digest substrate. True or False?
5. Emulsification of fats is accomplished by the acidity of the stomach. True or False?

The Urinary System

O V E R V I E W

In this unit you will observe both the macroscopic and the microscopic structure of the system which cleanses wastes from the circulating blood. You will then have the chance to examine urine, the product of this process, both chemically and microscopically.

O U T L I N E

Anatomy of the Urinary System

Quick Quiz 1

Physiology of the Urinary System

Anatomy of the Urinary System

Background

There are four major components of the urinary/excretory system: the kidney (where filtration of blood occurs), the ureters (which carry urine, the product of filtration, away from the kidneys), the bladder (which stores urine); and the urethra (which carries urine to the outside). Figure 21.1 illustrates these organs.

EXERCISE 1
The Kidney: Sheep Specimen and Human Model

Discussion

External Anatomy

The kidneys are paired organs located retroperitoneally in the lumbar area against the dorsal body wall. They are ensheathed in a fibrous connective tissue capsule, the re-

nal fascia, and surrounded by fat. The kidney is indented on the medial side at the hilus, and this produces a characteristic shape. The hilus opens into a hollow chamber of the kidney called the **renal sinus**, which is lined by a continuation of the fibrous capsule. It is through the hilus that all structures enter into and exit from the kidney: the renal artery and vein, nerves, lymphatics, and ureter.

Internal Anatomy

When the kidney is sliced through on a coronal plane, the interior anatomy becomes visible. The kidney consists of an outer granular "shell" area called the **cortex**, and an inner region, the **medulla**. The cortex contains most of the microscopic filtering units of the kidney, which are called **nephrons** and that ultimately drain toward the hilus. The medulla consists of many triangular **renal pyramids**, which are clusters of the **collecting ducts** that drain the nephrons into the cortex. Pyramids alternate with inward projections of the cortex, termed **renal columns**. The pyramid appears striated and ends in a point called the **renal papilla**, which empties into the **renal pelvis**, a saclike expansion of the ureter that

■ **FIGURE 21.1   The human urinary system**

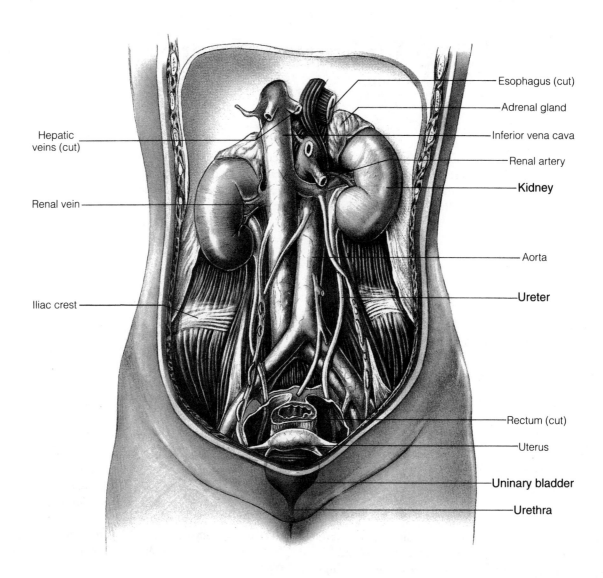

carries urine out of the kidney. It has many cuplike surfaces whose points drain the papillae. The points are the **minor calyces** (singular, calyx), which join into larger regions, the **major calyces**, which in turn empty into the pelvis itself. Thus the route of urine flow proceeds from collecting ducts, through openings in the papillae, to minor calyx, major calyx, renal pelvis, and ureter.

The blood supply of the kidney begins with the **renal artery**, a branch of the abdominal aorta, which enters the kidney at the hilus. The artery divides into **interlobar arteries**, which run between the pyramids. Where the medulla and cortex meet, the interlobar arteries branch into a series of incomplete arches, the **arcuate arteries**. These vessels in turn give rise to smaller, cortical **interlobular arteries**. Branches of these vessels, **afferent arterioles** lead to the microscopic **glomeruli**. Blood flows through the glomerulus and is filtered into the tip of the microscopic **kidney tubule**. Blood exits

from the glomerulus via the constricted **efferent arteriole**, which then goes on to serve the renal tubule as the **peritubular capillary** (the **vasa recta**). The venous circulation generally parallels the arterial. Ultimately the **renal vein** drains into the **inferior vena cava**. See Figures 21.1 and 21.2a and b.

Microscopic Structure

There are about one million nephrons in each kidney. About 80% of the nephrons are located in the cortex and are called **cortical nephrons**. The remaining 20% are located close to the medulla, and portions of them dip down deep into the medulla. These are termed **juxtamedullary nephrons**. See Figure 21.2b.

Each nephron consists of a **renal corpuscle** and a **renal tubule** and has its own blood supply. The renal

■ **FIGURE 21.2 The kidney**

(a) Longitudinal section — internal structure. (b) Longitudinal section — cortical and juxtamedullary nephrons.

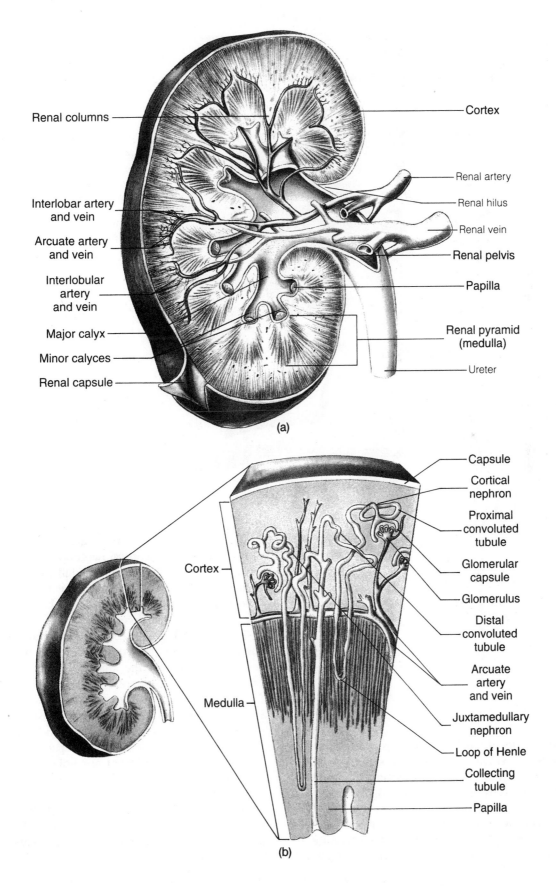

Renal columns

Cortex

Interlobar artery and vein

Renal artery

Renal hilus

Arcuate artery and vein

Renal vein

Interlobular artery and vein

Renal pelvis

Papilla

Major calyx

Minor calyces

Renal pyramid (medulla)

Renal capsule

Ureter

(a)

Capsule

Cortical nephron

Proximal convoluted tubule

Cortex

Glomerular capsule

Glomerulus

Distal convoluted tubule

Arcuate artery and vein

Medulla

Juxtamedullary nephron

Loop of Henle

Collecting tubule

Papilla

(b)

■ **FIGURE 21.3 The nephron**

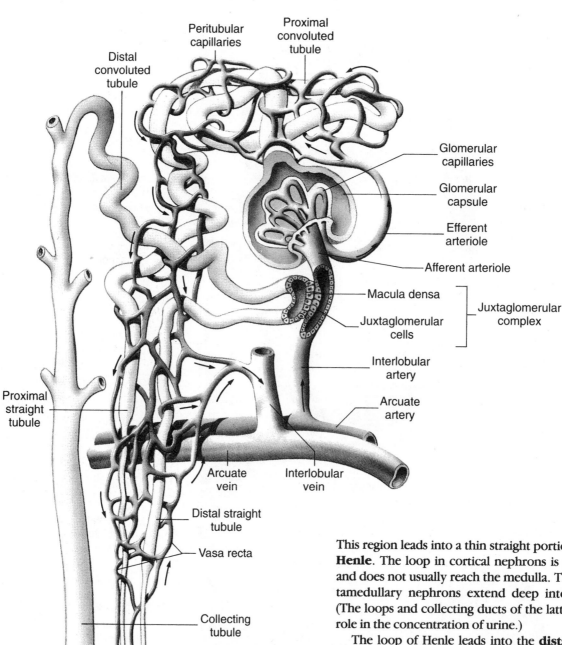

This region leads into a thin straight portion, the **loop of Henle**. The loop in cortical nephrons is relatively short and does not usually reach the medulla. The loops in juxtamedullary nephrons extend deep into the medulla. (The loops and collecting ducts of the latter play a major role in the concentration of urine.)

The loop of Henle leads into the **distal convoluted tubule**. A portion of the distal tubule, the macula densa, contacts the cells of the afferent arteriole at a region called the **juxtaglomerular apparatus (JGA)**. The afferent arteriolar cells release renin: (a) in response to low sodium delivery in the filtrate as it passes through the distal tubule, (b) in response to low blood pressure in the arteriole, and (c) in response to sympathetic stimulation to the cells of the afferent arteriole. Renin, in turn, through stimulating aldosterone secretion from the adrenal cortex, ultimately regulates sodium and water retention by the kidney. The distal convoluted tubule drains into the collecting duct. These are clustered in the renal pyramids, which empty into the renal pelvis and then into the ureter. See Figure 21.3.

corpuscle consists of a tuft of capillaries (the **glomerulus**), which sits within a capsule of modified epithelial cells, the glomerular capsule (**Bowman's capsule**). These epithelial cells are adapted for the filtration of blood.

Bowman's capsule leads into the renal tubule. The first tortuous portion is the **proximal convoluted tubule**. It is here that major sodium and water reabsorption occurs.

Materials

Diagrams and models of macroscopic and microscopic features of the kidney; injected sheep kidney; dissecting pan and instruments; microscope; c.s. slide of kidney.

Be sure to review safety precautions before handling fresh or preserved specimens.

Procedure

1. Observe the whole sheep kidney, noting renal capsule, fat, and hilus. Note blood vessels and ureter at the hilus.
2. Slice the kidney along a coronal plane. Observe the outer granular, reddish-brown cortex, and the inner, lighter medulla. Note also the saclike, white renal pelvis, which narrows into the ureter.
3. Observe the renal artery and its branches: the interlobar, arcuate, and interlobular arteries. The arcuate runs parallel to the surface of the kidney at the junction of the cortex and medulla. Identify renal veins that run alongside the arteries.
4. Note the striated renal pyramids, which taper to a point, the renal papillae. Identify the minor calyces into which the papillae empty. Observe the major calyces and the renal pelvis.
5. Identify the renal columns in between the pyramids.
6. Note the hollow renal sinuses around the concave cuplike surfaces of the pelvis. Pass a probe gently into the sinus toward the hilus.
7. Observe a model or diagram of microscopic structure of the kidney. Identify the afferent and efferent arterioles, the peritubular capillaries, and the vasa recta and the renal corpuscle (glomerulus and Bowman's capsule). Locate the proximal convoluted tubule, the loop of Henle (ascending and descending loops), the distal convoluted tubule, and the collecting duct.
8. Observe the slide of the kidney under the high-power lens of the microscope. Identify glomeruli and sections of the renal tubules.

Conclusions

- Look up the term "urolithiasis". How is it caused? Where do you see that it could cause a problem in the urinary tract?

  _____

  _____

  _____

  _____

- As you examine the nephron, note the difference in diameter of the afferent and efferent arterioles and explain this.

_____

_____

_____

_____

EXERCISE 2
The Urinary System: The Fetal Pig

Discussion

In the pig, the urinary and reproductive systems share a common tract to the outside (the urethra), and thus reference is commonly made to the "urogenital system." Reproductive structures are described separately in Unit 23 "The Reproductive System."

The basic structures to be described are the kidneys, the ureters, and the bladder and urethra, along with the blood supply. The **kidneys** are paired organs that lie on either side of the abdominal aorta and vena cava, and are retroperitoneal. Just medial and superior to the kidneys lie the **suprarenal** or **adrenal glands**.

The blood supply to kidney and adrenal glands branches off the aorta as the renal and adrenolumbar arteries, respectively. (Review Unit 17.) Drainage is via the renal and adrenolumbar veins, which feed into the vena cava. Renal vessels enter and leave the kidney at the **hilus** along with the ureter. The **ureter** passes caudally and retroperitoneally from the kidney to the bladder. The **bladder** is a muscular sac that empties into the **urethra**, which in turn passes to the outside. Since the urethra is joined by the **ductus deferens** in the male, and the **vagina** in the female, it opens to the exterior as the common **urogenital orifice**. See Figures 21.4 and 21.5.

Materials

A preserved and injected fetal pig; dissecting tray and instruments.

Be sure to review safety precautions before handling fresh or preserved specimens.

Procedure

1. Move the viscera of the abdominal cavity to one side. Observe the abdominal aorta and vena cava.
2. Locate the kidneys beneath the peritoneum along the dorsal body wall. Carefully remove the fat surrounding them and try to observe the narrow bandlike adrenal glands, superior and medial to the kidney.
3. Identify the renal artery and vein.
4. Locate the ureters leaving the kidney and trace their entrance to the urinary bladder which is attached to the ventral strip of the abdominal wall.

■ **FIGURE 21.4 The urogenital system of the male fetal pig**
Refer to the color photo gallery.

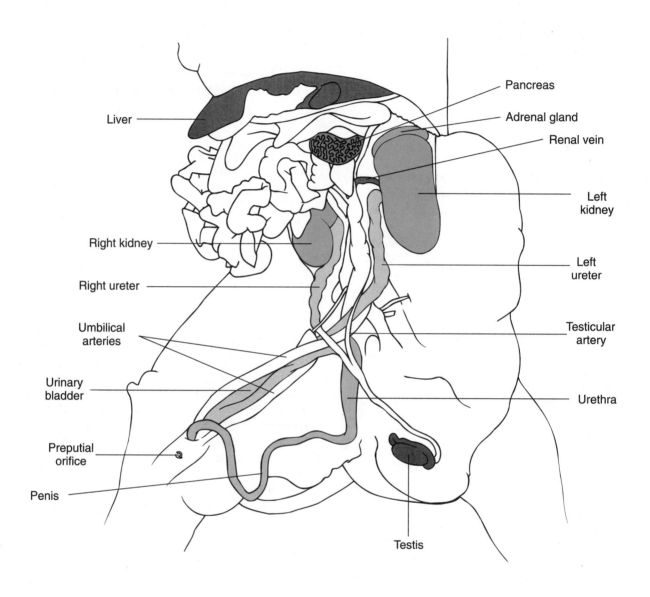

5. Locate the urethra leading out of the bladder. The pelvic cavity must be cut to expose the urethra. In the male pig the urethra will receive the ductus deferens as it passes along the dorsal wall to the caudal end of the body. Here it continues beneath the ventral surface of the skin as the penis which terminates at the preputial orifice or urogenital opening. In the female pig, the urethra is seen ventral to the body of the uterus and the vagina. The urethra and the vagina unite to form a common urogenital canal.

6. Return to one of the kidneys to examine its structure more closely. Make a longitudinal incision through the kidney. Note the outer renal capsule. Just beneath this observe the outer cortex and the inner medulla divided into renal pyramids. Locate the renal pelvis.

Conclusions

• Contrast the pathway of the urethra in the male and female pig.

Male: _____

_____

Female: _____

_____

- Explain what is meant by the term "retroperitoneal"?

_____

_____

- List the parts of the urinary system you observe in the fetal pig that are retroperitoneal.

_____

EXERCISE 3
The Urinary System: The Human

Discussion

The main difference between fetal pig and human anatomy in this system is that in the human female the urethra is totally distinct from the vagina. The urinary and reproductive systems are two separate systems. Human structures otherwise generally parallel those of the fetal pig (see Figures 21.4 and 21.5). In the human the adrenal

---

■ **FIGURE 21.5 The urogenital system of the female fetal pig**
Refer to the color photo gallery.

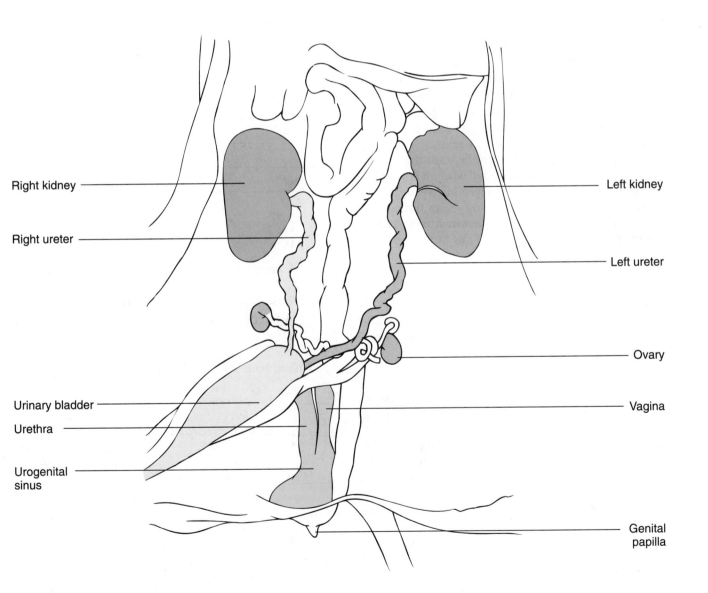

Right kidney

Right ureter

Urinary bladder

Urethra

Urogenital sinus

Left kidney

Left ureter

Ovary

Vagina

Genital papilla

glands sit on top of the kidneys, embedded in fat. The bladder has a capacity of 600 ml and a smooth triangular area on its internal floor, called the **trigone**. There are three openings in the bladder, one at each corner of the trigone. The two lateral, superior openings are where the ureters empty into the bladder, slanting downward. When the bladder contracts in emptying, these openings are compressed, thus preventing the refluxing of urine into the ureters. The inferior opening of the bladder drains into the **internal urethral orifice** and on into the urethra. The opening is surrounded by a thick circular mass of small muscle, the **internal sphincter muscle**.

The wall of the bladder consists among other tissues of bundles of *smooth* muscle fibers that are interlaced in all directions and depths. This muscle is called the **detrusor** muscle and is innervated by parasympathetic nerves from the sacral area of the spinal cord. These nerves function in controlling **micturition** (urination) by causing contraction of the detrusor muscle and relaxation of internal sphincter.

Surrounding the urethra is the **external urethral sphincter muscle**. This *striated* muscle exerts voluntary control over voiding. The female urethra is short in humans (about 3.8 cm) and longer in the male (about 20 cm). The male urethra is divided into three parts: prostatic, membranous, and spongy (or penile). The **prostatic urethra** is the segment just beneath the bladder that runs through the prostate gland, where it is joined by the **ejaculatory duct** of the reproductive tract. The **membranous urethra** is short and passes through the floor of the pelvis. The **spongy urethra** runs through the **corpus spongiosum** of the penis to the outside at the **urethral orifice**.

Materials

Charts and models of the human urinary system.

Procedure

1. Identify kidneys, adrenal glands, ureters, bladder, and urethra.
2. Identify the trigone and the openings into and out of it. Locate the internal and external sphincter muscles.
3. Locate the various portions of the male urethra and the urinary meatus.
4. Locate the separate urinary and vaginal orifices in the female.

Conclusions

- Why does enlargement of the prostate gland affect urination in the male?

- How is the backflow of urine from the bladder prevented during micturition?

- From what you know about the urinary tract, how can infections travel from bladder to kidney?

- Compare the fetal pig and human female urinary systems.

❖ Quick Quiz 1

Match the items in column A with those in column B.

	A	B
_____	1. leads away from glomerulus	(a) macula densa
_____	2. filters blood into Bowman's capsule	(b) afferent arteriole
		(c) glomerulus
		(d) vasa recta
_____	3. surrounds loops of juxtaglomerular tubules, carries blood	(e) efferent arteriole
_____	4. part of juxtaglomerular apparatus in distal convoluted tubule	
_____	5. secretes renin	

Physiology of the Urinary System

Background

The function of the urinary/excretory system is to cleanse the blood of wastes and to maintain water, electrolyte, and acid-base balance. Urine, the product of this cleansing, carries wastes in a regulated amount of water of variable pH. To form urine, the kidney filters blood at the renal corpuscle, and then carries out reabsorption and secretion of selected substances. Two hormones, ADH and aldosterone, regulate reabsorption of water. The kidneys respond to the osmolarity of body fluids, and to their electrolyte composition. When water intake is slowed, body fluids become concentrated. This increase in osmolarity is sensed by osmoreceptors in the hypothalamus of the brain. These cells secrete **antidiuretic hormone (ADH),** which causes a retention of water in the kidney tubules. Thus body fluid volume is preserved. Additionally the kidney itself releases the hormone renin as a response to low sodium delivery in the kidney tubules—either because body fluid sodium concentration is low or because delivery of blood to the kidney for filtration is low. In response to low sodium delivery, renin is released from the kidney. Renin activates angiotensinogen in the blood which goes on to form **angiotensin I** and then an active **angiotensin II.** Angiotensin II acts as a powerful vasoconstrictor (thus counteracting low blood pressure associated with fluid loss), and stimulates **aldosterone** release from the adrenal cortex. **Aldosterone** in turn causes sodium and water retention and potassium release at the kidney. Changes in body fluid volume and electrolyte composition can be caused, for example, by simple dehydration, burns, excessive sweating, and hemorrhage. Altering fluid and salt intake is another mechanism whereby changes can be effected.

EXERCISE 1
Examination of Urine

Discussion

The composition of urine varies according to the health and physiologic status of the individual. Thus fluid intake, dietary intake, cell metabolism, fluid losses, and physical activity all play a role in its ultimate makeup. Normally urine is a clear, pale yellow liquid that has a mild aromatic odor, a pH from 4.5-8.0, and specific gravity* in the range of 1.001-1.030. The characteristics and composition of urine are summarized in Tables 21.1 to 21.4.

Urine can be thought of as a solution in which water, the solvent (which accounts for 95% of urine), acts as a medium that carries various ions and molecules out of the body. The major organic wastes are generally: **urea** (from the deamination of amino acids), **uric acid** (from the breakdown of nucleic acids), and **creatinine** (from the metabolism of creatine). The **electrolytes** consist of ions like sodium, potassium, phosphate, sulfate, chlorides, bicarbonate, and ammonia. Amino acids and breakdown products of molecules like hemoglobin and bile salts are also present. Glucose, blood cells, and protein are usually not found in urine, and their presence may indicate certain pathologic conditions. However these substances may be found in nonpathologic conditions such as pregnancy, strenuous exercise, and dietary changes.

A. Physical and Chemical Characteristics

Materials

Urine samples in containers; graduate cylinders; Ketostix; Albustix, Clinistix, Hemastix, pH paper; glass stirring rods; medicine droppers; wax marking pencils; test tubes; rack and holder; 500-1000 ml beakers; water bath or hot plate; distilled water; Benedict's reagent; Fouchet's reagent; solutions of alkaline picrate, ammonium molybdate-nitric acid (see Appendix C); 30% acetic acid (v/v), dilute hydrochloric acid, 10% barium chloride, 3% silver nitrate, Bunsen burner; filter paper; small funnel; urinometer or refractometer; centrifuge; wooden applicator sticks

Note: Conduct all tests on student and unknown specimens; read all "stix" directions carefully before using.

Be sure to review safety precautions before handling fresh or preserved specimens.

Procedure

1. Obtain a "clean catch" of urine into a washed and dried (and sterilized if possible) container. Ideally the glans and urethral orifice in the male should be wiped clean with cotton moistened with warm water. The female should likewise wipe the urethral orifice from front to back. A small amount of urine is then voided and discarded, and the remainder

---

*Specific gravity (SG) is a measure of the weight of a given volume of a liquid, compared to the weight of an equal volume of distilled water. Specific gravity depends on both the *number* of particles in solution (concentration) and the chemical *nature* of these dissolved particles. An individual's ability to *concentrate* urine is related only to *number* of particles or osmolarity. Thus measuring the freezing point of urine with an osomometer is actually a better way to test urine-concentrating ability.

■ **TABLE 21.1   Physical Characteristics of Urine**

CHARACTERISTIC	NORMAL VALUES	COMMENTS
Color	Pale yellow	Because of the body's breakdown of hemoglobin, color varies according to concentration of solutes; color may be affected by blood, drugs, bile, and foods
Clarity	Transparent	May become cloudy on standing; high protein intake, vigorous exercise, or pregnancy may elevate urine protein levels temporarily (see Comments on Protein) and cloud the sample
pH	4.5–8.0 (average 6.0)	Whole wheat products, and high meat, egg, and cheese protein in the diet makes urine more acid; high vegetable protein intake makes urine more alkaline; bacterial infection may cause increase in pH; decrease in uncompensated acidosis, and increase in alkalosis
Specific gravity	1.001–1.030	A measure of ions and molecules dissolved in urine; SG decreases when a great deal of water is drunk, and vice versa; diuretics, chronic renal failure, and diabetes cause a dilute urine and thus a drop in SG; fever, dehydration, and kidney inflammation can cause an increase in SG; night urine is usually concentrated
Amount	0.6–2.5 liters per 24-hr period (average 1.41 liters)	Varies inversely with specific gravity (see above); when bladder holds about 200–250 ml, a feeling of a "full bladder" is experienced and the conscious desire to micturate occurs

■ **TABLE 21.2   Organic Constituents of Urine**

CHARACTERISTIC	NORMAL VALUES (TEST METHOD)	COMMENTS
CONSTITUENTS		
Urea	25 g/1500 ml	Result of amino acid deamination; normal constituent; represents almost half the solids in urine
Uric acid	1.5 g/500 ml	Result of nucleic acid catabolism; normal constituent
Creatinine	0.8 g/1500 ml (picric acid test)	Result of creatine catabolism; normal constituent
Glucose	None (Benedict's test or Clinistix)	Glycosuria or presence of glucose in urine indicates that the normal mechanism for reabsorbing glucose from kidney back to blood is overwhelmed; high dietary carbohydrate intake may produce this temporarily; pathologic glycosuria may indicate diabetes mellitus, which is not regulated because glucose that cannot be used by body cells pours out in urine
Protein	None (acetic acid or Clinistix)	Normally proteins are too large to be filtered from blood to urine; albumin is most abundant protein of blood and is usually the one detected; nonpathologic rise is seen in pregnancy, high dietary intake, strenuous exercise; pathologic increase may be seen in glomerulonephritis and hypertension
Ketone bodies	None (Ketostix)	Ketones are a by-product of fat metabolism; their presence in urine indicates an excess of fat use in place of carbohydrates for body energy (as in diabetes mellitus) or severe restriction of dietary carbohydrate intake; ketonuria is usually associated with acidosis
Hemoglobin	None (Hemastix)	Hemoglobin is released into blood when red blood cells hemolyze; when excessive hemolysis occurs, as in a transfusion reaction or in hemolytic anemia, hemoglobin appears in the urine
Bilirubin	None (Fouchet's test)	Bilirubin and biliverdin are bile pigments, the presence of which often indicates liver pathology (hepatitis, cirrhosis, obstructive jaundice); the test for bilirubin depends on oxidation of bilirubin to biliverdin; urine is usually greenish orange if bilirubin is present

TABLE 21.3 Inorganic Constituents of Urine

CONSTI-TUENTS[a]	NORMAL VALUES	TEST
Sodium	4.0 g/1500 ml	
Potassium	3.0 g/1500 ml	
Calcium	0.2 g/1500 ml	
Magnesium	0.1 g/1500 ml	Reflected in specific gravity
Iron	0.005 g/1500 ml	
Bicarbonate	1.5 g/1500 ml	
Ammonium	1.0 g/1500 ml	
Sulfate	1 g/1500 ml	Hydrochloric acid–barium sulfate
Phosphate	2 g/1500 ml	Nitric acid–ammonium molybdate
Chloride	9.5 g/1500 ml	Silver nitrate

[a]Normal constituents of urine; total concentration is reflected in specific gravity.

collected in a container. This minimizes collection of debris and secretions from the urinary tract.

2. Note the color, clarity, pH, and volume of the sample. Record in Table A.
3. Pour a portion of the urine into a urinometer until it is almost full. Place the float in the container by tapping the container and wait until the float has stopped bobbing and does not touch the sides of the container. Remove any bubbles from the sample. Take the specific gravity reading on the float stem at the urine-air interface. Or, place 1 or 2 drops of urine on the glass portion on the top of the refractometer. Read the specific gravity on the scale inside and record in Table A. Wash the instruments.
4. Test for **creatinine**. Pour 5 ml of urine sample into a test tube. Add 2 ml of alkaline picrate solution

carefully. Warm the mixture in a water bath, slowly. Record results in Table A. A red mixture is postive.

5. Test for **glucose**.

   a. Dip the Clinistix stick into the urine sample. Read results for glucose and protein (see step 6). Record results in Table A.
   b. Label two test tubes A and B. Add 5 ml of Benedict's solution to each of the two test tubes and follow with 8 drops of urine into A. Put 8 drops of distilled water into B. Bring both tubes to a boil, then let cool. Observe color and record in Table A. If sugar is present the color will be anywhere from blue-green (trace amount) to brick red (2% sugar).

6. Test for **protein.**

   a. Dip the Clinistix stick into the urine sample. Read and record in Table A.
   b. Filter a small amount of urine (about 5 ml) through filter paper in a small funnel, into a graduated cylinder. Mark two test tubes A and B. To A add 10 ml of urine; to B add 10 ml of distilled water. Mix well. Add 30% acetic acid by the drop and swirl until pH paper testing shows the solutions are slightly acid. Head just the top of the tubes in a Bunsen burner flame. If the heated portions become cloudy, add a few drops of acid and reheat. If the precipitate remains, proteins are present. Record results in Table A.

7. Test for **ketone bodies**. Dip a Hemastix stick into the sample. Read and record in Table A.

8. Test for **hemoglobin**. Dip a Hemastix stick into the sample. Allow to dry slightly. Read and record in Table A.

TABLE 21.4 Microscopic Constituents of Urine

CHARACTERISTIC	NORMAL VALUES (TEST METHOD)	COMMENTS
Red blood cells (hematuria)	None (microscopic examination)	Red blood cells are too large to be filtered into urine; presence usually indicates pathological conditions (e.g., infection or trauma)
White blood cells (pyuria)	None (microscopic examination)	White blood cells and/or pus indicate inflammation in the urinary tract
Casts	None (microscopic examination)	Casts represent hard fragments of protein formed in distal convoluted tubule and collecting duct of the urinary tract that dislodge and pass out in the urine (not to be confused with stones—precipitates of salts in urine); casts often contain trapped cells and granular debris and are classified as leukocyte casts, blood casts, and fatty casts; they often dissolve in dilute urine
Crystals	None (microscopic examination)	Not normally present in freshly voided urine but may precipitate as urine cools; generally not significant (except for urates in a patient with a history of gout, or oxalates in a history of stones, etc.); different crystals appear with varying shapes at changing pH value

9. Test for **bilirubin**. Label two test tubes A and B. To A add 10 ml of urine; to B add 10 ml of distilled water. Add a small amount of a 10% barium chloride solution to each test tube. After mixing, filter through filter paper and funnel and spread filter paper out on a paper towel. Add 1 drop of Fouchet's reagent to the precipitate formed on the filter paper. Greenish-blue color is positive. Record results in Table A.

10. Test for selected **inorganic ions**. Pour 5 ml of urine into each of three test tubes labeled S, (sulfate), P_1 (phosphate), and C_1 (chloride). Into tubes labeled S_2, P_2, and C_2 pour 5 ml of distilled water instead of urine. These will serve as controls.

   a. To test tubes S_1 and S_2 add a few drops of dilute HCl and 2 ml of a 10% $BaCl_2$ solution. If a white precipitate forms, sulfates are present in the urine. Record results in Table A.

   b. To test tubes P_1 and P_2 add 3 ml of ammonium molybdate-nitric acid solution. Note pale green color and light precipitate. Heat (in a hot water bath). Formation of a mustard yellow precipitate indicates phosphates are present. Record results in Table A.

   c. Add a few drops of silver nitrate to test tubes C_1 and C_2. If a white precipitate forms, chlorides are present. Record results in Table A.

Observations

■ **TABLE A**   **Urine Composition**

CHARACTERISTIC/ CONSTITUENT	STUDENT SPECIMEN	UNKNOWN
PHYSICAL		
Color		
Clarity		
pH		
Specific gravity		
Volume of sample		
ORGANIC		
Creatinine		
Glucose		
Protein		
Ketone bodies		
Hemoglobin		
Bilirubin		
INORGANIC		
Sulfate		
Phosphate		
Chloride		
MICROSCOPE		
Cells		
Casts		
Crystals		

B. Microscopic Examination

Materials

A microscope; clean slides; centrifuge and tubes; wax pencils; clock.

Be sure to review safety precautions before handling fresh or preserved specimens.

Procedure

1. Mix thoroughly what remains of your specimen. Pour about 12 ml into a conical centrifuge tube, marked with your initials. Place in centrifuge opposite another equivalent sample of blank (12 ml of water). Centrifuge at 1500–2500 rpm for 5 min. Carefully pour the supernatant off and discard. Resuspend the pellet in the test tube with a few drops of remaining urine by flicking the test tube with your fingers. Place a drop of the suspension on a slide and cover with a cover glass.

    a. Examine under low power for an overall view, then switch the objective to high power and cut the light to a minimum for good viewing of any casts.

    b. Examine the sample for cells (epithelial, red blood cells, white blood cells). A count of a few cells is normal; many cells indicate pathology. Record results in Table A. Use a (+) or a (−) to indicate presence or absence.

    c. Examine the sample for casts. These are usually oblong structures often with cells trapped in them. Record observations in Table A.

    d. Examine the sample for the presence of crystals. Record results in Table A.

2. Clean and sterilize all glassware.

Conclusions

● Did you find any abnormalities in your sample? Try to explain why they may have occurred.

_____

_____

_____

_____

● Would a molecule like myoglobin generally be detectable in your urine? Explain, based on your knowledge of the size of hemoglobin and myoglobin.

_____

_____

_____

_____

● Would gamma globulins? Explain.

_____

_____

_____

_____

● Look up kidney stones and gall stones. How do they differ in composition? Describe this, and their appearance.

_____

_____

_____

_____

The Endocrine System

OVERVIEW

In this unit, you will become familiar with microscopic and macroscopic features of the endocrine system and investigate the effects of the functioning of certain hormones.

OUTLINE

Some of the exercises in this unit use living specimens. Commercial audio-visual tapes may be substituted, if appropriate.

Anatomy of the Endocrine Glands

Background

The endocrine system is composed of a group of anatomically independent organs that help to regulate numerous metabolic processes. These organs can be thought of as a system because they all employ a common mode of functioning, the secretion of hormones. Because of their secretory activity, the endocrine organs are referred to as **glands**. Endocrine glands differ from other glands, such as sweat glands and sebaceous glands, in that their products (hormones) are released not through ducts, but rather into the bloodstream for distribution to other areas of the body. Endocrine glands are **ductless glands**.

United by a common mode of functioning, the pituitary, thyroid, parathyroid, and adrenal glands, the gonads, and the islets of Langerhans (pancreas) constitute the major endocrine glands illustrated in Figure 22.1. In addition, other hormone-secreting structures included in the endocrine system are the pineal and thymus glands as well as the glands of the digestive tract, which control the secretion of digestive juices and motility of the digestive tract. Many other organs secrete "local" hormones that act in or near the organ. These secretions are called **prostaglandins** and are derivatives of fatty acids.

■ **FIGURE 22.1 The human endocrine system**

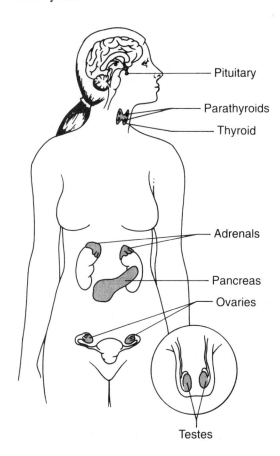

Pituitary

Parathyroids

Thyroid

Adrenals

Pancreas

Ovaries

Testes

■ **TABLE 22.1 Location and Gross Morphology of Human Endocrine Glands**

GLAND	LOCATION	GROSS MORPHOLOGY
Pituitary (hypophysis)	Sella turcica of sphenoid bone	Single rounded structure attached to base of brain by infundibulum; enveloped by the dura mater; consists of three regions that function as separate glands: adenohypophysis (pars distalis or anterior pituitary), intermediate pituitary (pars intermedia), and neurohypophysis (pars nervosa or posterior pituitary); the pars distalis and pars intermedia develop from an outgrowth of the roof of the mouth and are epithelial in nature; the pars nervosa develops as an extension of the hypothalamus and is therefore nerve tissue
Thyroid	Neck region anterior to the trachea	Single bilobed structure connected by a band of tissue; the isthmus, which lies just below the cricoid cartilage; encapsulated by a double layer of connective tissue
Parathyroids	Neck region, on each side of thyroid gland	Four tiny, oval bodies, isolated from the thyroid by a delicate connective tissue covering
Adrenals	Retroperitoneal; atop or near the superior border of each kidney	Paired, flattened and pyramidal; thick fibrous connective tissue capsule encloses each gland, which internally has two distinct regions: outer cortex (derived from mesoderm) and the inner medulla (derived from neural tissue)
Islets of Langerhans	Scattered throughout pancreas	Numerous clusters of cells within the pancreas set off from the other pancreatic tissue by a connective tissue covering
Pineal	Roof of posterior end of third ventricle of the brain	Single, conical structure covered by the pia mater
Thymus	Anterior mediastinum over the heart	Single flat, broad bilobed lymphoid organ enveloped by a layer of connective tissue; size varies with age, being largest during youth and regressing with age

EXERCISE 1

Macroscopic Anatomy of Endocrine Glands: Human and Fetal Pig

Discussion

Figure 22.1 and Table 22.1 summarize the location and gross anatomical features of the major endocrine glands. Since the ovaries and testes function primarily in the reproductive process, these glands are described separately in Unit 23 "The Reproductive System".

A. The Human

Materials

Models of the human torso and brain; charts and diagrams of the human endocrine system.

Procedure

1. Locate the endocrine glands housed in or near the brain.
2. Note their macroscopic appearance and relationship to other parts of the brain.
3. Locate the endocrine glands found in the neck and chest.
4. Observe their macroscopic appearance and relationship to other organs.
5. Locate the endocrine glands found in the abdominal and pelvic regions.
6. Note their macroscopic appearance and relationship to other organs. Record results in Table A.

B. The Fetal Pig

Materials

A preserved, injected fetal pig; dissecting tray and instruments.
   Be sure to review safety precautions before handling fresh or preserved specimens.

Procedure

Many of the endocrine glands have been observed during dissection of other systems.

1. Review the location and gross features of the thyroid gland, the thymus, and the pancreas. Note the large size of the thymus gland in the fetal pig.
2. Locate and observe the adrenal gland, which appears as a narrow band above each kidney.

Observations

◾ TABLE A The Human

AREA	ENDOCRINE GLAND APPEARANCE	OTHER SURROUNDING ORGANS
Brain		
Neck		
Abdomen/Pelvis		

◾ TABLE B The Fetal Pig

GLAND	ENDOCRINE GLAND APPEARANCE	OTHER SURROUNDING ORGANS
Thyroid		
Thymus		
Pancreas		
Adrenals		

Conclusions

- Are there significant differences between the organization of the endocrine systems of the fetal pig and human? Describe.

_____

_____

_____

_____

_____

_____

_____

- Is the overall location and appearance of the glands of the fetal pig the same as the human? Describe.

_____

_____

_____

_____

_____

EXERCISE 2

Microscopic Structure of Selected Endocrine Glands

Discussion

Like all other glandular tissue, the secretory components of endocrine glands are developed from epithelial tissue. The secretory cells of endocrine glands are arranged to make maximal contact with surrounding capillaries, since their secretions travel through the body via the blood.

Generally, the arrangement of cells is in cords and small clusters of cells, both of which are surrounded by blood vessels. Often, the small clusters of cells become "hollowed out," forming a ring of cells around a central area where secretory products are stored. This type of arrangement is called a **follicle**. Examples of these varying arrangements can be seen in the anterior pituitary, which has branching cords; the islets of Langerhans, which have clusters of cells; and the thyroid, which has **follicles**. Table 22.2 and Figure 22.2 illustrate and summarize the general microscopic arrangement and cell populations of selected endocrine glands.

Materials

A microscope; prepared slides of hypophysis, thyroid, adrenal glands, and pancreas.

Procedure

1. Examine the prepared slides of the pituitary gland. Using low power distinguish the three regions: pars distalis, pars intermedia, and pars nervosa. Using high power, observe the arrangement of cells in each region and identify the various cell types listed in Table 22.2.
2. Diagram in the Observations section the microscopic structure of the specified regions of the pituitary gland and label the cell types observable.
3. Examine a prepared slide of the thyroid gland. Using low power, observe the arrangement of the follicles and use high power to identify the acini, parafollicular cells and colloid-filled follicles.
4. Draw and label a diagram of the thyroid in the Observations section.
5. Examine a prepared slide of the adrenal gland. Use low power to distinguish the cortex and medulla. Using high power observe the adrenal cortex and distinguish the three zones. Observe arrangement of cells in each region.
6. Draw and label a diagram of the adrenal cortex in the Observations section. Indicate the three zones and illustrate the cellular arrangement.
7. Examine a prepared slide of the pancreas. Use low power to observe the distribution of islets of Langerhans and high power to distinguish the varied cell types.
8. Draw and label a diagram of the islets of Langerhans in the Observations section.

■ **TABLE 22.2 Microscopic Anatomy of Selected Endocrine Glands**

GLAND	DESCRIPTION	FUNCTION
PITUITARY		
Pars distalis	Glandular cells arranged in thick branching cords in close association with sinusoids of the pituitary portal system; three cell types are:	
	Acidophils (35%) Rounded cells with diameters of 15–19 mm concentrated in center of cords	Produce growth hormone and prolactin
	Basophils (15%) Shape varies from angular to oval and diameters range from 15 to 25 mm, located peripherally in cords	Produce thyrotropin and gonadotropins
	Chromophobes (59%) Small, light-staining rounded cells clumped together in center of cords	May be functionally undifferentiated precursors of other cell types
Pars intermedia	Not well developed in humans; consists of chromophobe cells in follicle arrangements with a few rows of deeply staining basophilic cells that extend into the pars nervosa	Secretes melanocyte-stimulating hormone
Pars nervosa	Comprised of neural tissue primarily; principal tissue consists of unmyelinated nerve fibers that comprise the **Hypothalamic-hypophyseal tract**	Conducts oxytocin and vasopressin from the hypothalamus to posterior pituitary
	Pituicytes are irregular cells found among the neural tissue; deeply staining masses of the neurosecretory products called **Herring bodies** can be observed in the nerve fibers	Modified glial cells that function as supporting elements
THYROID	Encapsulated by a fibroelastic connective tissue that partially penetrates and separates the gland into lobules; glandular tissue arranged in follicles called **acini**; lining of follicles is a simple epithelium that ranges from low cuboidal to squamous; centers of follicle are filled with colloidal material consisting of thyroglobulin and iodated amino acids; dispersed among follicular cells and between follicles are parafollicular cells, which are large and usually less deeply staining; follicle size and shape vary	Synthesis of thyroxine Synthesis of thyrocalcitonin
ADRENAL		
Cortex	Makes up bulk of gland and can be distinguished into three zones:	
	Zona glomerulosa (15%) Columnar type cells arranged in groups that are separated by a vascular connective tissue	Secrete corticosterone and aldosterone
	Zona fasciculata (75%) Polyhedral cells arranged in cords radiating outward from the core of the gland; cells, called **spongiocytes**, are highly vacuolated	Secrete corticosterone, cortisol, and androgens
	Zona reticularis (10%) Cells arranged in anastamosing cords with capillaries in spaces between the cords	Secrete corticosterone, cortisol, and androgens
Medulla	Glandular tissue arranged in cords and clusters of irregularly shaped cells; capillaries and veins run between the clusters; cells react strongly to chromium salts and are called **chromaffin cells**	Secretes epinephrine and norepinephrine
PANCREAS	Endocrine tissue arranged as clusters of cells scattered randomly throughout the organ; coiled cords of cells are separated from surrounding tissue by delicate reticular tissue; special histological technique allows three cells to be distinguished:	
	Alpha (α) cells With large dense granules enclosed by membrane	Secrete glucagon
	Beta (β) cells Smaller than α cells	Secrete insulin
	Delta (δ) cells Intermediate in size	Secrete somatostatin

■ **FIGURE 22.2 Microscopic anatomy of selected endocrine glands**
(a) Thyroid. (b) Adrenal cortex. (c) Pancreas.

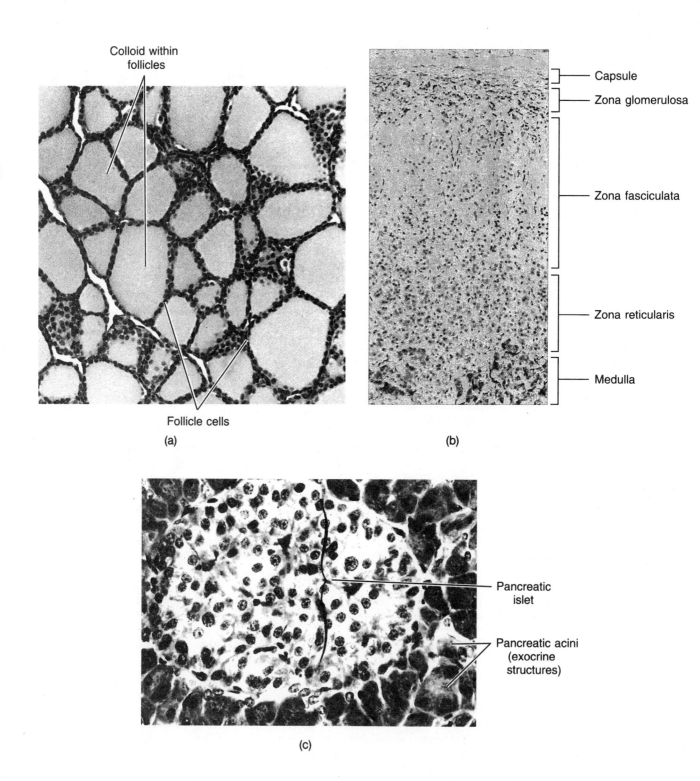

Colloid within follicles

Follicle cells

(a)

Capsule

Zona glomerulosa

Zona fasciculata

Zona reticularis

Medulla

(b)

Pancreatic islet

Pancreatic acini (exocrine structures)

(c)

Observations

```
┌─────────────────────────────┐
│                             │
│                             │
│                             │
│                             │
│        Pars nervosa         │
└─────────────────────────────┘
```

```
┌─────────────────────────────┐
│                             │
│                             │
│                             │
│                             │
│       Pars intermedia       │
└─────────────────────────────┘
```

```
┌─────────────────────────────┐
│                             │
│                             │
│                             │
│                             │
│        Pars distalis        │
└─────────────────────────────┘
```

```
┌─────────────────────────────┐
│                             │
│                             │
│                             │
│                             │
│        Thyroid gland        │
└─────────────────────────────┘
```

```
┌─────────────────────────────┐
│                             │
│                             │
│                             │
│                             │
│        Adrenal gland        │
└─────────────────────────────┘
```

```
┌─────────────────────────────┐
│                             │
│                             │
│                             │
│                             │
│          Pancreas           │
└─────────────────────────────┘
```

Conclusions

- How do the regions of the pituitary gland compare with regard to cell arrangement?

  _____

  _____

  _____

- What are the varied cellular arrangements found in endocrine glands and where are they found?

  _____

  _____

  _____

  _____

- Do the adrenal cortex and the adrenal medulla have different cell arrangements? Discuss.

  _____

  _____

  _____

  _____

• Do you see the endocrine and exocrine portions of the pancreas to be histologically distinct?

_____

_____

_____

_____

• Which is the more numerous cell tye of the pancreas, α or β cells?

_____

_____

_____

_____

❖ **Quick Quiz 1**

Match the areas in column A with the gland in column B.

A	B
_____1. neck	(a) ovary
_____2. "atop kidney"	(b) pituitary
_____3. brain	(c) islets of Langerhans
_____4. pelvis	(d) adrenal
_____5. abdomen	(e) thyroid

6. The thyroid gland has a follicle arrangement of secretory cells.   True or False?
7. Acidophils are found in the zona reticularis.   True or False?
8. "Cords" of glandular cells can be observed in the adrenal cortex.   True or False?

■ **TABLE 22.3 Major Endocrine Hormones**

HORMONE	SOURCE	TARGET	PRINCIPAL FUNCTION
Growth hormone [(GH); somatotropin (STH)]	Anterior pituitary	General body cells	Accelerates growth rate, particularly of muscle and bone; affects carbohydrate metabolism; elevates muscle glycogen
Thyrotropin [thyroid-stimulating hormone (TSH)]	Anterior pituitary	Thyroid gland	Stimulates synthesis and secretion of thyroid hormones
Corticotropin [adrenocorticotropic hormone (ACTH)]	Anterior pituitary	Adrenal cortex	Stimulates synthesis and secretion of glucocorticoids
Follicle-stimulating hormone (FSH)	Anterior pituitary	Ovary or testis	Stimulates development of ovarian follicles in females; promotes spermatogenesis in males
Luteinizing hormone (LH) [in males, interstitial-cell-stimulating hormone (ICSH)]	Anterior pituitary	Ovary or testis	Stimulates development of corpus luteum and secretion of its hormones in females; promotes development of sperm cells and secretion of testosterone in males
Prolactin [luteotropin, (LTH)]	Anterior pituitary	Corpus luteum and mammary glands	Stimulates progesterone secretion and lactation; unknown function in males
Melanocyte-stimulating hormone [(MSH) intermedin]	Pars intermedia	Melanocytes of skin	Stimulates melanin production and darkens skin but is insignificant in humans
Vasopressin [antidiuretic hormone (ADH)]	Produced in hypothalamus but stored and released by posterior pituitary	Cells of index: kidney distal tubules and collecting ducts of kidney	Stimulates water reabsorption
Oxytocin	Produced in hypothalamus but stored and released by posterior pituitary	Smooth muscle, particularly of uterus; mammary glands	Stimulates contraction of uterus in parturition; causes release of milk from mammary alveoli

Functional Aspects of Selected Hormones

Background

The regulatory effect of the endocrine system is achieved through the action of hormones on their target tissues. These hormones belong to several classes of chemical compounds: proteins (e.g., insulin), amines (e.g., epinephrine), steroids (e.g., testosterone), and fatty acids (prostaglandins). Despite the differences in chemistry, all hormones appear to have certain characteristics in common. First, hormones exert their influence at *low* concentrations. Second, they do not initiate reactions but rather regulate the rate of chemical reactions. Third, hormones are removed from the body continuously through metabolic degradation and excretion. Fourth, there appears to be a *specificity* of hormones for the target tissues. Hormones will only react with specific receptors within the membrane or cytoplasm of the target cell. Finally, the control of hormone secretion is achieved through **feedback** mechanisms. (See Figure 22.3.) These include: negative feedback control through *direct action*, as in the control by blood glucose levels of the secretion of insulin, and *indirect action*, as in the control of TSH or thyroxine secretion. The secretion of many pituitary hormones is regulated by "releasing factors" produced by the hypothalamus. These reach the pituitary gland through the portal system, and are a further example of an indirect control of hormone release.

Although each hormone has a specific function, as summarized in Table 22.3 the actions of hormones may be thought of as falling into one of three broad categories. Certain hormones, such as growth hormone, can be

■ **TABLE 22.3 Major Endocrine Hormones (continued)**

HORMONE	SOURCE	TARGET	PRINCIPAL FUNCTION
Thyroxine and Triiodothyronine (T_4 and T_3)	Thyroid gland	General; cells of body	Accelerates metabolic processes and oxygen consumption; stimulates glycogenolysis
Thyrocalcitonin (calcitonin)	Thyroid gland	Skeleton	Inhibits reabsorption of calcium and phosphorus from bone
Parathormone (PTH)	Parathyroid glands	Skeleton, kidney, and gastrointestinal tract	Promotes reabsorption from bone of calcium and phosphate; increases reabsorption of calcium and excretion of phosphates in urine; enhances absorption of calcium and phosphate across the gut
Mineralocorticoids (aldosterone)	Adrenal cortex (zona glomerulosa)	Renal tubules	Increases reabsorption of sodium ions
Glucocorticoids (cortisone, cortisol, and corticosterone)	Adrenal cortex (zona fasciculata)	General cells of body	Promotes liver gluconeogenesis and muscle glycogenolysis
Adrenal sex hormones (androgens; estrogen)	Adrenal cortex (zona reticularis)	General	Minor; perhaps augments gonadal hormones
Insulin	Islets of Langerhans, β cells	Liver, muscle, adipose cells	Enhances glucose and amino acid intake; regulates glucose metabolism and promotes protein synthesis
Glucagon	Islets of Langerhans, α cells	Liver	Stimulates glycogenolysis and gluconeogenesis
Estrogen	Ovary	Reproductive tract; mammary glands	Development and maintenance of reproductive tract; development of endometrium; promotion of secondary sex characteristics
Progesterone	Corpus luteum	Endometrium and mammary glands	Promotes the increase in size of the endometrium for implantation; prepares breasts to secrete milk
Testosterone	Interstitial cells of Leydig of testis	Reproductive tract	Promotes development and maintenance of reproductive tract and secondary sex characteristics; promotes sperm development

■ FIGURE 22.3   Generalized hormonal feedback control

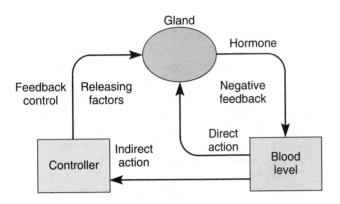

considered to be **morphogenic**, since they regulate the growth and development of an organ system or the whole organism. Other hormones (e.g., parathormone) may be regarded as **homeostatic** hormones, since they function in maintaining a relative constancy of a condition. Hormones such as adrenal epinephrine, which coordinate physiological events in stress, can be called **integrative** hormones.

The actions of hormones are conducted at the cellular level. The hormones which are proteins, peptide or catecholamine in origin act at the cell membrane to trigger the production of a secondary messenger, **cyclic AMP** within the cell. This secondary messenger is responsible for the resulting changes in cellular activity. Lipid soluble hormones do not act at the cell membrane, but through an intercellular mechanism. These function through the action of intracellular receptors that, when activated, alter gene activity that changes cellular activity.

The effects of the hormones studied in the following experiments appear to be mediated by cyclic AMP. For example, the action of vasopressin is mediated through the increased synthesis of cyclic AMP in the renal tubule cells. It is suggested that the increased cellular level of cyclic AMP makes the cells more permeable to water, thus increasing tubular reabsorption of water.

EXERCISE 1
The Effects of Vasopressin on the Frog

Discussion

Vasopressin or ADH has two major effects. It promotes the reabsorption of water by the renal tubules. This is referred to as an **antidiuretic action**. The second effect is an increase in blood pressure resulting from high levels of the hormone, causing vasoconstriction of arterioles. This action is referred to as a **pressor effect**. In this experiment you observe the antidiuretic effect of vasopressin.

Chemically, vasopressin is a polypeptide made up of eight different amino acids arranged cyclically with a small chain. It is quickly inactivated by the liver and kidneys; thus its effects occur quickly, but they also disappear rapidly.

Materials

Frogs, aquaria or water tanks; animal-weighing scales; Pituitrin solution (commercial preparation); 1-ml syringes, 25-gauge needles; amphibian Ringer's solution; paper towels.

Be sure to review safety precautions before handling fresh or preserved specimens.

Procedure

1. Inject the experimental frogs intraperitoneally with 0.5 ml of Pituitrin (Vasopressin) solution.
2. After injection, gently press the abdomen of the frog to expel urine from the bladder.
3. Blot the frog to remove excess fluid from the surface, and weigh the frog. Record this initial weight in Table C.
4. Place the frog in an aquarium so that the body surface is covered with water. (The eyes and nares should be above the water.)
5. Inject another frog (the control animal) in a similar fashion with 0.5 ml of amphibian saline; expel urine, blot dry, and weigh. Record the weight in Table C and place the frog in a separate aquarium.
6. Weigh the experimental and control animals at 30-min intervals for the next 2.5 hr, and again after 18 and 24 hr have elapsed. Record the weights in Table C.

Observations

Average weight change for:

a. Experimental _____

b. Control _____

■ **TABLE C Effects of Pituitrin/Vasopressin**

ANIMAL	INITIAL WEIGHT	MINUTES					HOURS	
		30	60	90	120	150	18	24
Experimental								
Control								
Change in weight								

Conclusions

- What was the effect of pituitrin/vasopressin on the weight of the frog?

  _____

  _____

- Explain the physiological mechanism that was operating in the animals.

  _____

  _____

  _____

  _____

  _____

- Were any other physical changes observed in the animals? Explain.

  _____

  _____

  _____

  _____

  _____

EXERCISE 2
The Effects of Thyroxine and Thiouracil on Tadpole Metamorphosis

Discussion

The active products of the thyroid gland are **thyroxine** (T$_4$) and **triiodothyronine** (T$_3$). In their formation, the follicular cells absorb iodide and oxidize it to iodine. Once iodine is formed, it is bound to the amino acid tyrosine, forming molecules of **monoiodotyrosine (MIT)** and **diiodotyrosine (DIT)**. This is done while the tyrosine is attached to the protein thyroglobulin. The next part of the sequence involves the coupling of DIT units together forming T$_4$ and the coupling of DIT and MIT units forming T$_3$. These reactions occur while the molecules are bound to thyroglobulin, to which they remain attached until they are secreted from the thyroid. When secretion of the hormones occurs, hydrolysis releases them from the thyroglobulin and the free hormones pass out of the cell and into interstitial fluid and ultimately into circulation. In the blood, T$_4$ is found circulating freely and also bound to certain plasma proteins. The proteins that bind T$_4$ are **albumin** and **thyroxine-binding globulin**. In addition, red blood cells bind T$_4$. Although the bound T$_4$ is inactive, it represents a circulating store of **thyroid hormone**.

The physiologic effects of free thyroid hormones are broad ranging and include the following:

1. Promotion of oxygen consumption by tissues, glycogenolysis and glycolysis, and a general increase in metabolic rate. Because this causes an increase in the production of body heat, this is referred to as **calorigenic action**.
2. Promotion of growth and differentation of tissues.
3. Increased carbohydrate absorption by the digestive tract.
4. Lowering of blood levels of cholesterol.

In this exercise, we examine the effects of thyroid hormone on growth and development.

It is known that thyroid hormone is involved in initiating the **metamorphosis** of certan amphibians. It is presumed that the triggering action of T$_4$ is accomplished by the activation of inherent genetic mechanisms within the cells yet to be elucidated. The process of metamorphosis in frogs varies, depending on the species.

Tadpoles normally hatch from the eggs within 3 weeks of fertilization. During the next 3 months, the tadpoles undergo extensive changes in both external and internal morphology. Among the major changes are reabsorption of the tail, transformation of the digestive tract, and development of lungs from gills. Since this process is largely regulated by thyroid hormones, blocking agents such as **thiouracil** can alter the normal developmental changes. In this experiment, *Rana pipiens* tadpoles about 6 weeks of age will be used so that metamorphic changes can be determined within a reasonable experimental period.

Materials

Seven tadpoles; seven water containers; air pumps and tubing; metric ruler, dissecting microscope; amphibian Ringer's solution; 1-thyroxine; thiouracil; spinach leaves; pH meter; 0.1 N NaOH; 0.1 N HCl; watch glasses, seven one-liter bottles; wax marking pencils.

Be sure to review safety precautions before handling fresh or preserved specimens.

Procedure

1. Measure out 7 separate liters of amphibian Ringer's solution. Place each liter in a separate bottle, and mark the bottles 1 through 7.
2. Add the reagents as follows.

BOTTLE/SOLUTION	REAGENT
1	—
2	Thyroxine, 10 mg
3	Thyroxine, 100 mg
4	Thyroxine, 300 mg
5	Thiouracil, 100 mg
6	Thiouracil, 300 mg
7	Thiouracil, 600 mg

3. Check the pH of each solution and adjust where necessary to a pH of 8.3 using 0.1 N NaOH and/or 0.1 N HCl.
4. Place a tadpole in a watch glass and observe under a dissecting microscope. Note the overall anatomical features, particularly fins, tail, mouth development, and legs. Measure the body length with a ruler.
5. Record the initial observations in Table D and then place the tadpole in the container filled with solution 1.
6. Repeat steps 4 and 5 six times more and place the tadpoles in separate containers filled with solutions 2 through 7, respectively.
7. Keep the containers at a temperature of about 23°C and aerate the solutions with the air pumps.
8. Feed the tadpoles pieces of spinach leaves, being sure to change the solutions periodically to keep the animals in a fresh environment.
9. Compare the size and morphological changes of the experimental and control animals on a 24-hr basis for at least 10 days.
10. Record all initial observations in Table D. Use a scale of 1 to 5 indicating minimal to maximal development.
11. Record changes in structures in Table E.

Observations

In Table E, indicate appearance/development/changes of the structures listed in Table D.

■ **TABLE D   Initial Observation of Tadpoles**

FEATURE OBSERVED	SOLUTION						
	1	2	3	4	5	6	7
Fins							
Tail							
Mouth							
Forelimbs							
Hind limbs							
Body length							

■ **TABLE E Effects of Thyroxine and Thiouracil**

MORPHOLOGIC CHANGES AFTER	SOLUTION						
	1	2	3	4	5	6	7
1 day							
2 days							
3 days							
4 days							
5 days							
6 days							
7 days							
8 days							
9 days							
10 days							

Conclusions

- Which experimental reagent speeded the rate of metamorphosis? Discuss.

_____

_____

_____

- What was the result of adding thiouracil to the tadpoles' environment?

_____

_____

_____

• Why did the thiouracil produce the observed results?

_____

_____

_____

• If thyroxine promotes growth, was its effect on the tadpole's tail expected? Discuss.

_____

_____

_____

EXERCISE 3
The Roles of Insulin and Epinephrine in Hypoglycemia

Discussion

Insulin is a protein containing two polypeptide chains linked by disulfide bridges. It is synthesized in the β **cells** of the islets of Langerhans of the pancreas and secreted into the bloodstream, where it circulates freely. One of the primary physiologic effects of insulin is to lower blood glucose levels. This is accomplished by accelerating the transport of glucose out of blood into muscle, liver, and kidney cells, and stimulating the intracellular use of glucose by these cells.

The major mechanism for the regulation of insulin secretion is the blood glucose level. Because the reserve of carbohydrate in neural tissue is low, the brain must be continually supplied with significant quantities of glucose for normal functioning. If blood glucose levels fall as in hypoglycemia or insulin overdose, the brain quickly shows evidence of this. If the condition is prolonged, irreversible brain damage can result, and even death.

Epinephrine, a catecholamine secreted by the adrenal medulla, has the opposite effect of insulin: it raises blood glucose levels by promoting muscle and liver glycogenolysis. In the liver, for example, epinephrine activates the conversion of ATP to cyclic AMP. Cyclic AMP in turn activates the production of a phosphorylase that catalyzes the breakdown of glycogen to glucose. Glucose diffuses out of the hepatic cells into circulation, thus raising the blood glucose levels.

A. Hypoglycemia in the Mouse

Materials

Five 1-ml tuberculin-type syringes and five 25-gauge needles; commercial insulin; 4 laboratory mice; 4 animal cages; 70% alcohol; solutions of 10% glucose and 10% fructose; epinephrine (1:1000); physiologic saline; protective gloves.

Be sure to review safety precautions before handling fresh or preserved specimens.

Procedure

Note: Read through the entire procedure before you begin.

1. Fast the mice for 24 hr before the experiment.
2. Prepare 4 syringes and needles for subcutaneous injection according to the chart in step 4.
3. Prepare a second set of syringes for subcutaneous injection according to the chart in step 6.
4. Inject each of the mice subcutaneously according to the following protocol.

ANIMAL	INITIAL SUBCUTANEOUS INJECTION
1	Saline (0.1 ml)
2	Insulin (0.1 ml)
3	Insulin (0.1 ml)
4	Insulin (0.1 ml)

5. After the administration of insulin or saline, observe the animal and check for convulsions. Hold the animal 2 to 3 in. above the table and drop it, checking for a righting reflex. Record results in Table F, as either "present" or "absent."
6. If convulsions appear in the experimental animals, give subcutaneous injections according to the following chart and record in Table F.

ANIMAL	ADDITIONAL SUBCUTANEOUS INJECTION
2	10% Glucose (0.05 ml)
3	10% Fructose (0.05 ml)
4	Epinephrine (0.1 ml)

7. Observe the results of this additional injection.

Note: If convulsions continue in *any* of the experimental animals, administer 0.1 ml of glucose solution again. Continue administration of glucose solution until convulsions stop.

8. Record the results in Table F.

Observations

■ **TABLE F   Hypoglycemia in Mice**

ANIMAL	INITIAL INJECTION	RIGHTING REFLEX INITIAL OBSERVATION	SECOND INJECTION	RIGHTING REFLEX OBSERVATION	ADDITIONAL GLUCOSE NECESSARY	RIGHTING REFLEX OBSERVATION
1	Saline		—			
2	Insulin		Glucose			
3	Insulin		Fructose			
4	Insulin		Epinephrine			

Conclusions

● What were the initial effects of insulin? Describe what was happening from a physiological perspective.

_____

_____

_____

● Which of the three substances (glucose, fructose, epinephrine) was able to counteract the effects of insulin? Discuss.

_____

_____

_____

● Why do you think a diabetic who is on an insulin regimen should carefully regulate physical activity such as exercise?

_____

_____

_____

● Why must a diabetic's meal schedule be regulated?

_____

_____

_____

❖ **Quick Quiz 2**

Match the item in column A with the gland in column B.

	A		B
_____	**1.** ADH	(a)	thyroid
_____	**2.** Triiodothyronine	(b)	adrenal cortex
_____	**3.** beta cells	(c)	neurohypophysis
_____	**4.** epinephrine	(d)	adrenal medulla
_____	**5.** mineralocorticoids	(e)	pancreas
		(f)	adenohypophysis
		(g)	thymus

6. Increased levels of vasopressin could increase overall body weight due to the accumulation of tissue fluids. True or False?

7. The presence of thiouracil would prevent reabsorption of a tadpole's tail.   True or False?

8. Epinephrine secreted by the adrenal gland has the same effect on glucose as insulin.   True or False?

UNIT 23

The Reproductive System

OVERVIEW

In this unit you will study both the microscopic and macroscopic features of the male and female reproductive system. Additionally, you will study the development of human gametes, i.e., egg and sperm, along with hormonal influences which regulate these activities.

OUTLINE

Anatomy of the Reproductive System

Background

To reproduce new, similar living organisms, the individuals of a species must pass on exact instructions to the succeeding generation. In the human species where both male and female participate in this process, the instructions are confined to the nucleus of sex cells (or gametes) in molecules called **DNA** (deoxyribonucleic acid). Nuclear material is divided into half the total amount of the *primitive* cells that will develop into sperm and egg through a process called **meiosis**. A father passes his half of the genetic material to his daughter or son in a **sperm**; a mother passes on the other half in an **ovum** or **egg**. At **fertilization** the sperm and egg unite. The fertilized egg now carries the complete complement of DNA. **Sperm development** in the male and **egg development** in the

female go on in the respective reproductive tracts. In addition, after fertilization, the female carries the developing fertilized egg in her tract.

EXERCISE 1
Microscopic Anatomy

Discussion

Male
In the male, the organ that produces **gametes**, or **sex cells (sperm)** is the **testis**. Within each testis there are hundreds of coiled tubules (the **seminiferous tubules**) in which **spermatogenesis**, or the development of sperm, occurs. A cross section of one of these tubules shows sperm in all stages of development, from the most primitive germ cells or **spermatogonia** to the more developed **spermatozoan**. The cells begin to develop at the periphery of the tubule and are finally released as

■ **FIGURE 23.1   The reproductive system of the human male**
(a) Gross anatomy – sagittal view. (b) Diagram of cross section of testis. (c) Diagram of cross section of seminiferous tubule.

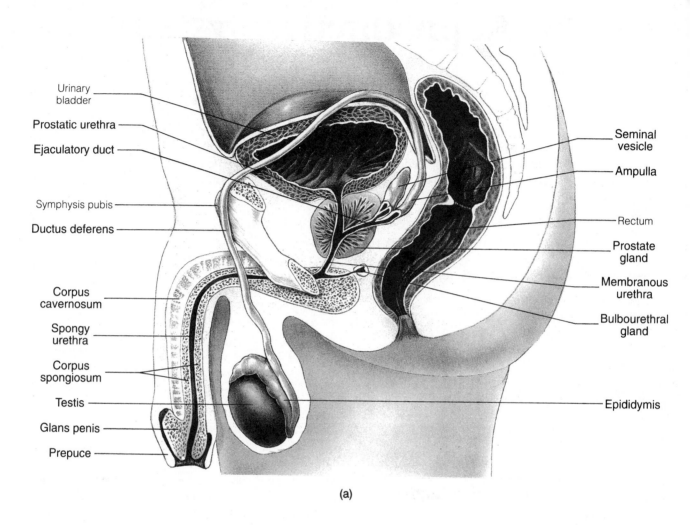

Urinary bladder

Prostatic urethra

Ejaculatory duct

Symphysis pubis

Ductus deferens

Corpus cavernosum

Spongy urethra

Corpus spongiosum

Testis

Glans penis

Prepuce

Seminal vesicle

Ampulla

Rectum

Prostate gland

Membranous urethra

Bulbourethral gland

Epididymis

(a)

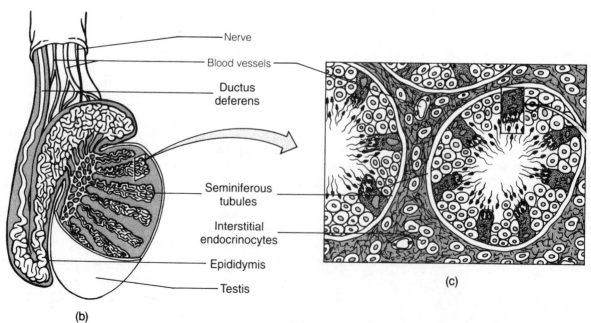

Nerve

Blood vessels

Ductus deferens

Seminiferous tubules

Interstitial endocrinocytes

Epididymis

Testis

(b)

(c)

mature cells at the central lumen. As they mature, developing sperm "tuck" their heads into the cytoplasm of supporting sustentacular cells, called **Sertoli cells** for nourishment. An additional important group of cells is found in **interstitial** tissues between tubules. Within the connective tissue area resides the **Leydig cells**, which are responsible for **testosterone** production. (See Exercise 1 under "Physiology of the Reproductive System" for details of sperm development.) See Figure 23.1.

Female

In the female, the organ that produces gametes or **sex cells (ova)** is the **ovary**. Within the ovary there is a demarcation of tissue; the inner medulla (consisting of large blood vessels, lymphatics, nerves, and scattered wisps of smooth muscle) and the outer cortex where ova develop. In an ovary from a sexually mature female, growing **follicles** (an immature egg and surrounding epithelial cells, the membrane granulosa) can be seen. The follicle in an ovary goes through different stages: from primary follicles with primitive eggs or **oögonia**, to developing follicles, to mature follicles ready to release an almost totally developed ovum. In a follicle that has begun growing, the egg, or ovum, projects as a mound into the fluid-filled cavity or **antrum** within. Additional follicle cells surround this egg and this enclosure is called the **corona radiata**. Also present in the ovary are the vascular **corpora lutea** (singular, **corpus luteum**). These bodies, the remains of ruptured follicles whose eggs have been released, secrete the hormones estrogen and progesterone

for a short time in that cycle. (See Exercise 2 under "Physiology of the Reproductive System" for the details of ovum development.) In pregnancy, a corpus luteum continues to grow for several months, and to release hormones. (If the ovum is not fertilized by a sperm, the corpus luteum degenerates into a white scar on the ovary, the **corpus albicans**.)

The **uterus** of the female reproductive tract is the organ where the fertilized egg implants and grows. There are three layers making up its walls: the outer **perimetrium** (as serosal layer); a middle **myometrium** (a muscle layer); and an inner **endometrium** (a mucosal layer of epithelium and underlying connective tissue, the lamina propria). The myometrium makes up the bulk of the uterine wall. It comprises bundles of long, smooth muscle fibers that are separated by connective tissue. There are three ill-defined layers of muscles: inner longitudinal fibers, a thick middle layer of circular and oblique fibers (with many blood vessels), and a thin outer longitudinal fiber layer. (Figure 23.2.)

The endometrium varies in appearance depending on what time of the menstrual cycle the section is made. In its basic form, the endometrium consists of simple columnar epithelium that is occasionally ciliated and sends glands deep into the connective tissue of the mucosa. During the four stages of the menstrual cycle (see Exercise 3 under "Physiology of the Reproductive System"), the endometrium undergoes the changes summarized in Table 23.1. Figure 23.3 indicates changes that occur in the menstrual cycle.

■ **TABLE 23.1 Stages of the Endometrium**

STAGE/PHASE	INFLUENTIAL HORMONE	DESCRIPTION
Proliferative (follicular)	Estrogen	Begins at the end of menstruation, see below: characterized by thickening of the endometrium as glands proliferate, lengthen, and become closely packed; many mitoses occur in lamina propria; glands become wide and convoluted toward end of this stage
Progestational (secretory)	Estrogen and progesterone	Endometrium continues to thicken as gland cells hypertrophy and edematous fluid accumulates; gland cells swell and release a thick glycogen-rich secretion; arteries grow toward surface and coil; upper portion of endometrium containing neck and convoluted portion of glands is the **functionalis** layer, lost in menstruation; bottom layer containing bases of glands is the **basalis** layer, not shed at menstruation
Ischemic (premenstrual)	Diminished estrogen and progesterone	Coiled arteries constrict and relax repeatedly; functionalis shrinks because of blood and oxygen deprivation
Menstrual	Diminished estrogen and progesterone	Functionalis dies and is shed as blood vessels near surface break; **necrotic** tissue, blood, and glandular secretions are discharged; basalis layer remains intact and repair begins from the remaining glandular epithelial cells

■ **FIGURE 23.2   The reproductive system of the human female**
(a) Gross anatomy – posterior view. (b) Sagittal section.

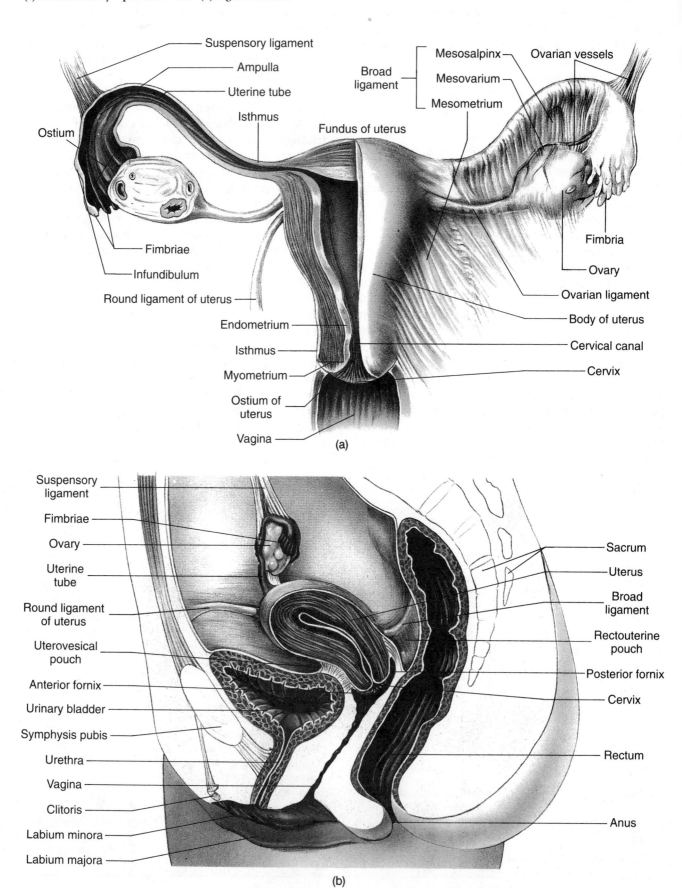

(a)

(b)

■ FIGURE 23.3 Cyclic changes in ovarian follicles

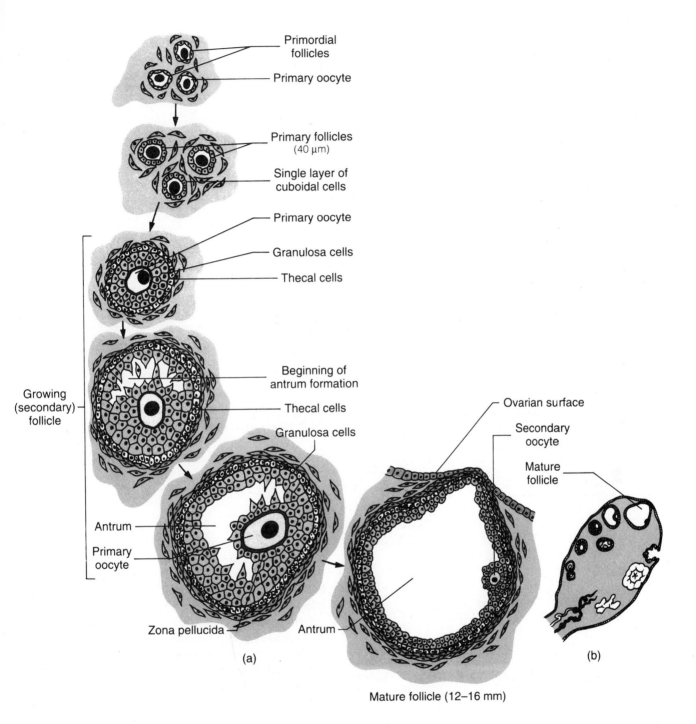

(a)

Mature follicle (12–16 mm)

(b)

A. The Testis

Materials

A microscope; slides of testis showing seminiferous tubule detail (c.s.).

Procedure

1. Oberve the slide of the testis under lower power.
2. Identify seminiferous tubules, mature sperm in lumina, developing sperm, and interstitial cell areas.
3. Observe the tubules under high power.

4. Identify cells in division by their dark staining chromosomes.
5. Identify heads of developing sperm, mature sperm, and flagella in the lumen.
6. Draw and label your observations in the space provided.

Observations

Seminiferous tubule

Conclusion

- How are the stages (immature to mature) of sperm development distributed in the seminiferous tubule cross section?

_____

_____

_____

_____

B. The Ovary

Materials

A microscope; slides of ovary with follicles at various stages during menstrual cycle.

Procedure

1. Observe the slide of the ovary under low power.
2. Identify the cortex region (surrounding the medulla) where follicles are present.
3. Observe the follicles under high power and observe the egg projecting into the center.
4. Draw a mature follicle in the space provided.

Observations

Ovary

Conclusion

- Are all follicles in an ovary in the same stage of development? Discuss.

_____

_____

_____

_____

EXERCISE 2
Macroscopic Anatomy: The Fetal Pig

Discussion

Male

The reproductive system of the male fetal pig consists of testis, epididymis, vas deferens (ductus deferens), urethra, and accessory glands: the paired seminal vesicles, the single prostate gland, and the paired bulbourethral (Cowper's) glands. Sperm are produced in the testis and travel through the duct system, where secretions from the accessory glands are added. Refer to Figure 23.4.

The **testes** develop inside the abdominal cavity and, shortly before birth, descend through an opening in the abdominal wall, the **inguinal canal**, into the **scrotal sac**. The sperm travel from the testes into the highly coiled **epididymis** on the surface of the testes and then into the **vas deferens**. The vas deferens passes through the inguinal canal along with blood vessels and nerves, making up the **spermatic cord**. The vas deferens loops around the ureters and passes behind the bladder and enters the **urethra**. Refer to Figure 23.5. After receiving the secretions of the **accessory glands**, the urethra passes into the **penis**. The penis terminates at the **preputial orifice (urogenital opening)** on the ventral abdominal wall just posterior to the umbilical cord.

■ **FIGURE 23.4 The reproductive system of the male fetal pig**
Refer to the color photo gallery.

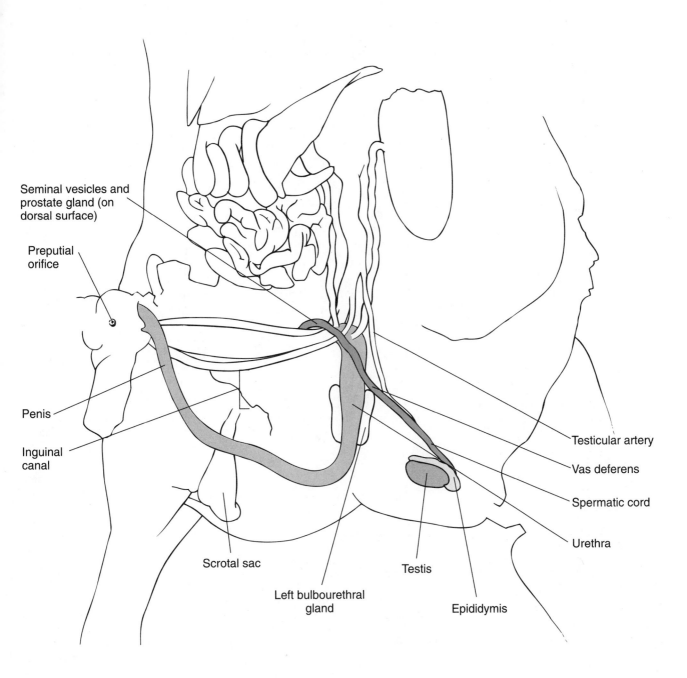

Seminal vesicles and prostate gland (on dorsal surface)

Preputial orifice

Penis

Inguinal canal

Scrotal sac

Left bulbourethral gland

Testis

Epididymis

Testicular artery

Vas deferens

Spermatic cord

Urethra

■ **FIGURE 23.5** **Dorsal view of the male fetal pig urethra and associated organs, greatly enlarged**

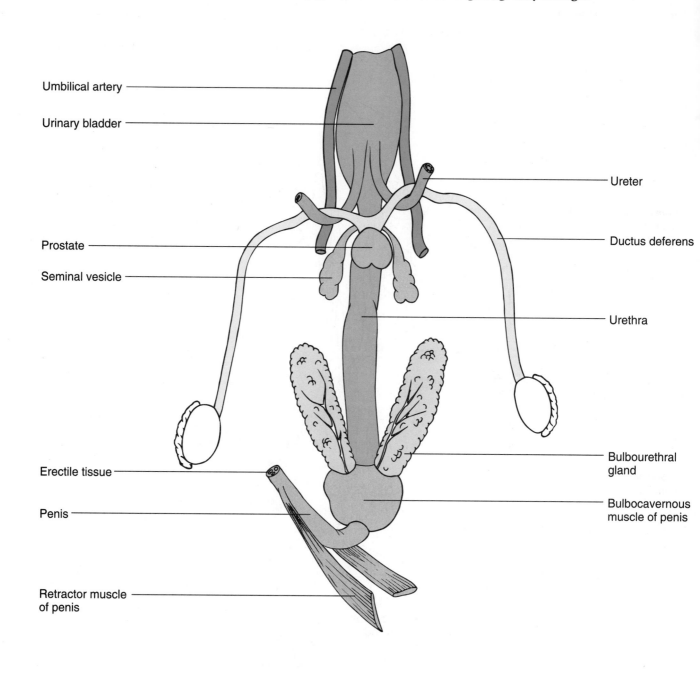

Umbilical artery

Urinary bladder

Prostate

Seminal vesicle

Erectile tissue

Penis

Retractor muscle
of penis

Ureter

Ductus deferens

Urethra

Bulbourethral
gland

Bulbocavernous
muscle of penis

Female

The reproductive system of the female fetal pig consists of the ovaries, uterine tubes (Fallopian tubes or oviducts), horns of the uterus, body of the uterus, vagina, and urogenital sinus (vulva). Eggs develop in the ovaries, travel through the duct system, and if fertilized, implant in the horns of the uterus. Refer to Figure 23.6.

Each oval shaped **ovary** lies posterior to the kidney within the abdominal cavity and is supported in a mesentery, the **broad ligament**. A thicker mesentery, the **round ligament**, crosses the broad ligament at right angles and attaches to the abdominal wall near the groin. On the dorsal surface of each ovary is a highly coiled **uterine tube (Fallopian tube, oviduct)**. The expanded

■ **FIGURE 23.6 The reproductive system of the female fetal pig**
Refer to the color photo gallery.

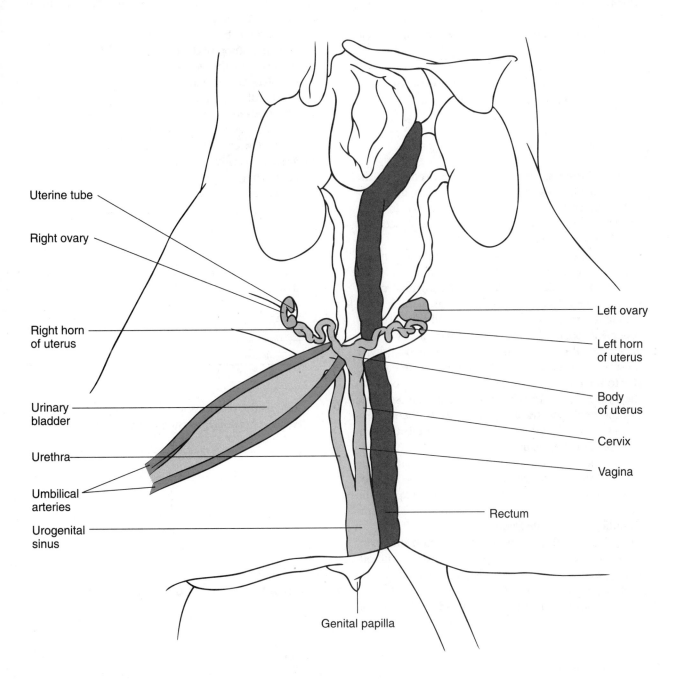

Uterine tube

Right ovary

Right horn
of uterus

Urinary
bladder

Urethra

Umbilical
arteries

Urogenital
sinus

Left ovary

Left horn
of uterus

Body
of uterus

Cervix

Vagina

Rectum

Genital papilla

end of the uterine tube is the **ostium** which has small fingerlike projections called **fimbriae**. These help to guide the ovum into the uterine tube which connects with the wider **horn of the uterus**. The ovum travels from the uterine tubes into the horn of the uterus where, if fertilized, it will implant. This extended space allows the pig to carry many young at once. The two horns of the uterus merge to form the **body of the uterus** which

lies dorsal to the urethra. The uterus leads into the **vagina**, which unlike in the human female, unites with the urethra from the bladder to form a common passageway, the **urogenital sinus (vulva)**. This passageway opens to the outside at the **urogenital orifice** at the location of the projection of tissue, the **urogenital papilla**. This is formed by the merging of the lateral folds called the **labia**.

A. The Male

Materials

A preserved and injected fetal pig; dissecting tray and instruments.

Be sure to review safety precautions before handling fresh or preserved specimens.

Procedure

1. Identify the scrotal sac under the skin ventral to the anus. Make a shallow incision into the sac and carefully remove the skin to expose the testis. In most specimens you will find that the testis has descended into the scrotal sac. See Figure 23.4.
2. On the dorsal surface of the testis, identify the epididymis and its continuation into the vas deferens.
3. Follow the vas deferens through the inguinal canal. Note the blood vessels that pass along with it comprising the spermatic cord. Gently pull the cord and observe its position in the abdominal cavity.
4. To expose the rest of the system the pelvic cavity must be opened. Observe the preputial orifice posterior to the umbilical cord. Observe the penis beneath the skin in the mid ventral area. Carefully separate out the penis and trace it posteriorly to the anal region. Carefully cut through the pelvic girdle and gently spread the legs apart.
5. Return to the vas deferens and follow its route in the abdominal cavity to where it enters the dorsal surface of the urethra. Gently turn the urethra to expose the dorsal surface. Note a pair of seminal vesicles surrounding the prostate gland.
6. Return the urethra to its normal position and continue to follow it. Observe the larger pair of bulbourethral glands (Cowper's glands) on either side of the urethra just before it enters the penis.
7. Retrace the penis to the preputial orifice (urogenital opening).

B. The Female

Materials

A preserved and injected fetal pig; dissecting tray and instruments.

Be sure to review safety precautions before handling fresh or preserved specimens.

Procedure

1. Identify the pair of oval shaped ovaries posterior to the kidneys. See Figure 23.6.

2. If not previously destroyed, observe the broad ligament supporting the ovary.
3. Locate the tiny coiled uterine tube (Fallopian tube, oviduct) curving around the ovary.
4. Observe the horn of the uterus.
5. Note the union of the two horns of the uterus to form the body of the uterus which lies dorsal to the urethra. The lower part of the body is the cervix which can be distinguished internally by its folded surface. This leads into the vagina.
6. Note the relationship of the urinary and reproductive tracts. Observe the merging of the urethra and the vagina to form the urogenital sinus (vulva) which opens to the external surface.
7. Observe the labia and genital papilla.

Conclusions

- Distinguish externally between the male and female fetal pig.

_____

_____

_____

_____

- Contrast the relationship of the urethra in the male and female reproductive systems of the fetal pig.

_____

_____

_____

_____

EXERCISE 3
Macroscopic Anatomy: The Human

Discussion

A comparison of Figures 23.1 and 23.2 with Figures 23.4 and 23.5 reveals many similarities between the fetal pig and the human. These will become apparent during this exercise.

Materials

Models and charts of the human reproductive system.

Procedure

1. Examine the models and identify the structures labeled in Figures 23.1 and 23.2.

2. Compare fetal pig and human structures by filling in
   Table A.

Observations

Record a (**+**) if present, and a (−) if absent.

TABLE A Comparison of Fetal Pig and Human Reproductive Structures

STRUCTURE	PIG	HUMAN	STRUCTURE	PIG	HUMAN
MALE			**FEMALE**		
1. Testis and scrotum			1. Ovaries in abdomen		
2. Epididymis			2. Broad and round ligaments		
3. Vas deferens			3. Fimbrae		
4. Vas deferens passes through inguinal canal			4. Oviduct		
5. Seminal vesicle			5. Horn of uterus		
6. Prostate gland			6. Body of uterus		
7. Juncture of vas deferens and urethra			7. Cervix		
8. Bulbourethral glands			8. Vagina		
9. Penis			9. Juncture of vagina and urethra		
10. Urogenital opening			10. Urogenital opening		

Conclusions

• List the differences between the fetal pig and human:

  a. Male structures   **b.** Female structures

  _____   _____

  _____   _____

  _____   _____

• What is the function of the seminal vesicles in the human male?

_____

_____

_____

_____

• What is the advantage of implantation in the horn of the uterus for the pig?

_____

_____

_____

❖ **Quick Quiz 1**

Match the statement in column A with the appropriate term in column B.

A		B	
_____ 1.	gamete	(a)	on surface of testis
_____ 2.	corona radiata	(b)	Sertoli cell
_____ 3.	preputial orifice	(c)	on external surface of female pig
_____ 4.	Cowper's glands	(d)	testis/ovary
_____ 5.	sustentacular cell	(e)	on external surface of male pig
_____ 6.	epididymis	(f)	sperm or egg
_____ 7.	genital papilla	(g)	surrounds egg
_____ 8.	gonad	(h)	bulbourethral glands

■ **FIGURE 23.7 Spermatogenesis**

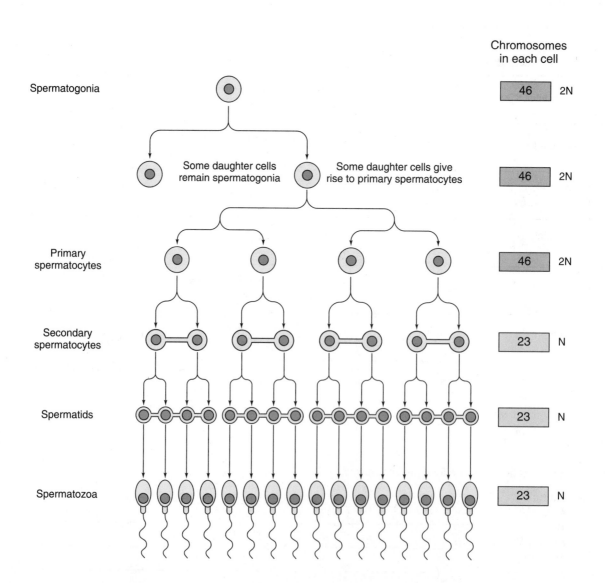

Physiology of the Reproductive System

EXERCISE 1
Maturation of Sperm

Discussion

Before reading this discussion, review the pertinent material of Exercise 1 of the preceding section. Refer to Figure 23.7 as you read.

The male reproductive tract produces sex cells called sperm. They develop from a primitive or germinating epithelium in the seminiferous tubules of the testis. These germinating cells have the normal body chromosome number of 46 chromosomes, or what is called the "2N" number of chromosomes. Under the influence of **follicle-stimulating hormone (FSH)** from the pituitary gland at puberty, these primitive cells divide. Some cells are kept in reserve for future sperm production while other cells go on to form mature sperm. Those that continue to develop are called **primary spermatocytes**. With a subsequent division, the chromosomal material of the developing sperm divides. The products of this reduction division, **secondary spermatocytes**, have half the chromosomal material of body cells: **23 chromosomes** or the "N" number. This process of **reduction division** is termed **meiosis**. The reduction division is followed by **mitosis** and produces four **spermatids** in a syncytial cluster. The entire process thus far has taken about 2 to 3 weeks, and is followed by **spermiogenesis**, requiring a further maturation time of about 8 weeks. Spermiogenesis ends in the production of four nonmotile, nonfertile spermatozoa.

The completion of maturation requires the hormone testosterone, produced by interstitial cells of Leydig upon stimulation of luteinizing hormone (LH) released from the pituitary gland. During development the sperm nestle into the Sertoli cells, presumably for nutrition. Sperm gain motility in the epididymis, and gain fertility in the female tract before fertilization.

At fertilization the sperm (1N) and the egg or ovum (1N) unite to form the fertilized egg (2N). This process reestablishes the normal somatic (body) cell number of chromosomes.

Mature sperm consist of a head, a midpiece, and a tail. The chromosomal material is condensed in the head, which is capped by an **acrosome** thought to contain the enzyme hyaluronidase. This enzyme aids in penetrating the cells that surround the egg at fertilization. The midpiece contains **mitochondria**, and the tail is a **flagellum** by which the sperm moves.

Materials

A microscope; slides of the seminiferous tubules and sperm smear (c.s.); electron micrograph (or diagram) of sperm.

Procedure

1. Observe the slide of the seminiferous tubule under high power.
2. Identify primary spermatogonia, Leydig cells, Sertoli cells, and nuclei of spermatocytes in division and mature sperm.
3. Observe the slide of sperm smear under the highest power of your microscope.
4. Identify the head region and flagellum of the sperm.
5. On the electron micrograph or diagram of a sperm, identify the head, neck, tail, and end piece, the acrosome, and the area of abundant mitochondria.
6. Label the accompanying diagram.

Observations

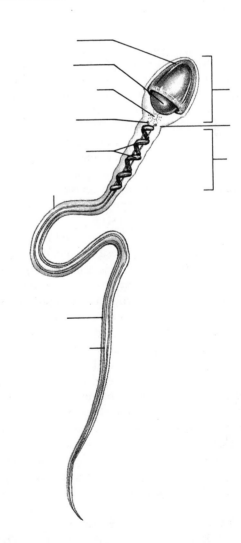

Conclusions

- Fill in the appropriate *N* number of chromosomes for each of the stages below.

CELL STATE	*N* NUMBER
Spermatogonium	_____
Primary spermatocyte	_____
Secondary spermatocyte	_____
Spermatid	_____
Spermatozoa	_____
Fertilized egg	_____

EXERCISE 2
Maturation of Ova

Before reading this Discussion, review the pertinent material of Exercise 1 of the preceding section. Refer to Figure 23.8 as you read.

Discussion

The female reproductive tract produces sex cells called ova. They develop from a primitive cell with 46 (the 2*N* number) chromosomes, called **oögonium**. The oögonia have been present in the human ovary since before birth and are surrounded by a cluster of epithelial follicle cells. Oögonia divide in the embryo and populate the developing ovary with **primary oöcytes**, cells that can develop to form **mature eggs**. During embryonic development primary oöcytes begin a reduction division or **meiosis** and stop half-way through the process. This division will be continued at puberty when FSH is secreted by the pituitary gland. As the egg develops and the follicle cells proliferate and form a sphere of cells around the egg (the **membrane granulosa**), the follicle cells and egg secrete the gelatinous **zona pellucida**, a covering around the egg. Additional follicle cells grow and surround the egg (the **corona radiata**), which sits on a mound (the **cumulus oöphorus**) within the fluid-filled space, the **antrum**. At this stage, the follicle is called a **Graafian follicle**. When the egg is ready to be released from the ovary the follicle bursts, expelling fluid and egg. The following cells remain behind as the **corpus luteum**, which disintegrates into a "scar" (the **corpus albicans**) after about 9 days if the egg is not fertilized. If the egg is fertilized, the corpus luteum survives the first few months of pregnancy to secrete the estrogen and progesterone before atrophying. These hormones support the uterus as it prepares for possible implantation of the **zygote** (fertilized egg).

The reduction division of the primary oöcyte continues when FSH is released from the pituitary. Two cells are formed: the **secondary oöcyte** (now with half the chromosomal content) and a small **polar body**, which degenerates. The cells of the follicle (now a Graafian follicle) release **estrogen**, a hormone that begins preparing the uterus for implantation. (See Table 23.1.) The secondary oöcyte is now released from the ovary at ovulation. It will divide again mitotically in the Fallopian tube when fertilization occurs to produce a mature ovum and a second degenerate polar body. The chromosome number of normal body cells (2*N*/46) is reestablished when the mature ovum (1*N*/23) is fertilized by a mature sperm (1*N*/23). The corpus luteum remains behind to continue secreting estrogen, and now progesterone too. These hormones continue to support the growth of the endometrium of the uterus until the corpus luteum dies. (See Figure 23.9 for a summation of hormonal changes.)

Materials

A microscope; slides of ovary with follicles.

Procedure

1. Review the slide of the ovary under low power under the microscope.
2. Observe follicles in different stages of development.
3. Identify follicle cells: antrum, zona pellucida, primary oöcyte, mature follicle, secondary oöcyte, primordial follicles.

Conclusions

- List the similarities/differences between egg and sperm formation with regard to the overall process, control, and timing.

  _____

  _____

  _____

  _____

  _____

- Compare the egg and sperm with respect to size, mobility, and genetic content.

  _____

  _____

  _____

  _____

  _____

  _____

■ **FIGURE 23.8 Oögenesis**

(a) Schematic representation. (b) Micrograph of a follicle.

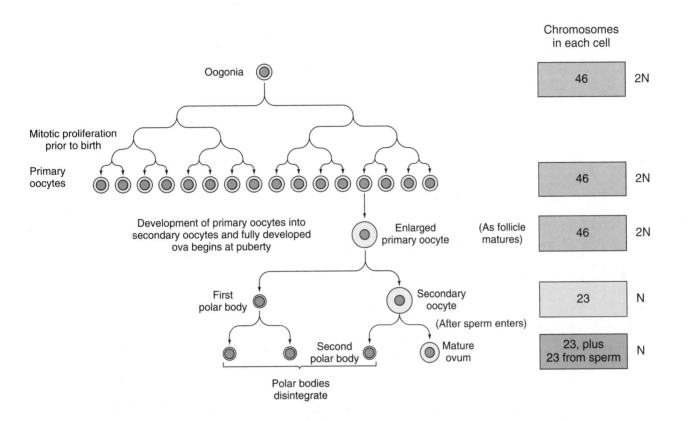

Chromosomes
in each cell

Oogonia

| 46 | 2N |

Mitotic proliferation
prior to birth

Primary
oocytes

| 46 | 2N |

Development of primary oocytes into
secondary oocytes and fully developed
ova begins at puberty

Enlarged
primary oocyte

(As follicle
matures)

| 46 | 2N |

First
polar body

Secondary
oocyte

(After sperm enters)

| 23 | N |

Second
polar body

Mature
ovum

| 23, plus
23 from sperm | N |

Polar bodies
disintegrate

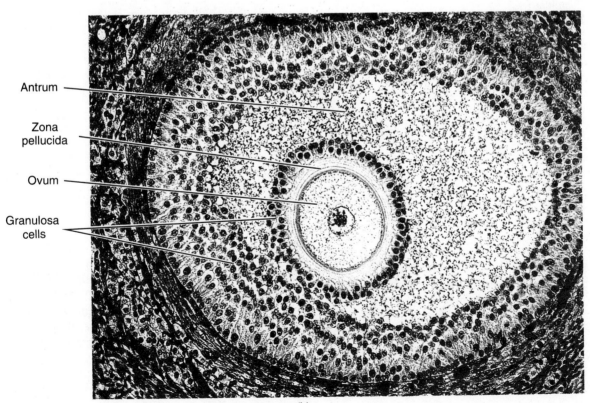

Antrum

Zona
pellucida

Ovum

Granulosa
cells

(b)

■ **FIGURE 23.9 Relation of hormones and ovum changes during the human female cycle**

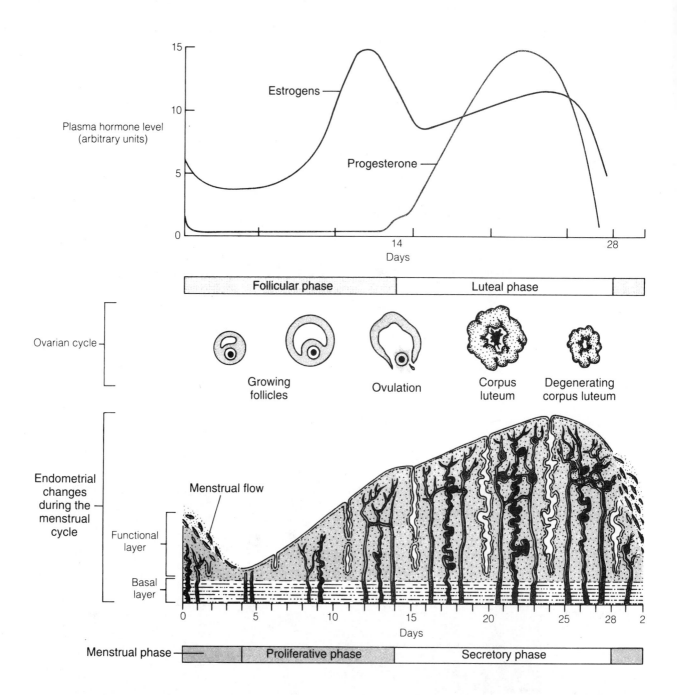

EXERCISE 3
Hormones of the Female Menstrual Cycle

Discussion

The ovary secretes hormones as the egg is developing, to prepare the uterus for implantation of that egg, should it be fertilized. These hormones are produced by the follicular cells.

Egg development is stimulated by FSH released from the pituitary gland. As the egg develops, the follicle cells will produce hormones that "turn off" FSH so that stimulation of additional egg development stops. When the follicle cells stop producing hormones they allow FSH to be released once again, thus making egg development a *cyclic* function (roughly 28 days). As the follicular hormones are effecting this **feedback inhibition** of pituitary hormones, they are also causing cyclic changes in the uterine lining.

As the egg develops, the follicle cells surrounding it produce estrogen. Estrogen does several things: (a) it stimulates the endometrium of the uterus to begin growing during the "**proliferative phase**" (see Table 23.1 and Exercise 1), (b) its levels surge about the twelfth day of the cycle and possibly cause LH to be released by the pituitary, and (c) as levels rise along with those of progesterone, estrogen and progesterone inhibit FSH secretion. LH secretion by the pituitary causes ovulation and supports the follicular cells left behind after ovulation, as the corpus luteum.

The corpus luteum continues secreting estrogen after ovulation. It also secretes progesterone. Together these two hormones cause continued development of the endometrium during the "**progestational phase**."

In addition, high levels of these two hormones effect a feedback inhibition on the pituitary gland and it stops secreting LH along with FSH. Support for the corpus luteum ceases and it begins to degenerate. It stops producing estrogen and progesterone and so uterine development is no longer supported. The "menstrual phase" of the uterus occurs. Finally, with low levels of estrogen and progesterone, the inhibition of FSH ceases and the cycle begins again.

Materials

Charts illustrating the relationships among pituitary, ovarian, uterine, and hormonal cycles; Figure 23.9.

Procedure

1. Examine the charts of the female cycle and identify pituitary hormones and ovarian hormones.
2. Identify the following events: FSH and LH secretion, growth of the follicle, development of the egg, estrogen secretion, ovulation, progesterone secretion, and the stages of uterine development.
3. Complete the diagram in the Observations section by: (a) drawing in the levels of FSH, LH, estrogen and progesterone and (b) identifying the phases including preovulatory, postovulatory follicular, luteal, menstrual, premenstrual, proliferative and secretory.

Observations

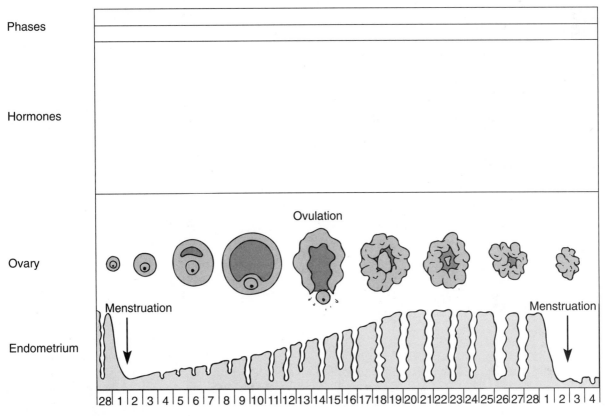

Relationships of female cycles

Conclusion

- Based upon your knowledge of the hormones in females, what is the rationale for the estrogen-progesterone birth control pill?

_____

_____

_____

_____

_____

_____

_____

_____

_____

_____

_____

_____

 Quick Quiz 2

Match the statement in column A with the phrase in column B.

	A		B
_____	1. maturation of spermatids	(a)	$2N$
		(b)	N
_____	2. structure found in male that contains enzyme to break down covering on egg	(c)	Fallopian tube
		(d)	polar body
		(e)	acrosome
		(f)	progesterone
		(g)	corpus albicans
		(h)	meiosis
		(i)	FSH
_____	3. stimulates spermatogenesis	(j)	spermiogenesis
_____	4. secondary spermatocyte		
_____	5. secreted by corpus luteum		
_____	6. "scar" tissue		
_____	7. site of fertilization		
_____	8. primary oöcyte		
_____	9. reduction division		
_____	10. formed in female to allow even distribution of DNA		

Basic Embryology and Genetics

O V E R V I E W

In this unit, you will study the early stages of embryological development. You will examine prepared slides and chick embryos which illustrate the early stages of development. You will also explore some basic concepts of inheritance.

O U T L I N E

Some exercises in this unit involve the use of living specimens. Commercial audio-visual materials may be substituted, if appropriate.

Early Embryological Development

Background

The gestation period for humans is about 38–40 weeks, and this developmental phase can be divided into three major subperiods: the period of the ovum (first two weeks), the period of the embryo (third through eighth week), and the period of the fetus (ninth week to parturition).

During the **period of the ovum**, mitotic division of the fertilized egg, transport of the developing egg to the uterus, and implantation in the uterine wall occur. The first two primary germ layers (precursors for various types of body cells) are also formed at this time. The **period of the embryo** is characterized by the formation of the third primary germ layer, the development of the placenta, and the establishment of the general body plan and internal organs. It is obvious that these first two periods are very critical, and disturbances during this time can result in serious malformation and dysfunction in an individual. Since the majority of the body organs and systems are established before the **period of the fetus**, which follows, this last part of gestation is characterized by the growth and development of the fetal organism.

EXERCISE 1
Fertilization

Discussion

Penetration of the egg by the sperm cell with the ultimate fusion of their nuclei to form a zygote is the process of

■ **FIGURE 24.1   Fertilization through cleavage: holoblastic pattern**

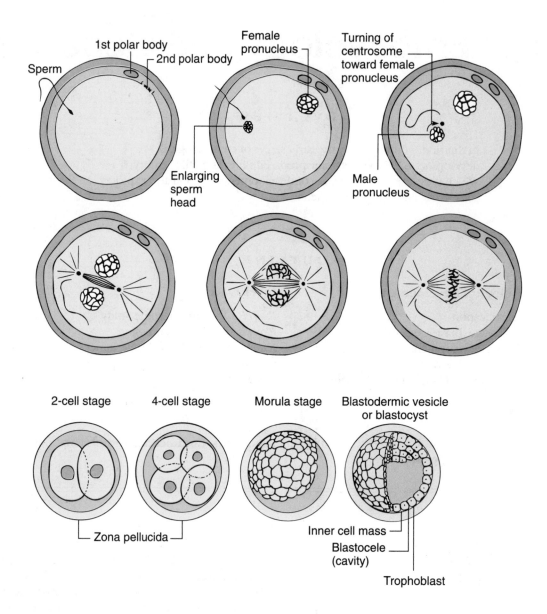

fertilization. In mammals, the **first maturation division** of the egg has preceded penetration by the sperm, occurring around the time of ovulation. The penetration of the egg by a sperm cell stimulates the **second maturation division**. During this time, the sperm head undergoes an increase in size inside the egg's cytoplasm, and is called a **pronucleus**. The egg nucleus is also called a pronucleus, and both bodies have migrated toward the center of the egg. In the center of the cell, the chromosomes of the two pronuclei intermix and group together around a mitotic spindle that has formed inside the cell. At this point, fertilization is completed and the cell is ready to undergo its first mitotic division. Fertilization serves to promote genetic variability through union of male and female pronuclei, from two different individuals, and to

supply the stimulus for further development and maturation of the zygote. See Figure 24.1.

The processes described above are not unique to humans, although they vary from species to species. Thus examples of fertilization in other animals can be used to illustrate the process in humans. In this exercise, slides of starfish are used to demonstrate the process of fertilization.

Materials

A microscope; prepared slides of starfish embryology including unfertilized eggs and fertilized egg with polar bodies.

Procedure

1. Using low power, examine slides of unfertilized star-fish egg; it can be recognized by the large, relatively clear nucleus with a prominent dark nucleolus.
2. Sketch the egg in the space provided in the Observations section.
3. Examine slides of the fertilized egg using low power. Locate the polar bodies and note whether sperm cells surround the egg.
4. Sketch the fertilized egg, polar bodies, and sperm cells, if present.

Observations

Unfertilized egg

Fertilized egg with polar bodies

Conclusion

- What are the relative sizes of the two types of germ cells?

EXERCISE 2
Cleavage, Blastocyst, and Germ Layer Development

Discussion

Immediately after fertilization the zygote undergoes a series of mitotic divisions. These divisions, referred to as cleavage, result in a closely packed ball of cells called the **morula**. The division of the cells up to this point has proceeded rapidly with no overall increase in the size of the zygote. The cells of the morula are quite small and are called **blastomeres**. As in fertilization, the pattern of division of the zygote varies depending on the type of egg. Eggs with a small amount of yolk evenly distributed throughout are called **isolecithal** and characterize human ova. Eggs containing large amounts of yolk concentrated in one area, such as a hen's egg, are called **telolecithal**. In a telolecithal egg, the yolk mass can alter the mode of cell division such that the area containing the yolk divides more slowly than the remainder of the cell. This results in an unequal division referred to as **meroblastic**. In isolecithal eggs mitosis is not affected, and thus the cell division results in daughter cells of equal size. This type of division, characteristic of human ova, is called **holoblastic.** Thus human ova are isolecithal and undergo holoblastic cleavage.

Very shortly after the morula is formed, a rearrangement of the blastomeres occurs. The solid mass of cells becomes "hollowed out" with the blastomeres forming an outer layer, the **trophectoderm**, which encloses a cavity called the **blastocoel**. There is a small cluster of cells on one side of the blastocoel called the **inner cell mass**. This area ultimately differentiates into germ layers of the embryo from which all tissues and organs develop. The developing zygote is now referred to as a **blastula** or **blastocyst**; it is about 4 to 5 days old and resides in the uterine cavity. It remains free in the uterine cavity until the seventh or eighth day, when it becomes implanted in the endometrium. After it comes in contact with the uterine lining, the blastocyst is surrounded by a double layer of trophectoderm cells that secrete proteolytic enzymes. The **proteolytic enzymes** digest the endometrium in the area around the blastocyst, eroding a small area of the endometrium into which the blastocyst sinks. The endometrium grows over the blastocyst and within a short time it is completely **implanted** in the uterus.

Following implantation, the inner cell mass undergoes development and differentiation, forming the three primary germ tissues: **ectoderm**, **mesoderm**, and **endoderm**. This sequence of changes is diagrammed in Figure 24.2. From the three primary germ layers, all the tissues and organs of the body develop. The structures produced by these germ layers are summarized in Table 24.1.

Since the starfish egg is isolecithal and undergoes holoblastic cleavage, it is used to illustrate the events of early embryological development (Figure 24.3).

Materials

A microscope; prepared slides of starfish embryology including whole mounts and/or sections of cleavage stages (2, 4, 8, and 16 blastomeres); slides of morula, blastula, and early gastrula stages.

■ **FIGURE 24.2 Human development: blastocyst through early germ layer formation**

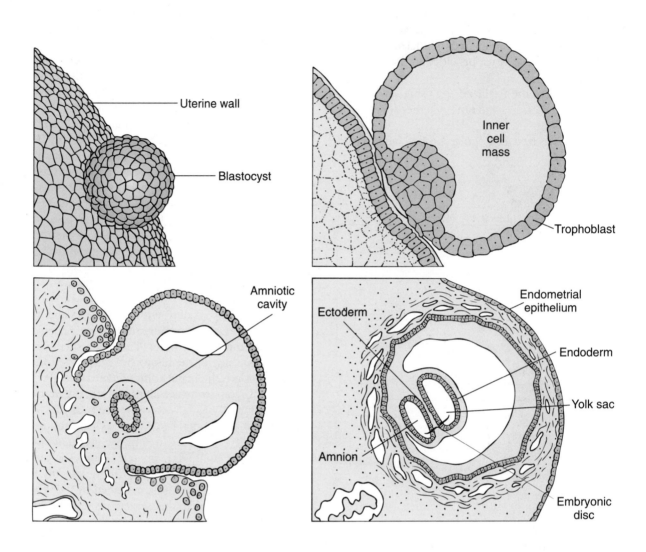

■ **TABLE 24.1 Primary Germ Layer Derivatives**

ECTODERM	MESODERM	ENDODERM
Epidermis of skin and its derivatives	Muscle tissue	Epithelial tissue of:
	Blood	Pharynx
Epithelial linings of	Cartilage	Larynx
nasal cavities	Bone	Trachea
sinuses	Lymphoid tissue	Lungs
mouth	Epithelial tissues of:	Thyroid
anus	Ovary and testis	Parathyroid
Nervous tissue	Kidney and ureter	Thymus
Neurohypophysis	Adrenal cortex	Digestive tube
Adrenal medulla	Lymphatic and blood vessels	Urinary bladder
		Urethra
		Vagina
		Adenohypophysis

Procedure

1. Examine slides of starfish cleavage. Locate the (2–16) blastomere stages. Notice how this holoblastic cleavage results in "tiers" of cells.
2. Sketch the various stages of cleavage in the space provided in the Observations section.
3. Locate a morula stage and sketch this stage also.
4. Examine blastula slides and identify the blastocoel, blastomeres, and trophectoderm cells. Refer to Figure 24.3. (You will observe that the peripheral cells are not uniform in thickness. At one side the cells are thicker, and these cells will invaginate in gastrula for-

mation. This is the simplest method of germ layer development and not like the more complicated process found in humans.)
5. Sketch the blastula stage and identify the structures cited in step 4.
6. Examine gastrula slides and identify the blastocoel, archenteron, ectoderm, mesoderm, and endoderm as shown in Figure 24.3.
7. Sketch the stages of gastrula formation and label the structures indicated in step 6.

■ **FIGURE 24.3 Early starfish embryology: holoblastic pattern**

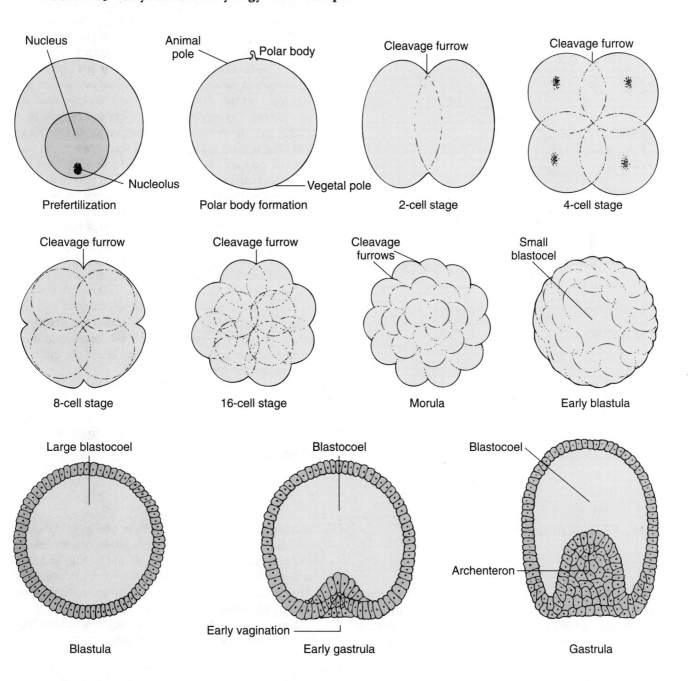

Observations

Morula	Blastula	Gastrula

Cleavage stages

Conclusions

• Are the blastomeres of a human ovum undergoing cleavage equal in all respects? Explain.

_____

_____

_____

• From which primary germ layer are the following derived?

a. Nails _____

b. Bones _____

c. Pancreas _____

d. Sweat glands _____

e. Blood vessels _____

f. Trigone _____

g. Cerebellum _____

EXERCISE 3
Early Chick Embryology

Discussion

It is beneficial to study the further embryological development of an organism that has a relatively short gestation period. Such an organism is the chick, which develops from a fertilized hen's egg. The egg you eat is an unfertilized hen's egg. Once it is released from the ovary, the egg travels down the oviduct where **albumin, shell membrane**, and **shell** are added. During this time, the process of cleavage and gastrula formation are completed. The hen's egg is laid about 18 hr after ovulation. The developing embryo in a freshly laid egg is seen as a white disc on the surface of the yolk and is called the **blastoderm**. The eggs used in the laboratory have been cooled; thus development has stopped at various stages depending how developed the egg was when it was laid. Once the eggs are incubated, however, development will resume and can be observed by removing part of the shell and covering the opening with a transparent "window."

After incubation for about 13 hr, two areas can be distinguished in the blastoderm. Figure 24.4 indicates the clear, central area called the **area pellucida**, from which embryonic structures will develop. This is surrounded by a mottled area called the **area opaca**, from which blood and blood vessels will develop. After 16 hr of incubation, an opaque band appears toward the center of the area pellucida, as shown in Figure 24.4. This is the **primitive streak**, the anterior end of which will develop into the embryo. At this stage the three germ layers are undergoing development and differentiation.

A. Preparation of Fertilized Eggs for Observation

Materials

Fertilized hen eggs; incubator; finger bowls; Locke's solution; chlorazine solution (for sterilization of instruments: see Appendix C); scissors; forceps and needle; diamond cutting knife; melted paraffin; 2.5-cm (1-in.) cover slips; artist's small paint brushes; cotton swab; sterile cotton; candling apparatus (see Appendix F).

Be sure to review safety precautions before handling fresh or preserved specimens.

■ **FIGURE 24.4 Chick embryology**
(a) An 18-hr embryo. (b) A 24-hr embryo. (c) A 33-hr embryo. (d) A 48-hr embryo. (e) A 72-hr embryo.

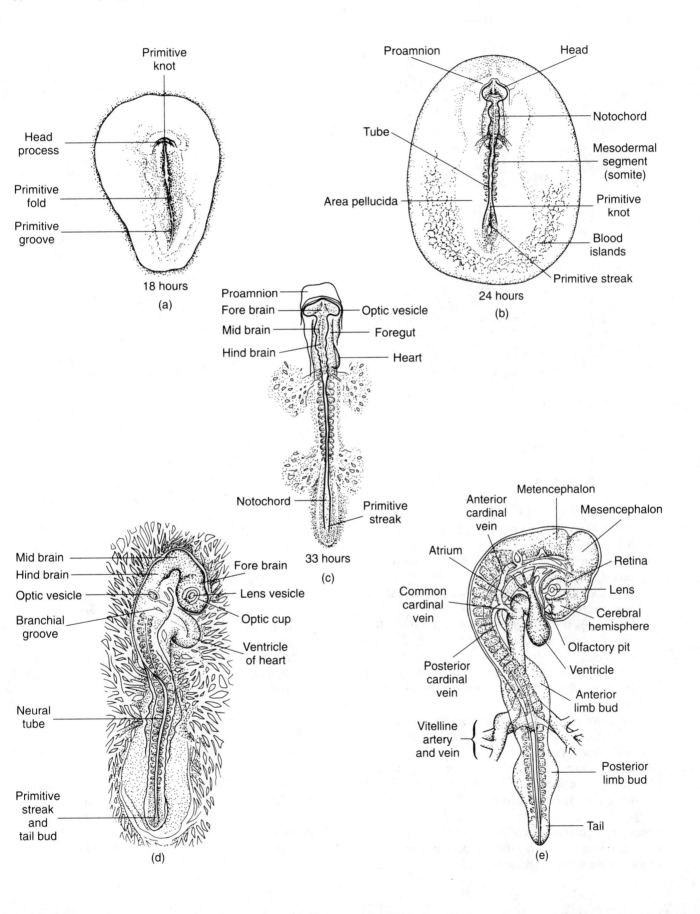

■ **FIGURE 24.5 Preparing the chick embryo for viewing**

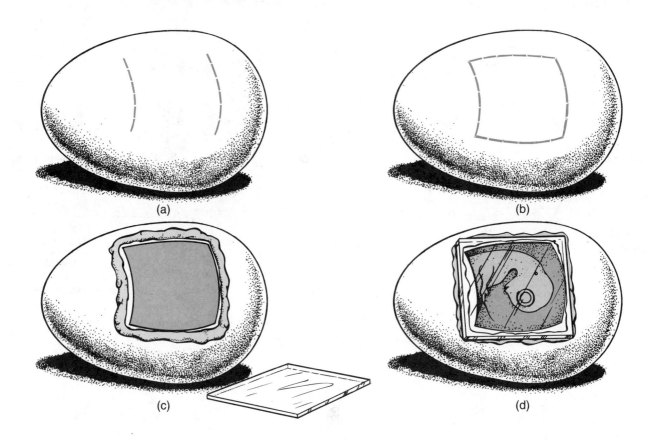

(a)

(b)

(c)

(d)

Procedure (Figure 24.5)

1. When the eggs arrive at the laboratory, rotate them gently every 24 hr to prevent adhesion of the membranes.
2. Prepare a bed of cotton in each of several finger bowls and place a fertilized egg in each of the bowls with the pointed end of the shell oriented slightly downward.
3. Let the egg rest in position for at least 5 min, allowing the blastodisc to settle and be observable upon candling.
4. Draw a 3/4 in. square (1.9 cm) on the shell over the blastodisc (generally slightly off center toward the blunt end of the egg).
5. Clean the top surface of the egg with a cotton swab soaked in chlorazine solution, then dry the surface with sterile cotton.
6. Using a sterile diamond cutting knife, gently saw through the shell along the marked square leaving the underlying shell membrane intact.

Note: Saw parallel sides of the square for the first two cuts and gently brush away shell particles as you saw. Complete the square, continuing to brush.

7. When sawing is complete, grasp the square and remove it from the egg, leaving the shell membrane intact.
8. Place a drop of Locke's solution on the shell membrane to moisten it. Then puncture the membrane in the center with a sterile needle and make a slitlike opening.
9. Using sterile scissors, cut away the shell membrane in the area where the shell has been removed.
10. Paint melted paraffin around the margins of the opening. Pass the cover slip through a flame to heat and sterilize it. Then place the cover slip over the shell opening and gently press it onto the melted paraffin so that the open area is sealed.

Note: Work quickly, to avoid contamination of the embryo.

11. Place the eggs in an incubator at 42°F, and after 24 hr move the egg slightly and slowly to prevent the embryo from sticking to the shell.
12. Replace the embryo in the same position in the incubator.

B. Observation of Developing Chick Embryos

Materials

A dissecting microscope; chick embryos of 18, 24, 33, 48, and 72 hr of age.

Be sure to review safety precautions before handling fresh or preserved specimens.

Procedure

1. Using embryology charts and Figure 24.4 as guides, examine chick embryos under a dissecting microscope.
2. Observe the different stages of development and identify the structures indicated in Figure 24.4.

Conclusions

- What are the identifying features of chick embryos of:

    a. 18 hr _____

    _____

    b. 24 hr _____

    _____

    c. 48 hr _____

    _____

    d. 72 hr _____

    _____

- At what hour of incubation can each of the following structures be observed?

    a. Eye _____

    b. Heart tube _____

    c. Blood _____

    d. Brain _____

    e. Blood vessels _____

    f. Ventricle of heart _____

    g. Limb buds _____

- How is the early development of a fertilized chick egg and fertilized human egg different?

    _____

    _____

    _____

    _____

Hormones of Human Fertilization

Discussion

In pregnancy, the placental trophoblast (cells involved in implantation) produces **human chorionic gonadotropin (HCG)** in very large quantities. This hormone is very similar to luteinizing hormone in its function. It stimulates the corpus luteum to continue secreting estrogen and progesterone. HCG reaches its peak at about the eighth week of pregnancy and then sharply declines until, at the twelfth week of gestation, the amount of hormone levels off to a constant value, which is maintained for the duration of the pregnancy. For the following experiments, it is necessary to concentrate the urine of a pregnant woman by centrifugation to ensure sufficient quantities of the hormone. The presence of HCG in the concentrate causes the male frog or toad to expel sperm within 2 to 4 hr after exposure to HCG. In the female toad, exposure to HCG causes ovulation within 24 hr.

A. Galli-Mainini Test (Male Frog)

Materials

Four male frogs *(Rana pipiens, Rana clamitans)* or male toads *(Bufo americanus)*; microscope and slides; morning urine specimens of pregnant and nonpregnant women; 2.5-cc syringe; test tubes.

Be sure to review safety precautions before handling fresh or preserved specimens.

Procedure

Refer to Figure 24.6 when injecting the frogs.

1. This procedure requires the use of three (or preferably four) frogs. Frog 1 is a negative control, injected with the nonpregnant urine; frog 2 is a positive control, injected with the pregnant urine; frogs 3 and 4 are to be injected with the test urines. These frogs may be used repeatedly, as long as a 3-day interval between tests is allowed.
2. Screen frogs for sperm production by taking frogs from container, drying them, and examining cloacal fluid under the microscope. This is accomplished by holding the rear legs in a flexed position against the frog's abdomen and placing the cloacal opening in contact with a glass slide. *No sperm* should be evident in the fluid of the animals to be used in the test.
3. With a 2.5-cc syringe, inject approximately 0.5 cc of concentrated positive test urine under the skin of the third frog's dorsal side as shown in Figure 24.6. (This will allow the urine to enter the frog's lymph sac.)

■ **FIGURE 24.6 Injecting lymph sacs of the frog**

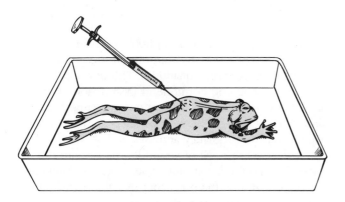

Procedure

4. Replace the frog or toad in a holding chamber at room temperature for 2 hr and repeat the procedure for collecting cloacal fluid described earlier.
5. If there are no sperm at 2 hr, the frog should be checked again at a later time, but before the 4-hr time limit.
6. Record results in the Observations section.
7. Repeat steps 2 to 6 on other test frogs with other urines.

Observations

Presence of Sperm

Frog 1 _____

Frog 2 _____

Frog 3 _____

Frog 4 _____

B. Hogben Test (Female Toad)

Materials

A female toad *(Xenopus laevis)*; morning specimens of pregnant and nonpregnant woman; test tubes; 2.5-cc syringe.

Be sure to review safety precautions before handling fresh or preserved specimens.

Procedure

1. In this test it is not necessary to check cloacal fluid.
2. Follow the procedure indicated by your instructor in injecting the concentrated urine into the abdominal area.
3. After 24 hr, check for ovulation by noting the release of eggs.
4. Record results in the Observations section.
5. Repeat steps 2 to 4 with other urines on other test frogs.

Observations

Presence of Expelled Eggs (Yes or No)

Frog 1 _____

Frog 2 _____

Frog 3 _____

Frog 4 _____

Conclusions

• Why are negative and positives controls necessary in the foregoing procedures?

_____

_____

_____

_____

• What can be concluded from the test you ran on the male specimens?

_____

_____

_____

_____

• What can be concluded from the test you ran on the female specimens?

_____

_____

_____

_____

Quick Quiz 1

Match the numbered statements with the lettered terms.

A		B
_____1. egg with small amount of yolk	(a)	isolecithal
	(b)	trophectoderm
_____2. results from cleavage of egg	(c)	morula
	(d)	holoblastic
_____3. division characteristic of human egg	(e)	telolecithal
	(f)	meroblastic
	(g)	blastula
_____4. encloses blastocoel		
_____5. typical of telolecithal egg		

Inheritance

Background

The branch of biology that is concerned with inheritance is known as **genetics** and can be traced as far back as Aristotle. Many of the principles of modern genetics, however, are based on the discoveries during the nineteenth century of an Austrian monk named Gregor Mendel. Although Mendel had no knowledge of DNA or what a gene was, careful experimentation with garden peas lead him to many fundamental concepts of heredity such as **dominance-recessiveness** and the principle of **independent segregation** of genetic characteristics. During the twentieth century, knowledge concerning the nature of genetic material, gene action and functions, and the genetic code has rapidly increased and provided a greater understanding of inheritance. This has promoted a field of investigation called medical genetics, which investigates patterns of inheritance as well as genetic and enzymatic abnormalities.

EXERCISE 1
Chromosome Study

Discussion

Every cell in the human body except the sex cells (germ cells) contains 46 chromosomes. Sex cells contain half of the number of chromosomes, such that a sperm cell has 23 chromosomes and an egg cell has 23. It is these chromosomes that determine heredity. For example, body type, sex, blood type, hair and eye color, and susceptibility to disease are all controlled at the start by the material of the chromosomes.

We know today that the chromosomal material is **DNA** and that the sequence of bases in the DNA determines specifically what a cell does. We can "see" the molecules of DNA only by chemical analysis. However, when a cell divides, the many miles of DNA in the nucleus coil up into separated segments that are visible to us as chromosomes. If mitosis is stopped chemically, these chromosomes become visible. Each chromosome contains enormous lengths of coiled DNA, and sections of this DNA "code for" proteins of our bodies. The functional sections of DNA molecules are called **genes** and contain 5000 to 30,000 base pairs. Each gene codes for one protein.

When cytogeneticists look at the chromosomes from body cells (usually white blood cells), they see 22 pairs from a male, plus 2 unpaired chromosomes, designated X and Y because of their shape. In the cells from a female they can see 23 pairs. The pairs can be grouped by size and shape and put into a pattern. This pattern, or **karyotype**, from a female is shown in Figure 24.7.

An irregularity in chromosome number and/or structure is called a **cytogenetic defect**, often manifested in

---

■ **FIGURE 24.7 Karyotype of a human female**

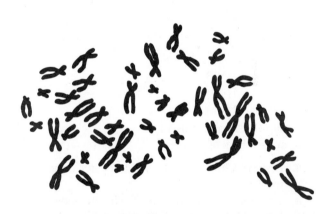

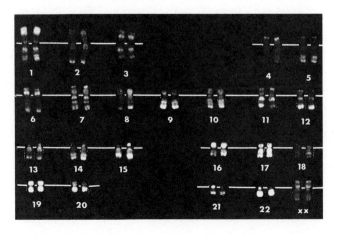

such abnormal features as retardation, abnormal body size, or of absence of sex development. For example, when there are 3 chromosomes of type #21, the fetal chromosome count is 47, not 46, and the baby is born with Down's syndrome.

A genetic counselor compiles a complete family medical history, including karyotypes where possible of three generations. He then informs the prospective parents of the status of the chromosome study, the history, and the chances of their having a normal child. Sometimes if the woman is already pregnant, a sample of the embryo's skin cells are withdrawn with the amniotic fluid (a process called **amniocentesis**), and a karyotype is done for the embryo.

Materials

Chromosome karyotype ("spreads") (following page); scissors; glue.

Procedure

1. Count the fetal number of chromosomes and record in the Observations section.
2. Cut out the chromosomes and sort them by size:

   Group A   3 largest pairs
   Group B   2 next largest pairs
   Group C   7 next largest pairs plus 2 X chromosomes
   Group D   3 pairs
   Group E   3 pairs
   Group F   2 pairs
   Group G   2 pairs (plus Y)

3. Next sort them by shape:

   a. Groups A and F constricted in center ("metacentric")
   b. Groups B, C, E, G, and Y chromosomes constricted off-center ("submetacentric")
   c. Groups D and G — the "short arm" is reduced to a tiny "satellite"

4. Pair and group the chromosomes according to size, shape, and position of centromere, as indicated above (see Figure 24.7), and paste them in place on the blank page provided.
5. Identify the 2 X (female) chromosomes; they are intermediate in size between 6 and 7,

                 or

6. Identify the Y (male) chromosome; it is like group G and usually slightly larger.
7. Identify any abnormalities in shape or number. Record in the Observations section.

Observations

a. Total chromosome count for the karyotype

_____

b. Abnormalities of the karyotype (if any)

_____

_____

_____

Conclusions

● What is the sex of the child from the karyotype you did?

_____

● Does this karyotype indicate any abnormality? Explain.

_____

_____

_____

● What would your personal karyotype look like with regard to sex chromosomes?

_____

_____

_____

_____

● What would a sample of amniotic fluid tell us about an unborn child? Explain.

_____

_____

_____

_____

● Based upon your knowledge of gamete development, suggest the ways abnormal chromosome numbers can result in a normal gamete.

_____

_____

_____

_____

_____

Chromosome spread

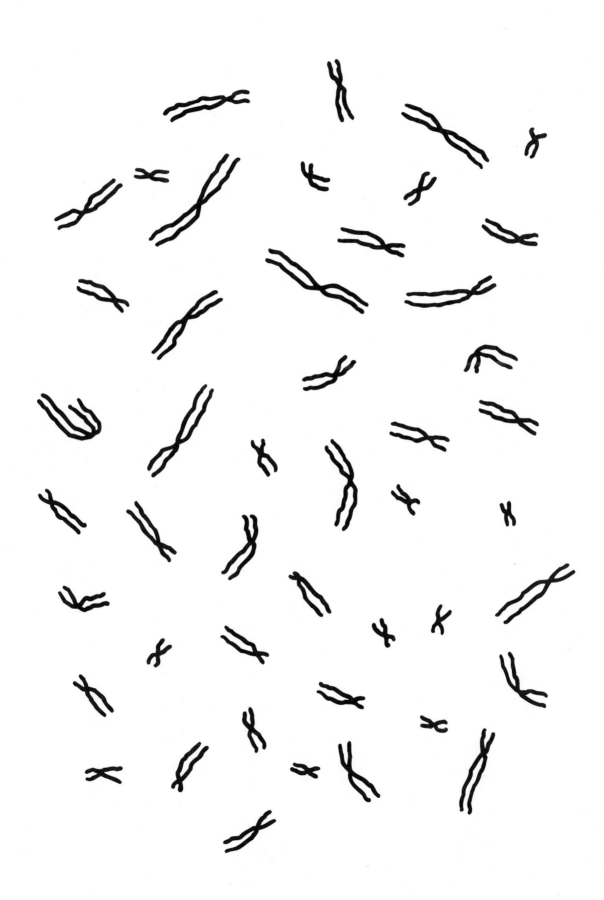

Karyotype

EXERCISE 2
Gene Expression

Discussion

Hereditary characteristics are determined by discrete, paired factors on the chromosomes called **genes**. One of each gene pair that an individual carries is inherited from each parent. The genetic makeup of an individual (the actual gene composition) is known as the **genotype**, and the bodily manifestation of that genetic composition is known as the **phenotype**.

If a zygote receives identical gene codes from each parent, it is **homozygous**; if it receives different gene codes it is **heterozygous**. When a person is heterozygous for a particular trait, often one of the genes is expressed phenotypically and is referred to as a **dominant gene**. The gene that is not expressed in this individual is called **recessive**. For example, in certain individuals a recessive gene causes failure to produce melanin, a pigment responsible for skin pigmentation. If an individual is homozygous recessive, there is no pigmentation—a condition called *albinism*.

There is a standard terminology employed to express genotype and solve genetics problems. A capital letter is used to denote the dominant trait and a lowercase letter indicates a recessive trait. Using albinism as an example, "A" would denote the gene for normal pigmentation and an "a" would indicate the gene causing lack of pigmentation. The genotype of an albino would be *aa*; the genotype of a normally pigmented person would be either *Aa* or *AA*.

A handy tool used in solving genetic problems is the **Punnett square**, illustrated in Figure 24.8. This can be used to show the different gametes produced by parents and the resulting offspring's genotypes from chance combinations at fertilization. Follow crosses 1 and 2 in Figure 24.8 as the steps in solving simple genetic crosses are enumerated.

Steps in doing genetics crosses:

1. Determine parent's genotype.
2. Determine the different types of gametes each parent can produce.
3. Draw a Punnett square and place the gametes of one parent on top of the square and the second parent on the side.
4. Fill in the square by combining the gametes listed across the top with those on the side.

The Punnett square is also used when more than one trait is considered. Figure 24.8 (cross 3) shows how to do a dihybrid cross (involving two traits). For example, one could determine the offspring's genotype in a cross of a homozygous normally pigmented person who has a crooked pinky (suppose heterozygous) with an albino

person whose pinky is normal. The crooked pinky is inherited as a dominant gene.

With the inheritance of some traits, the distinction between dominant and recessive is not always clearcut. There are cases of **co-dominance** or **incomplete dominance**. In humans, blood type is such an example. Persons who are type AB have both the A and B agglutinogens on the red blood cells. This is because neither the gene for A nor the gene for B is dominant; therefore both molecules are present on the red blood cell membrane.

Materials

A Punnett square (which you will devise).

A. Monohybrid Cross: One Trait

1. Individuals with alkaptonuria lack an enzyme that catalyzes the conversion of alkapton to carbon dioxide and water. As a result, the urine of these individuals turns black when exposed to air because of the accumulation of alkapton in the urine. Lack of the enzyme is a recessive trait.

   a. Use the symbol $\underline{A}$ to indicate production of the enzyme, and $\underline{a}$ to indicate lack of the appropriate enzyme.
   b. Set up a Punnett square showing the genotypes of both parents and the offspring, in a cross between two heterozygous individuals.

Observations

Punnett square

■ **FIGURE 24.8 Monohybrid cross**

Cross 1: Homozygous dominant—normally
pigmented male with

homozygous recessive—albino female
Male genotypes AA   Gametes A and A
Female genotype aa   Gametes a and a

	A	A	Male gametes
a	Aa	Aa	
a	Aa	Aa	

Female gametes (column label at left of a's)

All offspring will be heterozygous Aa in their genotype, and the phenotype will show normal pigmentation.

Cross 2: heterozygous male—normally pigmented
with
heterozygous female—normally
pigmented

Male genotypes Aa   Gametes A and a
Female genotype Aa   Gametes A and A

	A	a	Male gametes
A	AA	Aa	
a	Aa	aa	

Female gametes

Offspring can be:
homozygous dominant, AA—normally pigmented
heterozygous, Aa—normally pigmented
homozygous recessive, aa—no pigment, albino
Phenotypically the ratio would be 3-1 normal pigmented to albino.

Cross 3: Male homozygous normal pigmented—heterozygous crooked pinky with

Female albino—normal pinky
Male genotype AA Cc   Gametes Ac and AC
Female genotype aa cc   Gametes ac

	Ac	AC	Male gametes
ac	Aa cc	Aa Cc	
ac	Aa cc	Aa Cc	

Female gametes

Offspring could be heterozygous for normal pigmentation, with normal pinky (Aacc) and heterozygous for normal
pigmentation with crooked pinky (AaCc). If a person is homozygous dominant or heterozygous, the dominant gene
will cause normal pigmentation. For a recessive trait to be expressed, the individual must be homozygous for that
trait. Thus an albino has both recessive genes.

Conclusions

The following questions refer to alkaptonuria.

• What is the genotype of a heterozygous individual?

_____

• What is the genotype of a person with alkaptonuria?

_____

_____

• If a heterozygous man married a homozygous recessive woman, how many different genotypes are possible in their children?

_____

_____

• What are these genotypes?

_____

_____

• If two heterozygous individuals marry, how many different genotypes are possible in their children?

_____

• What are these genotypes?

_____

_____

_____

2. A cleft chin is a dominant characteristic.

   a. Use the symbol C for cleft chin and a c for no cleft.
   b. Set up a Punnett square showing the genotypes of both parents and the offspring in a cross between a homozygous recessive and heterozygous individual.

Observations

```
┌─────────────────────────────────────┐
│                                     │
│                                     │
│                                     │
│                                     │
│                                     │
│                                     │
│                                     │
│                                     │
│                                     │
│                                     │
│                                     │
│                                     │
│                                     │
│                                     │
└─────────────────────────────────────┘
```

Punnett square

Conclusions

These questions refer to the cleft chin trait.

• What is the genotype of an individual who is heterozygous for cleft chin?

_____

• What is the phenotype of a cc person?

_____

• If a heterozygous man marries a recessive woman, what are the possible genotypes of their children?

_____

_____

• If two heterozygous individuals marry, what are the possible genotypes of their children?

_____

_____

_____

B. Incomplete Dominance

Hypercholesterolemia is a condition in which individuals have an error in their system that regulates the body's production of cholesterol. The deficiency is a lack of receptors on the cell for low-density lipoprotein (LDL), which bind cholesterol on cell surfaces. Thus the LDL-cholesterol complex keeps circulating in blood. It does not bind to cells, and therefore intracellular cholesterol production is not turned off. Normal cholesterol levels range from 150 to 225 mg/100 ml of blood. Individuals who are heterozygous for hypercholesterolemia develop cholesterol levels of 300 to 550 mg/100 ml of blood and those who are homozygous for the trait have cholesterol levels of 650 to 1000 mg/100 ml of blood.

1. Use the symbol H to indicate normal numbers of LDL receptors, and h to indicate lower than normal numbers.
2. Set up a Punnett square showing the genotypes of both parents and offspring in a cross between two heterozygous individuals.

Observations

```
┌─────────────────────────────────────┐
│                                     │
│                                     │
│                                     │
│                                     │
│                                     │
│                                     │
│                                     │
│                                     │
│                                     │
│                                     │
│                                     │
│                                     │
│                                     │
│                                     │
└─────────────────────────────────────┘
```

Punnett square

Conclusions

The following questions refer to hypercholesterolemia.

- What is the genotype of a heterozygous individual?

  _____

- What is the phenotype of a heterozygous individual?

  _____

- Is more than one genotype indicative of hypercholesterolemia? Explain.

  _____

  _____

  _____

C. Dihybrid Cross: Two Traits

1. In humans, there is a dominant gene that provides an individual with the ability to roll the tongue in the shape of a "U." Another dominant gene results in the ability to taste the chemical phenyl thiocarbamide (PTC). The inability to roll the tongue, and nontasting of PTC, result from recessive genes.

   a. Use the symbols P for "tasters," p for "nontasters," R for tongue rolling, and r for nonrolling.
   b. Set up a Punnett square showing the genotypes of both parents and offspring in a cross between a homozygous recessive taster-tongue roller, and a homozygous dominant taster-tongue roller.

Observations

[blank box]

Punnett square

Conclusions

The following questions refer to the "tasting" and "tongue-rolling" traits.

- If a heterozygous taster and tongue roller marries someone who is heterozygous for tasting and cannot roll her tongue, how many different genotypes are possible in their children, and what are these genotypes?

  _____

  _____

  _____

- If two individuals who are heterozygous for each trait marry, what are the possible genotypes in their children?

  _____

  _____

2. Widow's peak, a downward point of hair in the center of the forehead, is dominant over the gene for a continuous hairline. In the condition called "blaze," a streak of white hair in the center of pigmented hair is dominant.

   a. Use the symbols W for widow's peak, w for no peak, B for blaze, and b for normal pigmentation.
   b. Set up a Punnett square showing the genotype of both parents and offspring in a cross between two heterozygous dominant widow's peak-blaze individuals.

Observations

[blank box]

Punnett square

Conclusion

• What possible genotypes would be expected in children whose mother and father were heterozygous for widow's peak and for blaze?

_____

_____

EXERCISE 3
Sex-Linked Inheritance

Discussion

Certain inherited characteristics are related to sex and sex chromosomes. In humans the sex chromosomes are the X and Y chromosomes. Since a portion of the X chromosome does not appear on the Y chromosome, genes that are carried on the nonhomologous portion cannot appear on the Y chromosome. These gene characteristics are called **sex linked**. Most of the sex-linked traits are recessive. A male (XY) carrying a recessive trait on the X chromosome will express the characteristic because there is no homologous dominant gene on the Y chromosome. Figure 24.9 illustrates the inheritance of hemo-

philia, which is transmitted by a sex-linked gene carried on the X chromosome. Other characteristics that are affected by sex are sex-limited and sex-influenced traits. **Sex-limited genes** are those that produce the phenotype in only one of the sexes. This may be because of hormone influences. These traits are generally secondary sex characteristics, such as beard development in men and breast development in women. **Sex-influenced genes** are those that appear to be dominant in one sex and recessive in the other. For example, baldness results in men who have at least one gene for the trait, whereas women must have two genes to be bald.

Procedure

Problem 1

1. Expression of genotype in sex-linked characteristics includes the sex chromosomes XY or XX, as well as capital letters for dominant trait and lowercase letters for recessive trait attached to the appropriate sex chromosome. For example, a hemophiliac male is X^hY while a normal female is $\underline{X^HX^H}$.

2. Set up a Punnett square showing the genotype of both parents and offspring in a cross section between a carrier female and nonhemophiliac male.

■ **FIGURE 24.9 Sex-linked inheritance**

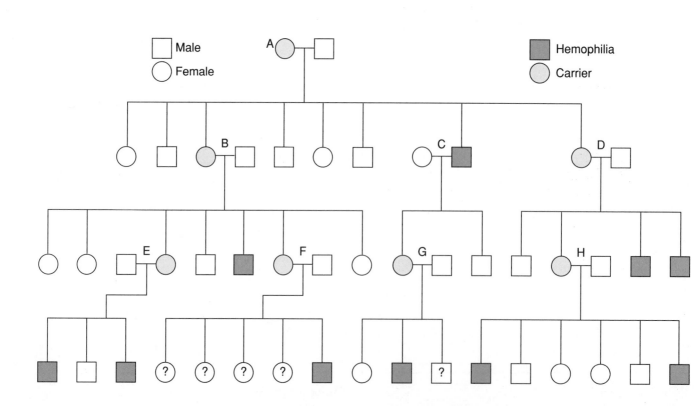

Observations

[blank box]

Punnett square

Conclusions

Complete the following questions concerning Figure 24.9.

- In the first cross (A), what is the genotype for the:

    a. Male _____

    b. Female _____

Observations

[blank box]

Pedigree for the Jones'

- In the third cross (C) what is the genotype for the:

    a. Male _____

    b. Female _____

- What is the genotype for the carrier females in crosses B, D, E, and F.

    _____

    _____

- What is the genotype for the male in crosses E, F, G, and H?

    _____

    _____

Problem 2

Color blindness is a sex-linked trait. A gene carried on the X chromosome plays a role in the formation of color-sensitive cells that distinguish red and green.

1. Use the symbols as shown in Figure 24.9 and the designation $\underline{C}$ for normal vision and $\underline{c}$ for color blind.
2. Draw a pedigree for the following family. John Jones, a color-blind man, marries Jane, a normal-visioned woman. They have two children, Kate who is a carrier, and Jim who has normal vision. Kate marries a normal-visioned man and they have four children: a normal daughter and a normal son, a carrier daughter, and a color-blind son. Jim marries a normal-visioned woman and they have four children, two daughters and two sons.

Conclusions

The following questions refer to color blindness.

● What was the genotype for:

   a. John _____

   b. Jane _____

   c. Kate _____

   d. Jim _____

● Will Jim's children have normal vision or be color blind? Explain.

_____

_____

● What are the genotypes of Jim's children?

_____

Problem 3

A sex-influenced gene affects the length of the index finger. The gene is dominant in men and recessive in women. Set up a Punnett square showing the genotypes of both parents and offspring in a cross between a woman with a short index finger and a man with a finger of normal size.

Observations

_____
| |
| |
| |
| |
| |
| |
| |
| |
| |
|_____|

Punnett square

Conclusions

● A woman with a short index finger marries a man who also has a short index finger. Will their sons have long or short index fingers? Explain.

_____

_____

_____

_____

● Will their daughters have long or short index fingers? Explain.

_____

_____

_____

_____

❖ Quick Quiz 2

Match the statement in column A with the term in column B.

	A	B
_____	**1.** often carried on the X chromosome	(a) phenotype
		(b) heterozygous
		(c) codominance
_____	**2.** bodily manifestation of trait	(d) genotype
		(e) sex-linked
_____	**3.** different genes for a trait in the same individual	
_____	**4.** actual genetic make-up	
_____	**5.** incomplete dominance	

Physiologic Self-Measurement

OVERVIEW

In this exercise you will familiarize yourself with the cyclic variations in certain body functions.

OUTLINE

Physiologic Self-Measurement

Background

There are variations in many physiologic functions in living organisms. Changes in hormone levels, DNA production, and blood pressure, for example, occur every day and so appear to have a specific biologic time structure. Through evolution, certain cycles have been "built into" human beings and have become "internalized." For example, there are **lunar cycles** (times with the periodicity of the moon), **circadian rhythms** (times with a 24-hr periodicity), and **annual rhythms** (yearly).

There are two qualifications of a biological rhythm. First, a rhythm must exist as a significant pattern for two or more cycles, after the environmental **synchronizer** (e.g., light and dark) has been eliminated. That is, it must be **"free-running."** Second, any abrupt change in the synchronizer must be responded to by a **gradual adjustment** in the rhythmic pattern of the cycle. For example, when you change time zones in travel, your regular "wakeup" time adjusts gradually to the new conditions.

Variability in physiologic and psychologic function has direct clinical effects. A doctor who measures blood pressure in the late afternoon (when blood pressure is usually at its low reading for an individual) may fail to pick up hypertension in a patient. Similarly, responsiveness to certain drug therapy appears to vary in laboratory animals, so administering medication for maximum effect on human patients may be a matter for consideration by the medical community.

EXERCISE 1
Measurement of Cyclic Phenomena

Discussion

This exercise introduces some physiologic measurements you can perform on yourself (e.g., pulse and body temperature). In carrying out these measurements, it is important to be aware of the following points: defining conditions; standardizing techniques; and the method of collecting and recording data.

1. *Conditions* should be constant. Therefore efforts to standardize conditions should be incorporated as follows:

   a. Sit down, away from distractions.
   b. Be sure the first measurement is taken within 15 min of rising, before daily activity starts.
   c. Always take the last measurement within 15 min of retiring.
   d. Avoid smoking, drinking of hot or cold beverages, eating, and exercise at least 30 min before taking measurements.

e. Note, under "Comments," any change in conditions (time, measurement device) or activity (exercise, alcohol; see code).

2. *To standardize* techniques, be sure to read the description of each test carefully, and understand it before you begin. Be sure you know how to perform and/or evaluate the test. Standardization of data and data *collection* can be achieved if the following procedures are used:

   a. The sequence of testing should be agreed on by individuals participating in the exercise.
   b. Only observed data should be recorded.
   c. Data should be plotted on predefined graph coordinates.
   d. Data should be recorded for 1 week before analysis, to overcome any "learning" that may occur in early stages of measurement taking, and to establish an "average" value for that parameter.
   e. You are responsible for the correctness of the scientific data you collect.

Some of the variables that can be measured by the group are as follows; decide as a group how often (e.g., hourly) to take these measurements each day.

1. **Oral Temperature** Shake the thermometer to below 96°F and place it either under your tongue or in your armpit for 5 min. Record the *variation* of the temperature as a + or − from the average temperature determined during the first week to the nearest tenth of a degree.
2. **Mood and Vigor** Rate yourself, using the following scale.

MOOD	RATING	PHYSICAL VIGOR
Depressed, "blue"	1	Inactive, tired
Somewhat depressed	2	Somewhat tired
Slightly less cheerful than usual	3	Slightly less active than usual
Usual state	4	Usual state
Slightly more cheerful than usual	5	Slighly more active than usual
Quite cheerful	6	Quite active
Happy, elated	7	Active, full of pep

3. **Time Estimation** This is an estimation by you of the length of a minute. Count evenly from 1 to 60, and have a stopwatch to check yourself. Record your *error*. If you take 65 sec to estimate a minute, your error is + 5, if you take 52 sec, your error is − 8.
4. **Pulse** Using the radial or carotid artery, count the number of heart beats in 1 min. Record as a + or − variation from the "average" number determined in first week.

5. **Eye-Hand Coordination** Time the following: touch your thumb to each fingertip of the same hand, counting 1, 2, 3, . . ., with each touch, as fast as possible. Start by touching the index finger and move in one direction only. Count to 25 and check for correctness: the thumb should be touching the tip of the index finger. Repeat immediately. Average the time the two sequences take and record the time as a + or a − variation from the "average" time determined in the first week.
6. **Adding Speed** Make up several columns of 50 single-digit numbers ahead of time. Time the following: add the numbers two at a time down the column (correcting errors as you notice them). For example, a column that reads 2, 6, 9, 2, 8 would be handled thus: 2 and 6, 6 and 9, 9 and 2, 2 and 8. Record the sums next to the column. Check for correctness and add 1 sec to your time for every incorrect answer. Record the time as a ± variation from the "average" time determined in the first week.
7. **Other Measurements** (The following examples are not included in the tables): blood pressure, "meter stick stop" for agility (hold the meter stick at a given mark toward the bottom, let it drop through your hand and note the level at which you stop it); and strength (use a hand grip and measure force of contraction). In addition to daily variations, monthly variations over the course of a semester may be done.

Materials

An oral thermometer; stopwatch; paper; pencil.

Note: The following abbreviations may be used in recording your data. These notations may be put directly on your graph where applicable.

Smoking[a]	S
Exercise	E
Strenuous exercise	E⁺
Anxiety	Ax
Anger	An
Alcohol	Al
Coffee, tea, other caffeinated beverages	C
Medicine (specify)	M
Weather—extreme changes (specify)	W
Temperature (specify) if over 100°F change to a fever thermometer; otherwise use a thermometer that reads to ± 0.15°F	T
Other (specify)	O

[a]Smoking, drinking, and eating, as well as exercise should be avoided for at least 30 min before taking reading. Otherwise, these activities should be noted in your data under "Comments," if they take place between data recordings.

■ SAMPLE TABLE Individual Data

TIME AFTER RISING	COMMENT	TEMPER-ATURE	MOOD	VIGOR	TIME ESTI-MATION	PULSE	EYE-HAND COORDI-NATION	ADDING SPEED
15 min, day 1								
3 hr, 15 min								
6 hr, 15 min								
9 hr, 15 min								
12 hr, 15 min								
15 hr, 15 min								
Final, day 1								
15 min, day 2								
3 hr, 15 min								
6 hr, 15 min								
9 hr, 15 min								
12 hr, 15 min								
15 hr, 15 min								
Final, day 2								
15 min, day 3								
3 hr, 15 min								
6 hr, 15 min								
9 hr, 15 min								
12 hr, 15 min								
15 hr, 15 min								
Final, day 3								
15 min, day 4								
3 hr, 15 min								
6 hr, 15 min								
9 hr, 15 min								
12 hr, 15 min								
15 hr, 15 min								
Final, day 4								
15 min, day 5								
3 hr, 15 min								
6 hr, 15 min								
9 hr, 15 min								
12 hr, 15 min								
15 hr, 15 min								
Final, day 5								

Procedure

1. Decide on time period and testing cycle for the group.
2. Set up graph axes for the recordings and appropriate time periods for your readings.
3. Run through each test so that procedure and technique are clear.
4. Carry out tests 1 to 6 in the description section for the given time period. Allow a 1-week period of running the tests before you begin taking data, during which you should establish "average" values. Record in the Observations section.
5. Record all data in tables and then plot on graphs (see following examples). Note "Comments" in table and code for them at points on graph.

Observations

All data should be graphed on separate lines, set up for each comparison as in the sample data lines on the next page. Data should be compared to those taken by the group. Interpretations of data from just one person must not be used to draw conclusions because they may be the result of individual variation. After you have taken into account special situations (illness, excitement, etc.), examine the group data for patterns in the 24-hr periods. Collect data from the group and complete Table A.

Personal Averages

Temperature	_____
Mood	_____
Vigor	_____
Time estimation	_____
Pulse	_____
Eye-hand coordination	_____
Adding speed	_____

Conclusions

• What might be the considerations a doctor would take into account in administering two medications to a patient?

_____

_____

_____

_____

_____

_____

_____

• What types of cycle did you observe (individual and group)? What other cycles might also exist? Give examples, where possible.

_____

_____

_____

_____

_____

a. Individual _____

b. Males _____

c. Females _____

■ **TABLE A** Summarized Group Data[a]

MEASUREMENT	VARIATION (12 HR, CIRCADIAN, ETC.)	TIME OF VARIATION	SHOWING PATTERN		
			% OF GROUP	% OF MALE	% FEMALE
Temperature					
Mood					
Vigor					
Pulse					
Eye-hand coordination					
Adding speed					

[a]Size of group _____ % Male _____ % Female _____

■ **SAMPLE DATA GRAPH (INDIVIDUAL, GROUP)**

Skill/Measurement

temperature   x -------------------x
mood   o -------------------o
vigor   □----------------□

Legend for Comments
(for individual data)

e_1 = exercise, jogging
a_1 = alcohol (glass of wine, etc.)

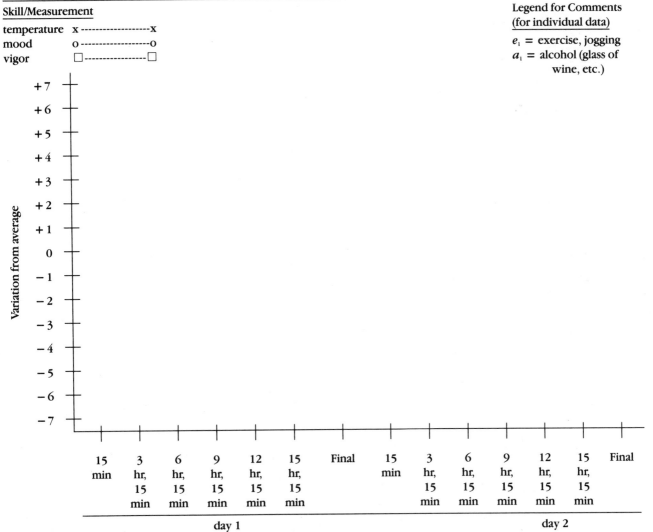

❖ Unit 1 Test: Terminology and Body Regions

1. A cut that slices across the length of the esophagus is a:
   - (a) longitudinal cut
   - (b) posterior cut
   - (c) transverse cut
   - (d) sagittal cut
2. Which set of items is **improperly** paired?
   - (a) navel—umbilical region
   - (b) lower tip of sternum—hypogastric region
   - (c) ribs—thoracic region
   - (d) intestine—abdominopelvic cavity
3. Which of these is located in the thoracic region?
   - (a) heart and lungs
   - (b) trachea
   - (c) breasts
   - (d) all the above
4. Which of these sections/cuts would give nearly identical superficial halves?
   - (a) coronal
   - (b) frontal
   - (c) midsagittal
   - (d) transverse

5. Which of these is the most superficial structure?
   - (a) stomach
   - (b) skin of the abdomen
   - (c) colon
   - (d) kidney

Match the structure in A with the cavity or region of B.

	A		B
_____	6. hip	(a)	ventral cavity
_____	7. lower ribs	(b)	iliac region
_____	8. thyroid	(c)	cervical region
	gland	(d)	hypochondriac region
_____	9. brain	(e)	dorsal cavity
_____	10. small		
	intestine		

◆ Unit 2 Test: Measurements and the Compound Microscope

Select the best answer for each of the questions. For questions 1 through 6, use the following data and field view. A student looks through the ocular (15x) of a compound microscope with the high-power objective in place (100x) and sees in the field:

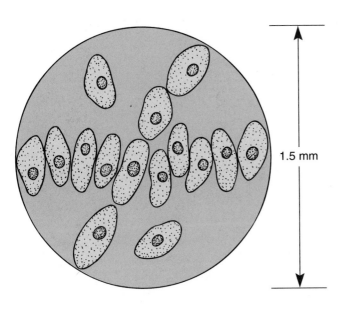

1.5 mm

1. What is the total magnification?
   (a) 50x
   (b) 100x
   (c) 150x
   (d) 1500x
2. What is the width of each cell in micrometers?
   (a) 50 μm
   (b) 150 μm
   (c) 1500 μm
   (d) none of these
3. What is the width of each cell in angstroms?
   (a) 15 Å
   (b) 150 Å
   (c) 1500 Å
   (d) none of these

4. What is the field diameter?
   (a) 0.15 μm
   (b) 15 μm
   (c) 150 μm
   (d) 1500 μm
5. How many cells can fit lengthwise in this field?
   (a) 10
   (b) 8
   (c) 4
   (d) 2
6. What is the length of each cell?
   (a) 150 μm
   (b) 375 μm
   (c) 500 μm
   (d) 1500 μm
7. The resolving power of a microscope:
   (a) is the ability to clearly separate two points
   (b) depends on the wavelength of light and numerical aperature
   (c) both a and b
   (d) neither a nor b
8. The function of the iris diaphragm on the microscope is to:
   (a) act as a source of light
   (b) control the amount of light hitting the specimen
   (c) magnify
   (d) hold the specimen
9. Which of these functions is carried out by the objective lens?
   (a) acts a source of light
   (b) controls the amount of light hitting the specimen
   (c) magnifies
   (d) changes working distance
10. If a student sees the cell membrane as an indistinct line at 1500x, what can be done to improve the resolution?
   (a) use a more powerful ocular lens
   (b) use a more powerful objective lens
   (c) use an electron microscope
   (d) use more light

❖ Unit 3 Test: The Cell

1. Which of the following is not a major region of the cell?
   - (a) cytoplasm
   - (b) cell membrane
   - (c) ribosomes
   - (d) nucleus

2. Which of the pairs below is improperly matched?
   - (a) centrioles–centrosome
   - (b) ribosomes–energy production
   - (c) RNA–nucleolus
   - (d) Golgi body–glycoprotein production

3. Which of the following organelles are not membrane bound?
   - (a) lysosomes
   - (b) mitochondria
   - (c) ribosomes
   - (d) vacuoles

4. Duplication of DNA takes place during:
   - (a) prophase
   - (b) metaphase
   - (c) telophase
   - (d) interphase

5. Chromatid pairs are attached at the:
   - (a) centromere
   - (b) centrosome
   - (c) chromatin
   - (d) none of these

6. Which of the following is a characteristic of cellular membranes?
   - (a) composed of phospholipids
   - (b) contains protein and some carbohydrate
   - (c) porous
   - (d) all the above

7. The movement of a substance from an area of higher to lower concentration is called:
   - (a) active transport
   - (b) diffusion
   - (c) pinocytosis
   - (d) filtration

8. Which of the following processes requires the presence of a carrier molecule within the cell membrane?
   - (a) filtration
   - (b) phagocytosis
   - (c) active transport
   - (d) none of these

9. If cells are placed in a hypertonic solution, containing a nondiffusable solute, the result will be a:
   - (a) net diffusion of water into the cell
   - (b) net diffusion of water out of the cell
   - (c) slight movement of solute out of the cell
   - (d) net diffusion of both water and solute into the cell

10. Which of the following processes results in the expenditure of energy by the cell?
    - (a) osmosis
    - (b) diffusion
    - (c) filtration
    - (d) phagocytosis

11. How can osmosis be explained as diffusion of water?

    _____

    _____

12. Describe how active transport is occurring in the following situations:

    (a) The thyroid gland

    _____

    (b) The establishment of the sodium/potassium gradient at the membrane

    _____

    (c) The absorption of glucose in the small intestine

    _____

1. Which of the following is not true of epithelial tissue?
   (a) It forms thin layers between organs and holds them together
   (b) It carries on secretion and absorption
   (c) It has modifications such as cilia and microvilli
   (d) It is anchored to a basement membrane and then connective tissue.

2. From your observations, stratified squamous:
   (a) is composed of many layers
   (b) contains numerous goblet cells
   (c) lines the urinary bladder
   (d) is multinucleate

3. Cartilage:
   (a) is composed of cells suspended in fluid
   (b) lacks a direct blood supply within the tissue
   (c) is a specialized loose connective tissue
   (d) is all of the above

4. Which one of the following connective tissues which you have observed exhibits a regular pattern of its fibers?
   (a) areolar tissue
   (b) reticular tissue
   (c) hyaline cartilage
   (d) tendons

5. A chondrocyte is found in:
   (a) bone
   (b) adipose
   (c) areolar
   (d) cartilage

6. Which of the following serves to bind epithelial tissues to connective tissues?
   (a) microvilli
   (b) intercalated discs
   (c) basement membrane
   (d) all of the above

7. From your observations which of the following is (are) correctly paired with respect to muscle?
   (a) skeletal–striated
   (b) cardiac–intercalated discs
   (c) smooth–no pattern obvious
   (d) all of the above

8. Where would simple cuboidal epithelium be found?
   (a) lining the digestive tract
   (b) smooth muscle
   (c) kidney tubules
   (d) none of these

9. Which of these would be affected if the innermost lining of the digestive tract were destroyed?
   (a) epithelium
   (b) connective tissue
   (c) both a and b
   (d) neither a nor b

10. Which of these is a characteristic of simple columnar epithelial cells?
    (a) flat shape
    (b) single centrally located nucleus
    (c) may secrete mucous
    (d) all of these

11. Why would a person receiving chemotherapy be likely to suffer from disorders of the digestive system?

    _____

    _____

12. Why are cancers of the skin and gut heard of more than of muscle?

    _____

    _____

13. How do the adaptations you observed in the columnar cells of the trachea aid in airway hygiene?

    _____

    _____

Unit 5 Test: Skin

1. The dermis is:
   (a) stratified squamous epithelium
   (b) loose connective tissue
   (c) dense connective tissue
   (d) none of these
2. Where are most of the derivatives of skin found?
   (a) the dermis
   (b) the subcutaneous region
   (c) the stratum lucidum
   (d) the stratum granulosum
3. The stratum germinativum:
   (a) contains pigment-producing cells
   (b) is actively mitotic
   (c) is the deepest layer of the epithelium
   (d) all of these
4. What kind of tissue does the skin "sit on"?
   (a) dense connective tissue
   (b) loose connective tissue
   (c) epithelial tissue
   (d) reticular tissue
5. Where are depositions of adipose tissue generally located in relation to the skin?
   (a) papillary region
   (b) reticular region
   (c) stratum basale
   (d) subcutaneous region
6. The stratum corneum is:
   (a) living
   (b) dead
   (c) producing melanin
   (d) all of these

7. Describe how the various tissue layers are affected by these disorders:

   (a) Decubitus ulcers

   _____

   _____

   (b) Second and third degree burns

   _____

   _____

   (c) Acne

   _____

   _____

   (d) Psoriasis

   _____

   _____

   (e) The three skin cancers

   _____

   _____

   _____

   _____

   _____

   _____

❖ Unit 6 Test: The Skeletal System

1. Which of the following statements is *false?*
   (a) The major inorganic components of bone are calcium phosphate and calcium carbonate.
   (b) The diaphysis is composed of compact bone.
   (c) Cancellous bone is made of a loose arrangement of trabeculae.
   (d) The Haversian systems of cancellous bone are composed of concentric lamellae.

2. Which of the following bones is not considered to be in the axial skeleton?
   (a) zygomatic
   (b) hyoid
   (c) patella
   (d) sphenoid

3. Which of the following would be a short bone?
   (a) sternum
   (b) cuneiform
   (c) vertebrae
   (d) fibula

4. Which of the following markings do the thoracic vertebrae use for articulation with the ribs?
   (a) tuberosity
   (b) condyle
   (c) sulcus
   (d) facet

5. Which of these facial bones is singular?
   (a) maxillary
   (b) vomer
   (c) conchae
   (d) palatine

6. Which of the following statements is *true?*
   (a) The orbit of the eye is formed by the frontal, sphenoid, maxillary, zygomatic, lacrimal, ethmoid, and palatine bones.
   (b) The mandible articulates with the temporal bone by means of the coronoid process.
   (c) The maxillary branch of the trigeminal nerve passes through the foramen ovale.
   (d) The temporal bone has a paranasal sinus.

7. Which of the following statements is *false?*
   (a) Union of the temporal and parietal bones results in the squamosal suture.
   (b) The union of the lamina and pedicle results in the lambdoidal suture.
   (c) The sacrum results from fusion of five sacral vertebrae.
   (d) The atlas articulates with the occipital condyles of the skull.

8. Which of the following pairs are incorrectly matched?
   (a) lumbar–5
   (b) cervical–transverse foramen
   (c) thoracic–bifid spinous process
   (d) sacrum–wedge shape

9. Which of the following statements is *true?*
   (a) Rib pairs 11 and 12 have no ventral attachment.
   (b) The sternal angle is the point at which the manubrium and body of the sternum join.
   (c) True ribs attach directly to the sternum via costal cartilage.
   (d) All statements are true.

10. Which statement is *true?*
    (a) The clavicle forms a joint with the scapula at the acromial process of the scapula.
    (b) The humerus articulates with the scapula at the coracoid process.
    (c) The radius articulates with the humerus at the trochlea of the humerus.
    (d) Metacarpal I articulates with the hamate bone.

11. The acetabulum is:
    (a) a round depression in the innominate bone
    (b) the point at which the ilium, ischium, and pubis fuse
    (c) the articulation point for the femur
    (d) all the above

12. At its proximal end the fibula articulates with the:
    (a) tibia
    (b) patella
    (c) femur
    (d) all the above

13. Which of these bones does not have a styloid process?
    (a) temporal
    (b) mandible
    (c) radius
    (d) ulna

14. Which is *true* concerning the tibial tuberosity?
    (a) It is on the posterior surface.
    (b) It is located inferiorly in the bone.
    (c) Both a and b.
    (d) Neither a nor b.

15. Which of the following bones is a tarsal?
    (a) hamate
    (b) navicular
    (c) triquetrum
    (d) trapezoid

16. The greater sciatic notch is associated with the:
(a) os coxae
(b) carpals
(c) sternum
(d) none of these

17. The coronoid fossa is found on which bone?
(a) radius
(b) humerus
(c) femur
(d) fibula

18. The medial malleolus is found on which bone?
(a) temporal
(b) fibula
(c) tibia
(d) femur

19. Which of the following pairs is correct?
(a) tarsals–7 bones each foot
(b) phalanges–56 bones total
(c) carpals–16 total
(d) all of these

20. Which foramen is associated with the pelvic girdle?
(a) mental
(b) lacerum
(c) obturator
(d) spinosum

21. What is the distribution of spongy and cancellous bone tissue in a typical long bone?

_____

_____

_____

22. List the differences between the male and female pelvis.

_____

_____

_____

_____

❖ Unit 7 Test: Articulations

1. The joint between the navicular and cuboid bones is an example of which type of joint?
   (a) condyloid
   (b) pivot
   (c) gliding
   (d) hinge
2. The type of movement exhibited at a pivot joint is:
   (a) adduction
   (b) flexion
   (c) rotation
   (d) all of the above
3. The epiphyseal plate is an example of which type of joint?
   (a) synchondrosis
   (b) symphysis
   (c) synovial
   (d) none of these
4. Turning the sole of the foot inward is an example of:
   (a) pronation
   (b) supination
   (c) inversion
   (d) eversion
5. The articulation between the medial and proximal phalanges is an example of which type of joint?
   (a) gliding
   (b) hinge
   (c) condyloid
   (d) saddle

6. A pivot joint is:
   (a) nonaxial
   (b) uniaxial
   (c) biaxial
   (d) triaxial
7. Movement of an extremity away from the midsagittal plane is:
   (a) abduction
   (b) adduction
   (c) inversion
   (d) none of these
8. A classification of joints based on degree of mobility would **not** include:
   (a) amphiarthrosis
   (b) gomphosis
   (c) diarthrosis
   (d) amphiarthrosis
9. Synovial fluid:
   (a) is secreted by the synovial membrane
   (b) nourishes the articular cartilage
   (c) both a and b
   (d) neither a nor b
10. The pubic symphysis is:
    (a) an example of a diarthritic joint
    (b) held together by hyaline cartilage
    (c) a synchondrosis
    (d) none of these
11. Label the diagram on the following page.

Unit 9 Test: Muscles

Match the descriptions in column A with the muscles in column B.

A	B
_____ **1.** covers most of the dorsal thigh of the pig	(a) trapezius
	(b) gluteus maximus
_____ **2.** most superior of the muscles listed	(c) rectus femoris
	(d) latissimus dorsi
	(e) supraspinatus
	(f) biceps femoris
_____ **3.** part of the quadriceps femoris	(g) rectus abdominis
	(h) gracilis
_____ **4.** covers most of the rump	(i) semimembranosus
	(j) masseter
_____ **5.** part of the hamstrings	
_____ **6.** lies over the scapula	
_____ **7.** superficial back muscle	
_____ **8.** abdominal muscle which has fibers running parallel to midline	
_____ **9.** medial thigh muscle	
_____ **10.** covers the most area of the muscles listed	

A	B
_____ **11.** in between the ribs	(a) rhomboideus
	(b) external oblique
_____ **12.** chest muscle	(c) masseter
_____ **13.** flexes the foot	(d) digastric
	(e) cutaneous
_____ **14.** V-shaped muscle	(f) pectoralis major
	(g) intercostals
_____ **15.** group of deep back muscles	(h) gastrocnemius
	(i) sartorius
_____ **16.** extends the forearm	(j) triceps brachii
_____ **17.** outermost abdominal muscle	
_____ **18.** cheek muscle	
_____ **19.** tailor's muscle	
_____ **20.** found in the pig and not in the human	

21. Use the terms introduced in Unit 1 to thoroughly describe the location of the following muscles on the pig body, and their relation to muscles next to them:

(a) Rectus femoris _____

_____

_____

_____

(b) Triceps brachii _____

_____

_____

_____

(c) Internal oblique _____

_____

_____

_____

(d) Serratus ventralis _____

_____

_____

_____

(e) Sternomastoid _____

_____

_____

_____

❖ Unit 10 Test: Muscle Physiology

Match the definitions in column A with the terms in column B.

	A		B
_____	**1.** fused contractions with no apparent relaxation	(a)	treppe
		(b)	ARP
		(c)	complete tetany
		(d)	single muscle twitch
_____	**2.** staircase effect showing increasing strength of contraction	(e)	incomplete tetany
_____	**3.** period which must be allowed to pass in order to elicit a second contraction		
_____	**4.** constant contraction showing small relaxation curves		
_____	**5.** simplest muscle contraction		

6. If muscle responds in an All-or-None way to a stimulus, how do you explain the phenomenon of recruitment?

_____

_____

7. What is the difference in the response of the muscle when you compare treppe to tetany? _____

_____

_____

8. Explain why the muscle does not totally relax from the stimuli leading to tetany.

_____

_____

9. In the preparation with the nerve attached, how does the onset of fatigue compare when the nerve is stimulated versus the muscle being stimulated directly?

_____

_____

Why? _____

_____

_____

❖ Unit 11 Test: Anatomy of the Nervous System

1. Which of the following is the functional unit of the nervous system?
   (a) nerve
   (b) neuron
   (c) neuroglia
   (d) none of these

2. The part of the neuron that transmits the impulse toward the cell body is the:
   (a) chromophilic substance
   (b) axon
   (c) node of Ranvier
   (d) dendrite

3. Which of the following is not part of the brain stem?
   (a) medulla
   (b) pons
   (c) cerebrum
   (d) midbrain

4. Which of the following can be seen from a dorsal view of the brain?
   (a) olfactory bulb
   (b) pons
   (c) corpora quadrigemina
   (d) trigeminal nerve

5. The connective tissue covering which is wrapped around a fascicle is the:
   (a) neurilemma
   (b) endoneurium
   (c) perineurium
   (d) epineurium

6. Which of the following separates the cerebrum from the cerebellum?
   (a) longitudinal fissure
   (b) septum pellucidium
   (c) transverse fissure
   (d) aqueduct of Sylvius

7. Which of the following is not associated with the spinal cord?
   (a) central canal
   (b) cauda equina
   (c) basal ganglia
   (d) none of these

8. The dorsal root of a spinal nerve transmits what kind of impulse?
   (a) afferent
   (b) motor
   (c) sensory
   (d) mixed

9. Which of the following can be identified in a cross section of the spinal cord?
   (a) central canal
   (b) anterior medial fissure
   (c) horns
   (d) all of these

10. Label the diagram below.

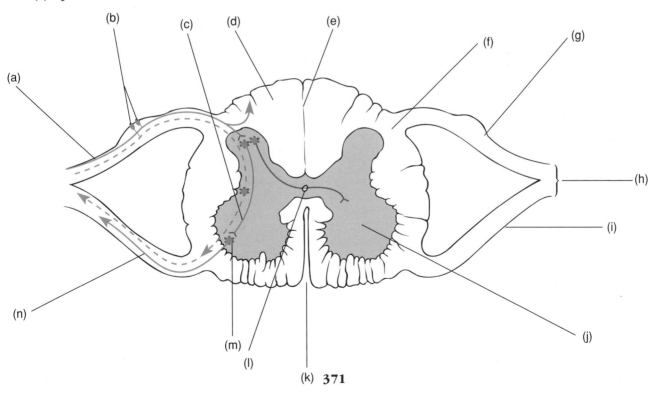

Unit 12 Test: Selected Functions of the Nervous System

1. Which of the following terms can not be appropriately applied to the detection of the coin on the forearm?
   (a) exteroception
   (b) cutaneous sensation
   (c) nocioception
   (d) tactile sense

2. Which of the following statements is true about cutaneous sensations, based upon your observations?
   (a) The distribution of receptors is uneven throughout the body.
   (b) The receptors stimulated by ice cubes are rapidly adapting.
   (c) both a and b
   (d) neither a nor b

3. The terms thermoreceptor, pressoreceptor or chemoreceptor are based upon a classification which indicates:
   (a) location
   (b) adaptation time
   (c) modality
   (d) none of these

4. Reflexes are:
   (a) carried out at the subconscious level
   (b) able to be modified by overriding cerebral activity
   (c) generally consistent and predictable
   (d) all of these

5. Which of the following is an example of a simple spinal reflex?
   (a) patellar
   (b) corneal reflex
   (c) both a and b
   (d) neither a nor b

6. Outline in proper sequence the components in a simple two neuron reflex.

   _____

   _____

   _____

   _____

   _____

   _____

❖ Unit 13 Test: Special Senses

1. Which of the following terms may be used to describe the eye?
   (a) photoreceptor
   (b) chemoreceptor
   (c) mechanoreceptor
   (d) thermoreceptor
2. Of which layer of the eye is the ciliary body a modification?
   (a) sclera
   (b) cornea
   (c) uvea
   (d) retina
3. The area of the eye that contains only cones is the:
   (a) optic disc
   (b) tapetum lucidum
   (c) macula lutea
   (d) fovea centralis
4. Changing the shape of the lens so that light rays from varying distances converge on the retina is called:
   (a) summation
   (b) condensation
   (c) accommodation
   (d) near point
5. The organ of Corti is:
   (a) ossicles
   (b) located on the basilar membrane
   (c) housed in the semicircular canals
   (d) none of these
6. Information regarding dynamic equilibrium is provided by the:
   (a) utricle and saccule
   (b) semicircular canals
   (c) tectorial membrane
   (d) otoliths

7. Nerve deafness can result from damage to which cranial nerve?
   (a) I
   (b) II
   (c) VIII
   (d) XI
8. Which statement concerning the sense of taste is true?
   (a) The majority of taste receptors are located on the tongue.
   (b) To be tasted, a chemical must be in solution.
   (c) Olfaction can have an effect on gustatory sensations.
   (d) All statements are true.
9. Where are the taste receptors for *bitter* taste concentrated on the tongue?
   (a) the front tip
   (b) along the sides
   (c) at the back
   (d) in the middle
10. When you spin around, the eye movements that occur are:
    (a) in the same direction as the head movement
    (b) in the opposite direction of the head movement
    (c) in varying directions
    (d) none of these

❖ Unit 14 Test: Blood

1. Which of the following sets of values would be normal for the human?
   (a) total RBC 5 million/mm^3, total WBC 8000/mm^3, hematocrit 43, hemoglobin 11.2 g/100 ml
   (b) total RBC 6 million/mm^3, total WBC 4000/mm^3, hematocrit 47, hemoglobin 16.0 g/100 ml
   (c) total RBC 5 million/mm^3, total WBC 6000/mm^3, hematocrit 44, hemoglobin 14.6 g/100 ml
   (d) total RBC 5.5 million/mm^3, total WBC 7500/mm^3, hematocrit 42, hemoglobin 17.6 g/100 ml

2. Which formed element is involved in the initiation of clotting?
   (a) erythrocyte
   (b) neutrophil
   (c) lymphocyte
   (d) thrombocyte

3. A relatively large phagocytic cell with a bean-shaped nucleus is a
   (a) lymphocyte
   (b) monocyte
   (c) basophil
   (d) eosinophil

4. A decrease in number of erythrocytes could
   (a) raise the hematocrit
   (b) lower the hemoglobin content
   (c) alter blood type
   (d) none of the above

5. Band cells are immature forms of
   (a) erythrocytes
   (b) thrombocytes
   (c) neutrophils
   (d) basophils

6. If a sample of blood agglutinates in anti-B but not in anti-Rh, the blood type is
   (a) B positive
   (b) AB negative
   (c) B negative
   (d) A positive

7. Fibrin is derived directly from
   (a) thrombin
   (b) thromboplastin
   (c) platelets
   (d) fibrinogen

8. Which of the formed elements is responsible for antibody production?
   (a) erythrocytes
   (b) thrombocytes
   (c) lymphocytes
   (d) monocytes

9. Which of the following pairs is correctly matched?
   (a) sedimentation rate (Wintrobe) — 6 mm/hr
   (b) bleeding time — 2 min
   (c) coagulation time — 8 min
   (d) all of these

10. Calcium ions are required for normal
    (a) hemoglobin levels to be maintained
    (b) clotting to occur
    (c) both a and b
    (d) neither a nor b

1. The right atrium:
   (a) receives blood from the body
   (b) receives blood from the lungs
   (c) contains oxygenated blood
   (d) two of the above
2. The base of the heart:
   (a) is located in an inferior position
   (b) is also known as the apex
   (c) has the great vessels associated with it
   (d) is all the above
3. The blood in the coronary sinus:
   (a) is deoxygenated
   (b) is on its way to the myocardium
   (c) enters the left atrium
   (d) is all the above
4. The endocardium:
   (a) lies on the outer surface of the heart
   (b) lines atria and ventricles
   (c) is continuous with the adventitia
   (d) is all the above
5. The right atrioventricular valve:
   (a) is tricuspid
   (b) allows deoxygenated blood through it
   (c) is anchored by chordae tendineae
   (d) is all the above
6. The moderator band runs:
   (a) from tricuspid valve to ventricular myocardium
   (b) within the semilunar valve
   (c) across the left atrium
   (d) across the right ventricle
7. Intercalated discs run:
   (a) parallel to the sarcolemma
   (b) perpendicular to striations
   (c) perpendicular to the long axis of the cell
   (d) none of these
8. Left atrium and left ventricle:
   (a) comprise a functional syncytium
   (b) are separated by the mitral valve
   (c) carry deoxygenated blood
   (d) send blood to the lungs

9. The vessels returning blood to the heart from the lungs are the:
   (a) pulmonary arteries
   (b) pulmonary veins
   (c) systemic arteries
   (d) none of these
10. The left ventricle:
    (a) is thicker-walled than the right ventricle
    (b) is separated from the left atrium by the tricuspid valve
    (c) is separated from the right atrium by the interventricular septum
    (d) none of the above
11. Explain the purpose of the:

    (a) chordae tendineae _____

    _____

    (b) thick wall of the left ventricular myocardium

    _____

    _____

    (c) semi-lunar valves _____

    _____

    (d) interventricular septum _____

    _____

    (e) auricles _____

    _____

1. The predominant effect of an increase in calcium concentration around the heart in the laboratory appears to be a(n):
   (a) increase in the rate of contraction
   (b) increase in the strength of contraction
   (c) decrease in the rate of contraction
   (d) decrease in the strength of contraction

2. Increasing the temperature of fluid which bathes the heart:
   (a) increases the rate at which calcium diffuses out of the SR
   (b) increases the rate of active transport of calcium out of the SR
   (c) increases the rate of calcium being pumped out of the muscle cell
   (d) all of these

3. Which system is a β-adrenergic blocker interfering with?
   (a) parasympathetic
   (b) cholinergic
   (c) sympathetic
   (d) none of these

4. If K⁺ outflow were increased for the heart muscle cell what would occur?
   (a) heart rate would increase
   (b) heart rate would decrease
   (c) heart rate would be unaffected
   (d) none of these

5. What do you call the result of tying a knot around the area of the heart between the atrium and ventricle?
   (a) interventricular block
   (b) interatrial block
   (c) right-left block
   (d) atrioventricular block

6. If you compare the rate at which the atria *alone* contract to the rate at which the ventricle *alone* contract, what would you see?
   (a) atria are slower than ventricles
   (b) ventricles are slower than atria
   (c) both are the same
   (d) none of these

7. If you counted R-R waves to obtain the heartbeat, would it be any different rate than if you counted P-P waves?
   (a) yes
   (b) no

8. What causes the sounds of the systolic blood pressure reading?
   (a) blood pounding against the arterial wall
   (b) blood being squeezed by the arterial wall
   (c) the stethoscope creating pressure on the artery
   (d) all of these

9. Which of these would give similar blood pressure values? (a) ankle, standing; (b) wrist, standing; (c) top of foot, standing
   (a) a and b
   (b) b and c
   (c) a and c
   (d) all the same

10. Which of these is correctly paired?
    (a) Bundle of His — normal initiator of hearbeat
    (b) AV node — delay of impulse
    (c) triangle of heart — outline of excitatory tissue
    (d) chordae tendinae — carry impulses into myocardium

11. Outline the pathway of the impulse from the sinoatrial node into the ventricular myocardium, including rapid and delayed pathways.

    _____

    _____

    _____

    _____

    _____

    _____

    _____

    _____

    _____

    _____

Unit 17 Test: Blood Vessels and Fetal Circulation

Match the items in column A with those in column B.

A   **B**

_____ 1. carries blood (a) superior vena cava
to the liver   (b) external iliac vein

_____ 2. runs along   (c) ductus ateriosus
the trachea   (d) foramen ovale

_____ 3. branch off   (e) anterior mesenteric
the aorta   artery

_____ 4. drains the leg (f) brachiocephalic artery

_____ 5. drains blood (g) hepatic artery
from the   (h) carotid artery
head, neck,   (i) subclavian vein
and upper   (j) renal artery
extremity

_____ 6. supplies
blood to the
small
intestine

_____ 7. allows blood
to bypass the
lungs in fetal
circulation

_____ 8. supplies
blood to the
kidney

_____ 9. drains the
arm

_____ 10. opening in
the fetal
heart

11. The vessel which is built for handling strong pressures is the vein.   True or False?

12. The endothelial cell is a simple cuboidal cell adapted for absorption from, and secretion into the capillary. True or False?

13. The left and right subclavian arteries branch off the aortic arch in a symetrical way.   True or False?

14. The statement that arteries carry well oxygenated blood holds true only in adult circulation. True or False?

15. The umbilical vein carries poorly oxygenated blood. True or False?

16. Explain from what organs, and through which vessels, the blood gets carried to the adult liver.

_____

_____

_____

_____

17. Explain how the most highly oxygenated blood is obtained in the fetus, and how it then is routed through the heart.

_____

_____

_____

_____

1. The thoracic duct drains most of the lymphatic channels of the body. True or False?
2. The major difference between the veins in the circulatory system and lymphatic vessels is the presence of valves. True or False?
3. As lymph flows through a lymph node, antibodies may be added to the fluid. True or False?
4. The saclike enlarged origin of the thoracic duct is called the trabeculae carneae. True or False?
5. The brachiocephalic artery is the vessel which receives lymph from the right lymphatic duct and thoracic duct. True or False?

6. Lymphatic capillaries are characterized by:
   (a) one way permeability
   (b) endothelial cells
   (c) blind endings
   (d) all of these
7. Particles found in blood will be filtered out at which structure?
   (a) lymph nodes
   (b) spleen
   (c) both a and b
   (d) neither a nor b
8. Accumulation of fluid in the right arm might indicate a blockage in which of the following:
   (a) cysterna chyli
   (b) thoracic duct
   (c) both a and b
   (d) neither a nor b

9. Which of the following pairs is/are correct?
   (a) red pulp — reticular tissue
   (b) white pulp — lymphatic tissue
   (c) both a and b
   (d) neither a nor b
10. Which of the following would not be considered part of lymphatic system?
    (a) macrophages
    (b) lamina propria
    (c) tonsils
    (d) thyroid gland

11. Explain why lymph nodes are often removed when a mastectomy is done to remove cancer.

   _____

   _____

   _____

   _____

12. Based upon your understanding of the circulation of lymph, why could damage to the right lymphatic duct result in edema in the right hand?

   _____

   _____

   _____

   _____

❖ Unit 19 Test: The Respiratory System

1. Which below is not characteristic of the respiratory system of the fetal pig?
   (a) 5 lobes of the lung
   (b) nasopharynx
   (c) alveoli
   (d) diaphragm of muscle

2. Which of these muscles are used in a normal human inspiratory effort?
   (a) scalenes
   (b) pectorals
   (c) intercostals
   (d) sternocleidomastoid

3. Which of these muscles attach to the ribs or rib cage?
   (a) scalenes
   (b) sternocleidomastoid
   (c) intercostals
   (d) all the above

4. Which of these would be found at the respiratory bronchiole level?
   (a) goblet cells
   (b) cartilage
   (c) smooth muscle
   (d) columnar epithelium

5. Which set is correctly paired?
   (a) conchae–nasal cavity
   (b) lingual tonsil–nasopharynx
   (c) adenoids–laryngopharynx
   (d) all the above

6. Which is characteristic of the larynx?
   (a) the large thyroid cartilage
   (b) the cricoid cartilage, which sits on top of the arytenoid cartilages
   (c) a small unpaired corniculate cartilage
   (d) all the above

7. Which is properly located in the larynx?
   (a) epiglottis: superior to vestibule
   (b) false vocal cords: superior to true vocal cords
   (c) stratified squamous epithelium: superior portion of the larynx
   (d) all the above

8. The nerve that causes contraction of the diaphragm
   (a) travels within the carotid sheath
   (b) originates from the pons
   (c) exits from the spinal cord at the cervical level
   (d) none of these

9. Which of these would be decreased in asthma?
   (a) VC
   (b) $FEV_{1.0}$
   (c) MEFR
   (d) all the above

10. Which of these is true concerning the pleura?
    (a) The parietal pleura lines the chest wall.
    (b) The visceral pleura surrounds the lung.
    (c) There is normally fluid in the pleural cavity.
    (d) All the above.

11. Explain the day-to-day control over respiration.

    _____

    _____

    _____

    _____

12. Explain what disease process would limit expiratory flowrate.

    _____

    _____

    Why?

    _____

    _____

13. Explain what disease process would limit inspiratory capacity.

    _____

    _____

    Why?

    _____

    _____

❖ Unit 20 Test: The Digestive System

1. Which layer of the wall of the digestive tract contains Auerbach's plexus?
   (a) mucosa
   (b) lamina propria
   (c) muscularis
   (d) serosa

2. Haustra characterize:
   (a) small intestine of the fetal pig
   (b) esophagus of the fetal pig
   (c) fundus of human stomach
   (d) colon of the human

3. The pancreas secretes exocrine products into the _____ via the _____.
   (a) jejunum, sphincter of Oddi
   (b) duodenum, ampulla of Vater
   (c) hepatic duct, round ligament
   (d) ileum, ligamentum teres

4. The liver:
   (a) is suspended from the underside of the diaphragm
   (b) is divided into right and left lobes by the falciform ligament
   (c) empties bile via the hepatic duct
   (d) is all the above

5. Which is incorrectly matched?
   (a) cardiac sphincter — stomach
   (b) ileocecal valve — entrance to small intestine
   (c) pyloric sphincter — entrance to duodenum
   (d) all

6. Which of the following enzymes functions in an acid environment?
   (a) ptyalin
   (b) amylopsin
   (c) rennin
   (d) pepsin

7. Trypsinogen is activated to trypsin in the:
   (a) stomach
   (b) small intestine
   (c) mouth
   (d) pancreas

8. Which enzyme is secreted by the pancreas and functions in the small intestine?
   (a) rennin
   (b) lactase
   (c) steapsin
   (d) ptyalin

9. Bile is produced by:
   (a) the gall bladder
   (b) hepatic cells
   (c) zymogenic cells
   (d) crypts of Lieberkühn

10. Which of the following enzymes is proteolytic?
    (a) trypsin
    (b) carboxypeptidase
    (c) maltase
    (d) all the above

11. Compare the digestive system of the fetal pig and human from an anatomical perspective.

    _____

    _____

    _____

    _____

    _____

12. You are eating just the white (protein) of a hard boiled egg. List the organs in which chemical digestion would occur and specifically indicate the enzymes which would be involved.

    _____

    _____

    _____

    _____

    _____

Match descriptions of A with structures in B. Structures in column B may be used more than once or not at all.

	A		**B**
_____	**1.** carries urine to outside	(a)	kidney cortex
_____	**2.** contains trigone	(b)	ureter
		(c)	urethra
_____	**3.** collecting ducts converge here	(d)	kidney medulla
		(e)	bladder
_____	**4.** does most of filtering of blood		
_____	**5.** is common to male urinary and reproductive tracts		

6. Which is an abnormal constituent of urine?
  (a) ketone bodies
  (b) sodium
  (c) urea
  (d) creatinine

7. Which of the following values is normal?
  (a) pH = 8.3; SG = 1.001
  (b) pH = 7.8; SG = 1.000
  (c) pH = 6.0; SG = 1.020
  (d) pH = 3.0; SG = 1.025

8. Which of these conditions would cause renin to be released?
  (a) high Na⁺ in blood
  (b) high blood pressure
  (c) sympathetic stimulation
  (d) all the above

9. Which of these is affected by ADH?
  (a) proximal convoluted tubule
  (b) Bowman's capsule
  (c) loop of Henle
  (d) collecting ducts

10. Which of these is a normal constituent of urine?
  (a) hemoglobin
  (b) glucose
  (c) ketone bodies
  (d) all of these

11. Explain the difference between the two types of
  (a) water reabsorption in the kidney:

  _____

  (b) What are they called?

  _____

  (c) What is the difference between them?

  _____

  _____

  (d) What controls each of them?

  _____

  _____

Unit 22 Test: The Endocrine System

1. Which of the following is/are **true** concerning endocrine glands?
   (a) Endocrine glands secrete hormones through ducts into circulation.
   (b) Hormones which regulate growth and development are called morphogenic hormones.
   (c) All hormonal action is mediated by cyclic AMP action.
   (d) all of the above.

2. Thyroid hormones have physiologic effects which:
   (a) have a calorigenic effect
   (b) inhibit differentiation of tissues
   (c) both a and b
   (d) neither a nor b

3. The major mechanism for the regulation of insulin secretion is:
   (a) blood epinephrine level
   (b) blood glucose level
   (c) neither a nor b
   (d) both a and b

4. Epinephrine is:
   (a) secreted by the adrenal medulla
   (b) opposite in effect from insulin
   (c) a catecholamine
   (d) all of the above

5. The thyroid gland:
   (a) is bilobed
   (b) secretes PTH
   (c) has a "cord" arrangement
   (d) is regulated directly by blood level of thyroxine

6. Which of the following is **not** released by the pituitary?
   (a) TSH
   (b) PTH
   (c) ACTH
   (d) STH

7. Which of the following hormones has a diuretic effect?
   (a) ADH
   (b) oxytocin
   (c) aldosterone
   (d) none of these

8. Which of the following pairs is correctly matched?
   (a) glucagon–glycogenolysis stimulation
   (b) thyroxine–metamorphysis stimulation
   (c) insulin overdose–hypoglycemia
   (d) all of these

9. Pars distalis is a term associated with the:
   (a) zona glomerulosa
   (b) beta cells
   (c) adenohypophysis
   (d) acini

10. Which of the following glands is located near the larynx?
    (a) pituitary
    (b) thyroid
    (c) pancreas
    (d) adrenals

11. For each of the glands or regions listed below indicate which type of cellular arrangement (cord or follicle) is found?

    (a) thyroid _____

    (b) pars distalis _____

    (c) adrenal medulla _____

    (d) zona fasciculata _____

12. For each of the hormones listed below indicate at least one physiological effect.

    (a) thyroxine _____

    (b) insulin _____

    (c) epinephrine _____

    (d) ADH _____

1. Which of the following cells has the 2*N* number of chromosomes?
   (a) spermatozoan
   (b) primary spermatocyte
   (c) secondary oöcyte
   (d) all of the above

2. Mobile spermatozoa are found in the:
   (a) seminiferous tubule
   (b) epididymis
   (c) both a and b
   (d) neither a nor b

3. Which of the following structures is correctly paired for males?
   (a) testis – reproductive function only
   (b) urethra – urinary function only
   (c) prostate gland – urinary function only
   (d) bladder – reproductive function only

4. Which of the following would be found in the female fetal pig?
   (a) cervix
   (b) vagina
   (c) urogenital sinus
   (d) all of these

5. Which of these is a muscle layer of the uterus?
   (a) endometrium
   (b) myometrium
   (c) perimetrium
   (d) none of these

6. Which of the following hormones would peak just prior to ovulation?
   (a) LH
   (b) FSH
   (c) progesterone
   (d) all of these

7. Which of the following is/are incorrectly paired?
   (a) Leydig cell – testosterone production
   (b) Sertoli cell – FSH production
   (c) spermatozoan – gamete
   (d) all of these

8. Which of the following is true regarding the menstrual cycle?
   (a) The luteal phase is preovulatory.
   (b) FSH triggers rupture of the follicle.
   (c) Rising estrogen levels along with progesterone inhibit FSH.
   (d) None of the statements are true.

9. Which of the following is/are associated with the Fallopian tube?
   (a) fimbriae
   (b) infundibulum
   ((c) both a and b
   (d) neither a nor b

10. Which of the following terms could be associated with a mature sperm cell?
    (a) fimbriae
    (b) bicornuate
    (c) acrosome
    (d) none of these

11. For each of the hormones listed below indicate the specific site/cell which is responsible for its production.
    (a) testosterone _____
    (b) progesterone _____
    (c) leutinizing hormone _____
    (d) estrogen _____
    (e) FSH _____

12. List the physiological processes which are affected by the following hormones.
    (a) FSH _____
    _____
    (b) LH _____
    _____
    (c) testosterone _____
    _____

❖ Unit 24 Test: Basic Embryology and Genetics

1. Eggs with large amounts of yolk concentrated in one region are referred to as:
   (a) isolecithal
   (b) meroblastic
   (c) telolecithal
   (d) holoblastic
2. Digestion of an area of endometrium by the developing zygote results from release of enzymes produced by which cells?
   (a) endometrial
   (b) trophectodermal
   (c) mesodermal
   (d) entodermal
3. Hair and nails are derivatives of which primary germ layer?
   (a) trophectoderm
   (b) endoderm
   (c) mesoderm
   (d) ectoderm
4. Which of the following pairs is correctly matched?
   (a) hen's egg – isolecithal, meroblastic cleavage
   (b) human ova – telolecithal, holoblastic cleavage
   (c) hen's egg – telolecithal, meroblastic cleavage
   (d) no pairs are correctly matched
5. The developing embryo forms from the cells of the:
   (a) trophectoderm
   (b) inner cell mass
   (c) blastocoel
   (d) archenteron
6. Which of the following statements is/are true?
   (a) In mammals, the first maturation division precedes penetration by the sperm.
   (b) In mammals, the second maturation division occurs after penetration of sperm.
   (c) Fertilization promotes genetic variability.
   (d) All the above.
7. Mesoderm ultimately gives rise to which of the following?
   (a) brain
   (b) thyroid follicular cells
   (c) skeleton
   (d) alveoli
8. HCG is produced by the:
   (a) developing embryo
   (b) endometrium
   (c) both a and b
   (d) neither a nor b
9. Chromosomes:
   (a) are made of DNA
   (b) contain the genetic code
   (c) come in pairs (except X and Y)
   (d) all the above

10. A karyotype:
    (a) groups chromosomes by size and shape
    (b) provides information concerning the sex of an individual
    (c) can indicate abnormal fetal development
    (d) all the above
11. If the parental genotypes are BB and bb, the offspring will be:
    (a) heterozygous
    (b) half homozygous, half heterozygous
    (c) homozygous dominant
    (d) homozygous recessive
12. In the heterozygous state, expression of both traits is known as:
    (a) dominance
    (b) recessiveness
    (c) codominance
    (d) none of the above
13. If two individuals who are heterozygous for a trait produce offspring, the offspring will show a predicted ratio of:
    (a) 1:1
    (b) 2:1
    (c) 1:2:1
    (d) 1:1:1
14. The genotype for a female carrier for hemophilia is:
    (a) X^hX^h
    (b) X^Hx^h
    (c) X^hY^H
    (d) X^HY^h
15. If a trait appears to be dominant in one sex and recessive in the other sex, it is referred to as:
    (a) sex influenced
    (b) sex limited
    (c) sex linked
    (d) none of these
16. List the major events that occur in each of the following developmental periods:

    ovum _____

    _____

    embryo _____

    _____

    fetus _____

    _____

17. What are somites? _____

    _____

397

18. Having dimples is a dominant trait in humans. Based upon this information solve the following:

    (a) If 25% of the children in a family do not have dimples and 75% of the children do, suggest the possible genotype for the parents? Explain.

    _____

    _____

    _____

    (b) If someone with dimples marries an individual with no dimples can they have children with no dimples? Explain.

    _____

    _____

    _____

19. What is the difference between the genotype and phenotype?

_____

_____

_____

1. A circadian rhythm repeats every:
   (a) 12 hr
   (b) 24 hr
   (c) week
   (d) month
2. When a synchronizer has been removed from a biological rhythm:
   (a) it "free-runs"
   (b) there is a gradual adjustment in the pattern of the cycle
   (c) both a and b
   (d) neither a nor b
3. Which of these could be considered an environmental synchronizer?
   (a) light and dark
   (b) phases of the moon
   (c) electromagnetic radiation
   (d) all the above

4. For a pattern to qualify as a "biological rhythm" it must:
   (a) exist as a significant pattern for at least 1 cycle
   (b) "free-run" when the environmental sychronizer is removed
   (c) exhibit a 24-hr variation pattern
   (d) remain in the exact pattern once the synchronizer is removed
5. Sarah has a cold. Her oral temperature is higher at night than when she wakes up in the morning. This could be because of:
   (a) cyclic variations in body temperature
   (b) a hot meal ingested just before measurement
   (c) variations in the effectiveness of her medication
   (d) all the above

Measurements

METRIC UNIT	METRIC EQUIVALENT	UNIT ENGLISH EQUIVALENT
LENGTH		
1 meter (m)	—	1.09 yards; 39.37 in.
1 centimeter (cm)	0.01 m $(1 \times 10^{-2}$ m$)^a$	0.394 in.
		1 in. = 2.54 cm
1 millimeter (mm)	1×10^{-3}	0.0394 in.
1 micrometer (μm)	1×10^{-6} m	3.94×10^{-5} in.
1 nanometer (nm)	1×10^{-9} m	3.94×10^{-8} in.
1 angstrom (Å)	1×10^{-10} m	3.94×10^{-9} in.
MASS		
1 kilogram (kg)	1000 g	2.205 lb.
1 gram (g)	—	1 pound (lb) = 453.6 g
		1 ounce (oz) = 28.35 g
VOLUME		
1 liter (L) or (l)	1000 ml	1.057 quarts (qt)
		33.81 fluid oz
		946 ml = 1 qt
1 milliliter (ml)	0.001 liter	0.034 fluid oz
		30 ml = 1 fluid oz
		5 ml = 1 teaspoon (tsp)
1 cubic centimeter; cc; (cm³)	0.999972 ml	0.0338 fluid oz
1 cubic millimeter (mm³)	0.001 cc	—

[a]The mode of expression using a 10 plus as an exponent, is called scientific notation.

Anatomical Landmarks

HEAD

Parotid gland

Anterior to ear lobe above ramus of mandible; **cranial nerve VII** runs through this gland

NECK

Cervical vertebra (7)

Midline of posterior surface of the neck; this and thoracic vertebra 1 can be palpated as the first elevations on the back of the neck because they are prominent

Hyoid bone

Directly posterior to protrusion of mandible

Thyroid cartilage
Cricoid cartilage

Midline of anterior surface of the neck; can be palpated in the ventral midline of the neck

Carotid artery

Just lateral to **thyroid cartilage**; pulse can be easily palpated here

CHEST

Sternal angle (angle of Louis)

Manubrium joins the body of the sternum; starting point for counting ribs because **second costal cartilage** joins sternum laterally; exists at level of the body of **thoracic vertebrae 4 and 5**; level at which **trachea** bifurcates

Nipple

Located at the level of the **fourth intercostal space** in males

Mediastinum

Space between the lungs, which contains **trachea, heart, great vessels**, and **thymus gland** and serves as a passage for **esophagus, vagus**, and **phrenic nerves**; just posterior to the manubrium are **innominate veins** and **superior vena cava**

ABDOMEN

Rectus abdominis muscles

On each side of midline; lateral borders demarcated by curved depression called the **semilunar line**

Quadrants

If an imaginary line is drawn perpendicular to the midline across the abdomen at the highest level of the iliac crest, the abdomen is divided into four quadrants:
Upper right quadrant contains gall bladder and much of the liver
Lower right quadrant contains appendix and ascending colon
Upper left quadrant contains spleen and stomach
Lower left quadrant contains descending and sigmoid colons

Solutions

 ## Unit 14: Blood

White Blood Cell Diluting Fluid

Use commercially prepared solution or a 2 to 3% aqueous solution of acetic acid to which gentian violet has been added until solution is pale blue-violet.

Heller-Paul-Wintrobe Mixture

Mix 6 mg of ammonium oxalate and 4 mg of potassium oxalate.

 ## Units 16 and 17: The Cardiovascular System and Blood Vessels

Adrenalin and Acetylcholine

Use commercially available preparations.

Ringer's with Excess K⁺

Use 1 ml of a 10% K^+ solution and 99 ml of Ringer's solution.

Ringer's With Excess Ca²⁺

Use 1 ml of 10% Ca^{2+} solution, and 99 ml of Ringer's solution (stronger solutions may be required).

Cholinergic Blocker

For example, atropine; use 0.4 mg of atropine sulfate per liter of water.

Adrenergic Blocker

For example, propanolol (Inderal); use 1 mg of propano-hydrochloride per milliliter of water.

Other Cardiac Drugs

Digitalis (0.2–2% solution); Pilocarpine (05–5% solution).

 ## Unit 21: The Urinary System

Saturated Picric Acid Solution

Dry picric acid is explosive and so should always be kept damp. Place an excess of picric acid in an amber glass bottle that contains distilled water. Shake occasionally. After a few days decant the solution into a second amber glass bottle. Add a small amount of the solid to maintain a saturated solution.

Alkaline Picrate Solution

Just before using: to 10 ml of saturated picric acid solution add 2 ml of a 10% w/v solution of sodium hydroxide. Mix. The alkaline solution should be only about twice as dark as the acid solution.

Fouchet's Reagent

Dissolve 25 g of trichloracetic acid in 100 ml of distilled water. Then add 10 ml of a 10% (w/v) ferric chloride solution.

Ammonium molybdate – Nitric Acid Reagent

Dissolve 3.5 g of ammonium molybdate [$(NH_4)_2 MoO_4$] in 75 ml of distilled water. Add 25 ml of concentrated nitric acid.

Unit 24: Basic Embryology and Genetics

Locke's Solution

Add to 100 ml of distilled water: 0.9 g of sodium chloride, 0.914 g of anhydrous calcium chloride, 0.04 g of potassium chloride, and 0.01 g of sodium bicarbonate.

Chlorazine Solution

95% alcohol to which a few drops of 1% iodine has been added.

SOLUTIONS

SOLUTION NAME	HOW TO PREPARE[a]
PERCENT	
Weight to volume (% w/v)	To make a 10% w/v NaCl solution, dissolve 10 g of NaCl in a volume of distilled water that (with the salt) comes *up to* 100 ml
Volume to volume (% v/v)	Used for mixing two liquids. To make a 10% v/v ethyl alcohol solution, combine 10 ml of ethyl alcohol with distilled water until the combined volume comes *up to* 100 ml
Weight to weight (% w/w)	Used for two liquids or solids in a liquid. To make a 5% w/w glucose solution, dissolve 5 g glucose in distilled water until *total weighs* 100 g
CONCENTRATION	
Molar *(M)*	To make a 1 *M* solution of Na_2SO_4, dissolve 152 g of Na_2SO_4 [or 1 gram molecular weight (gmw) or 1 mole] in distilled water until the combined solution comes *up to* 1 liter.
Molal *(m)*	To make a 1 *m* solution of Na_2SO_4, dissolve 152 g (1 gmw or 1 mole) *in 1 liter* of distilled water
Normal *(N)*	For this solution, the gram equivalent weight is used. For simple acids, bases, and solids, divide the gmw by the larger valence in the molecule to obtain the gram equivalent weight (gew). To make a 1 *N* solution of Na_2SO_4, use 76 g of Na_2SO_4 (152 g/2), and proceed as for a molar solution; that is, dissolve 76 g of Na_2SO_4 in distilled water until the combination of salt and water comes *up to* 1 liter

[a]If no solvent is given, use water as the dissolving medium. The material that is dissolved, the solute, varies.

Normal Laboratory Values for Body Fluids

Normal Blood Values

Bleeding time	Less than 4 min
Clotting time	
Capillary method	Up to 4 min
Venous blood	4–10 min
Cell counts	
Erythrocytes	
Males	5–6 million/mm^3
Females	4.5–5 million/mm^3
Leukocytes	
Neutrophils	3000–5800/mm^3 (54–62%)
Lymphocytes	1500–3000/mm^3 (25–33%)
Monocytes	285–500/mm^3 (3–7%)
Eosinophils	50–250/mm^3 (1–3%)
Basophils	15–50 mm^3 (less than 1%)
Platelets	150,000–450,000/mm^3
Hematocrit	
Males	40.0–54.0/100 ml
Females	37.0–47.0/100 ml
Hemoglobin	
Males	14.0–18.0 g/100 ml
Females	12.0–16.0 g/100 ml
Glucose	70–100 mg/100 ml
Lactic acid	6–16 mg/100 ml
Urea [blood-urea-nitrogen (BUN)]	10–20 mgl/100 ml

Normal Serum Values*

Bilirubin	0.3–1.1 mg/100 ml
Calcium	4.5–5.0 mEq/liter
Carbon dioxide (pCO$_2$)	
Arterial	40 mm Hg
Venous	46 mm Hg
Chloride	100–106 mEq/liter
Iron	75–175 mg/100 ml

Lipids (total)	450–850 mg/100 ml
Cholesterol	150–280 mg/100 ml
Phospholipids	9–16 mg/100 ml
Fatty acids	190–420 mg/100 ml
Neutral fats	0–200 mg/100 ml
Oxygen (pO_2)	
Arterial	95 mm Hg
Venous	40 mm Hg
Potassium	3.5–5.0 mEq/liter
Proteins (total)	6.0–8.0 g/100 ml
Albumin	3.5–5.5 g/100 ml
Globulin	1.5–3.0 g/100 ml
Sodium	136–145 mEq/liter
Uric acid	3.0–6.0 mg/100 ml

Normal Cerebrospinal Fluid Values

Cells	Less than 5 mm^3 mononuclear
Chloride	120–130 mEq/liter
Gamma globilin	1.3–4.7 mg/100 ml
Glucose	50–75 mg/100 ml
Protein	15–45 mg/100 ml
Pressure	75–180 mm H_2O

Normal Urine Values

Aldosterone	6–16 mg/14 hr
Ammonia	20–70 mEq/liter
Calcium	Less than 250 mg/24 hr
Creatine	Less than 100 mg/24 hr
Creatinine	0.4–1.8 g/24 hr
Hemoglobin	0
Protein	0–less than 30 mg/24 hr qualitative
Sugar	0
Urea	25–35 g/24 hr

*1 mEq equals 1 gram equivalent weight (e.g., calcium: 40 ÷ valence: 2).

Apparatus Setups

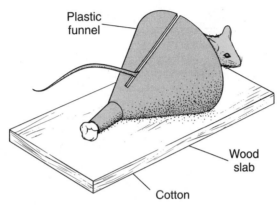

Mouse injection apparatus

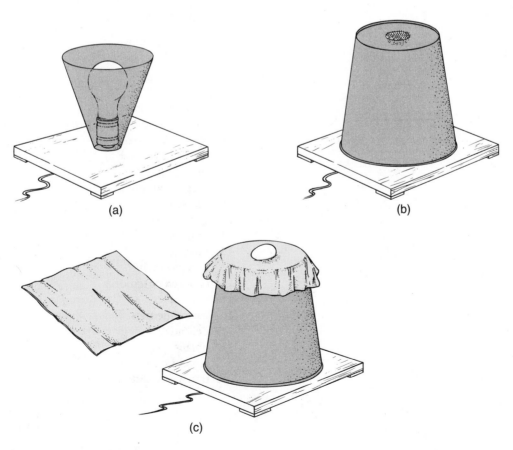

Egg candling apparatus

Quick Quiz Answers

 Unit 3

Quick Quiz 1

1. b
2. c
3. a
4. d
5. b

 Unit 6

Quick Quiz 1

1. e
2. j
3. a
4. d
5. h
6. i
7. g
8. b
9. f
10. c

Quick Quiz 2

1. d
2. e
3. a
4. h
5. g
6. c
7. f
8. c
9. b
10. a

 Unit 9

Quick Quiz 1

1. d
2. c
3. d
4. a
5. b

 Unit 10

Quick Quiz 1

1. latissimus dorsi
2. frontalis
3. orbicularis oris
4. sternocleidomastoid, external intercostals etc.
5. sternocleidomastoid
6. pectoralis major
7. biceps brachii
8. rectus abdominus
9. masseter
10. gluteus maximus
11. trapezius
12. sartorius
13. triceps brachii
14. quadriceps femoris
15. deltoid
16. gastrocnemius
17. deltoid
18. tibialis
19. gluteus maximus
20. transverse abdominus

❖ Unit 11

Quick Quiz 1

1. d
2. c
3. a
4. b
5. e

Quick Quiz 2

1. d
2. e
3. c
4. b
5. a
6. h
7. g
8. f

❖ Unit 12

Quick Quiz 1

1. c
2. e
3. a
4. d
5. b

Quick Quiz 2

1. c
2. d
3. e
4. e
5. b

❖ Unit 13

Quick Quiz 1

1. b
2. a
3. b
4. c
5. b
6. d
7. f
8. c

Quick Quiz 2

1. c
2. d
3. b
4. e
5. a

❖ Unit 17

Quick Quiz 1

1. c
2. b
3. a
4. b
5. d

Quick Quiz 2

1. c
2. d
3. c
4. b
5. d

Quick Quiz 3

1. a
2. b
3. a
4. c
5. d

Unit 18

Quick Quiz 1

1. lymph
2. one way
3. sinuses
4. white
5. thoracic duct

 Unit 19

Quick Quiz 1

1. d
2. d
3. d
4. d
5. d

Quick Quiz 2

1. e
2. d
3. c
4. a
5. f

 Unit 20

Quick Quiz 1

1. c
2. d
3. b
4. a
5. d

Quick Quiz 2

1. True
2. True
3. True
4. False
5. False

 Unit 21

Quick Quiz 1

1. e
2. c
3. d
4. a
5. b, e

 Unit 22

Quick Quiz 1

1. e
2. d
3. b
4. a
5. c
6. True
7. False
8. True

Quick Quiz 2

1. c
2. a
3. e
4. d
5. b
6. True
7. True
8. False

Unit 23

Quick Quiz 1

1. f
2. g
3. e
4. h
5. b
6. a
7. c
8. d

Quick Quiz 2

1. j
2. e
3. i
4. b
5. f
6. g
7. c
8. a
9. h
10. d

❖ Unit 24

Quick Quiz 1

1. a
2. c
3. d
4. b
5. f

Quick Quiz 2

1. e
2. a
3. b
4. d
5. c

NOTES

NOTES

NOTES

NOTES

NOTES

Color
Photo
Gallery

■ **FIGURE 8.1 External ventral views of posterior region of male and female fetal pig**

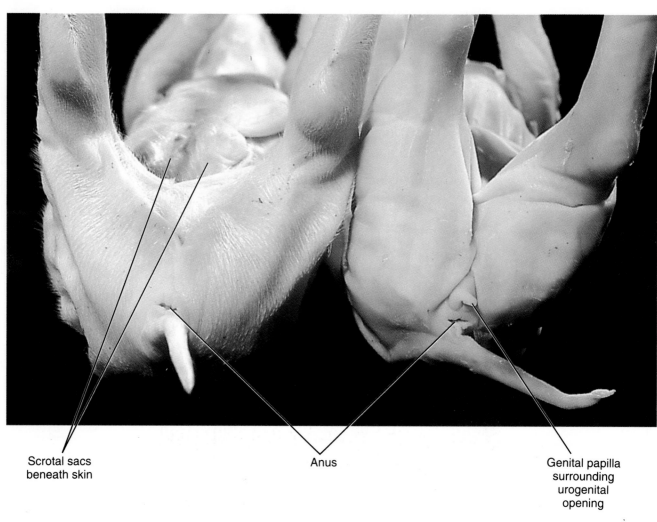

Scrotal sacs
beneath skin

Anus

Genital papilla
surrounding
urogenital
opening

Male

Female

■ **FIGURE 9.2   Superficial muscles of the right ventral thigh of the fetal pig**

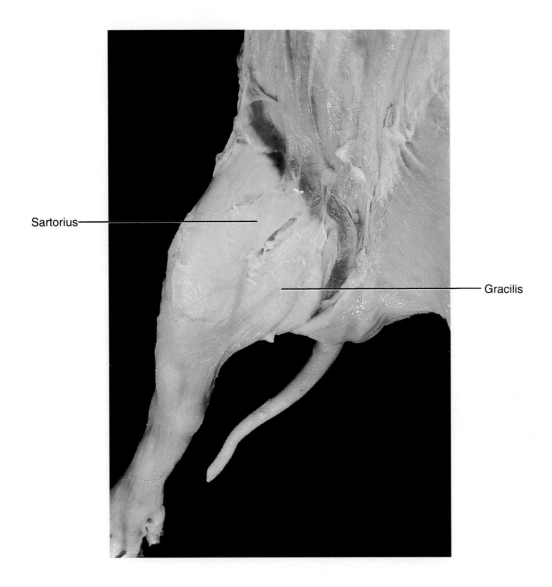

■ **FIGURE 9.3 Deep muscles of the right ventral thigh of the fetal pig**

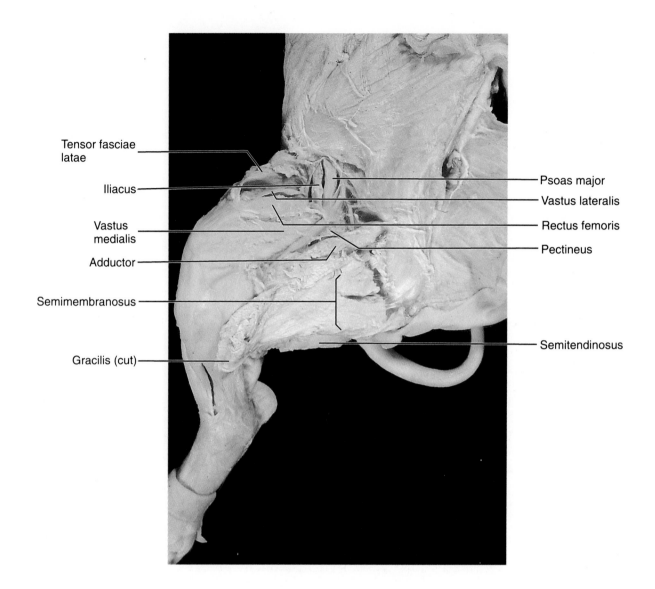

Tensor fasciae latae

Iliacus

Vastus medialis

Adductor

Semimembranosus

Gracilis (cut)

Psoas major

Vastus lateralis

Rectus femoris

Pectineus

Semitendinosus

■ **FIGURE 9.4 Selected muscles of the right lower hind limb of the fetal pig**

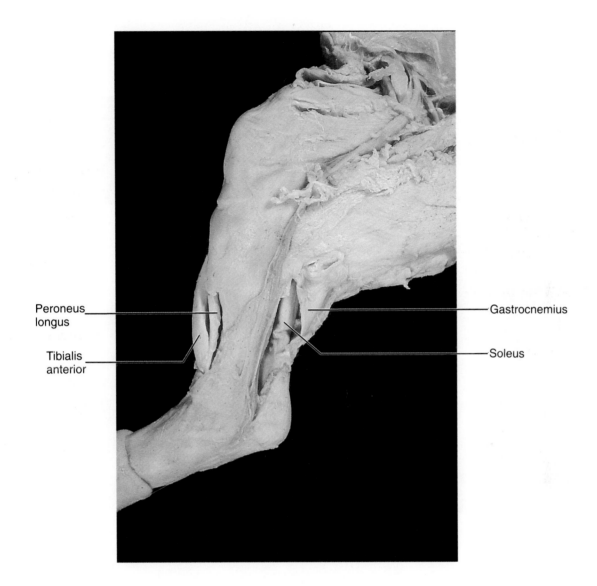

Peroneus longus

Tibialis anterior

Gastrocnemius

Soleus

■ **FIGURE 9.5 Superficial muscles of the right buttocks and dorsal thigh of the fetal pig**

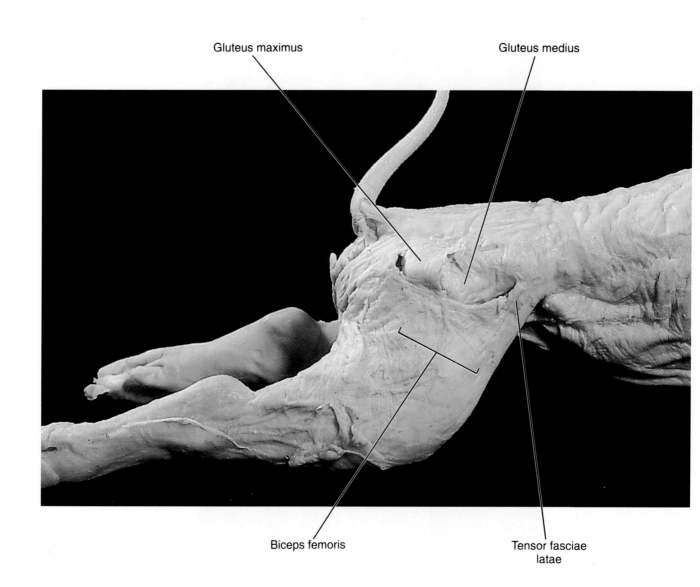

Gluteus maximus

Gluteus medius

Biceps femoris

Tensor fasciae latae

■ **FIGURE 9.6 Superficial muscles of the right back and shoulder of the fetal pig**

Brachiocephalic

Deltoid

Latissimus dorsi

Trapezius

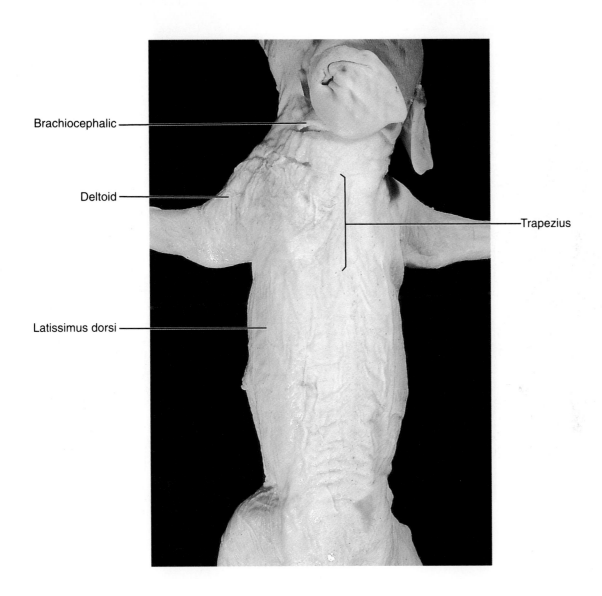

■ **FIGURE 9.7   Selected right deep muscles of the back of the fetal pig**

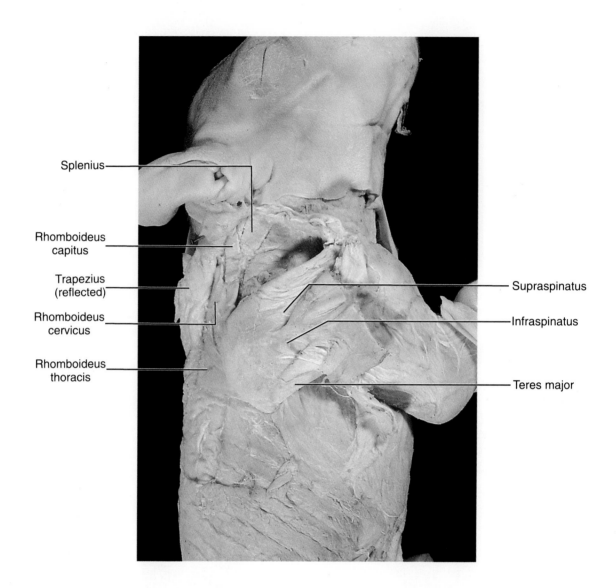

Splenius

Rhomboideus
capitus

Trapezius
(reflected)

Rhomboideus
cervicus

Rhomboideus
thoracis

Supraspinatus

Infraspinatus

Teres major

■ **FIGURE 9.8 Selected muscles of the right forelimb of the fetal pig**

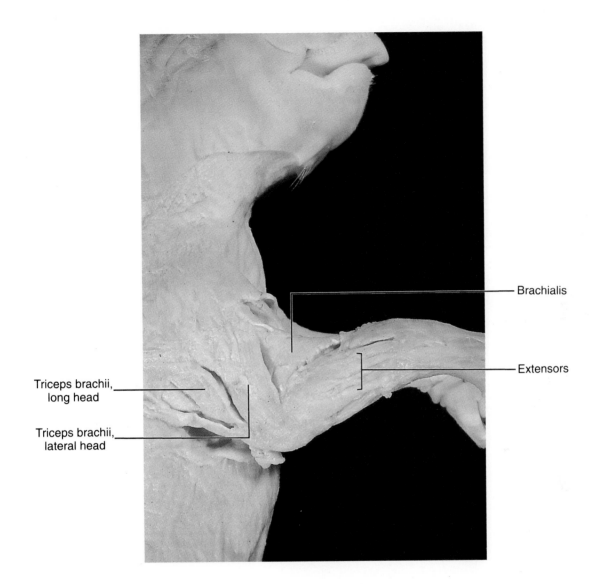

Brachialis

Extensors

Triceps brachii,
long head

Triceps brachii,
lateral head

■ **FIGURE 9.9   Selected muscles of the chest of the fetal pig**

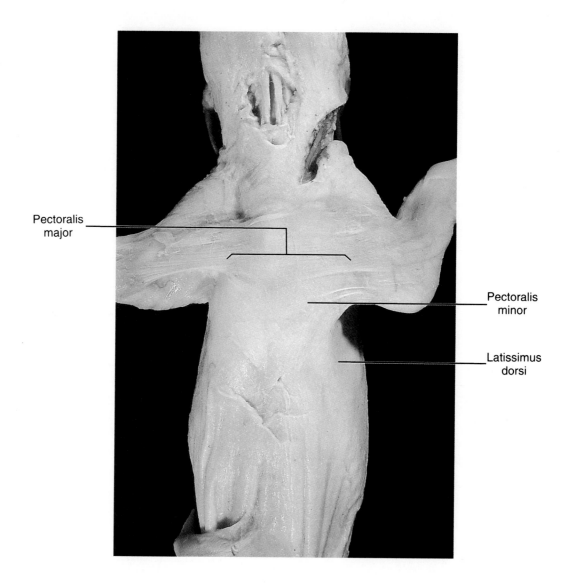

Pectoralis
major

Pectoralis
minor

Latissimus
dorsi

■ **FIGURE 9.10 Selected muscles of the neck of the fetal pig**

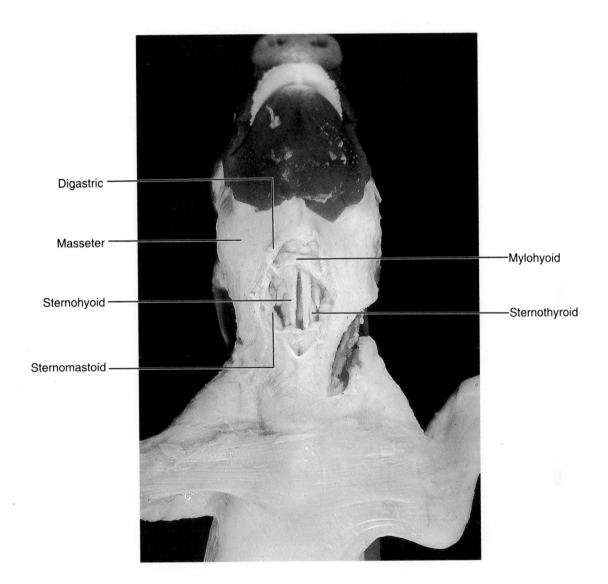

Digastric

Masseter

Sternohyoid

Sternomastoid

Mylohyoid

Sternothyroid

■ **FIGURE 9.11   Selected deep muscles of the ventral chest of the fetal pig**

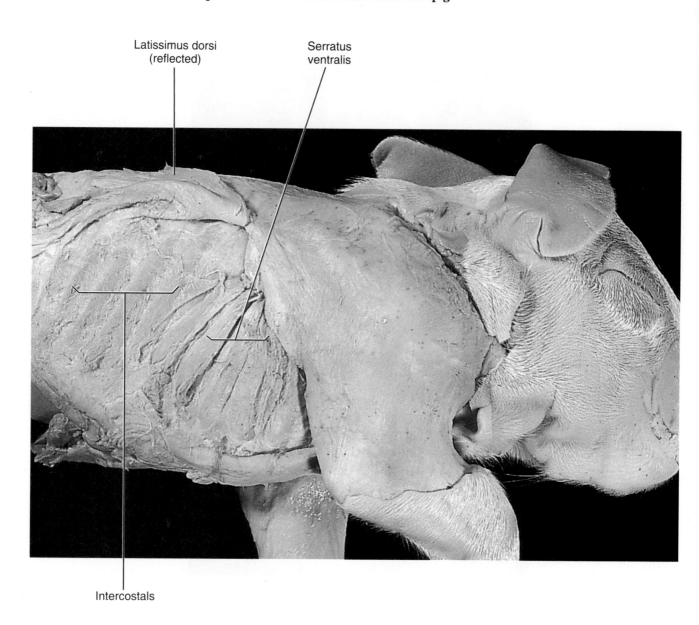

Latissimus dorsi
(reflected)

Serratus
ventralis

Intercostals

FIGURE 9.12 Muscles of the abdomen of the fetal pig

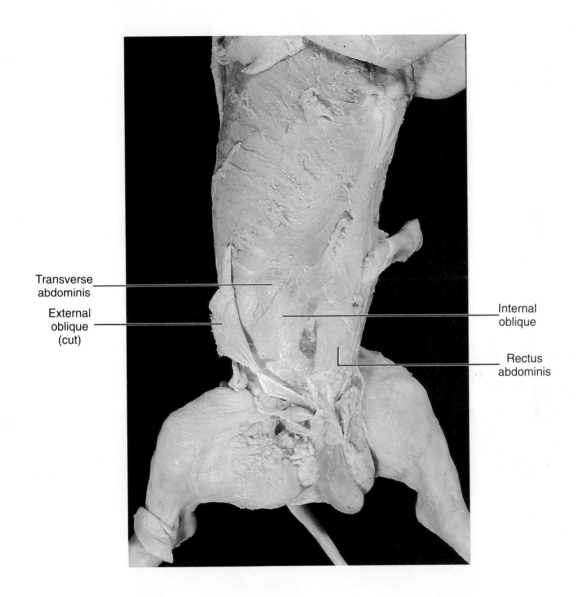

Transverse abdominis

External oblique (cut)

Internal oblique

Rectus abdominis

■ **FIGURE 11.5 The sheep brain: dorsal view**

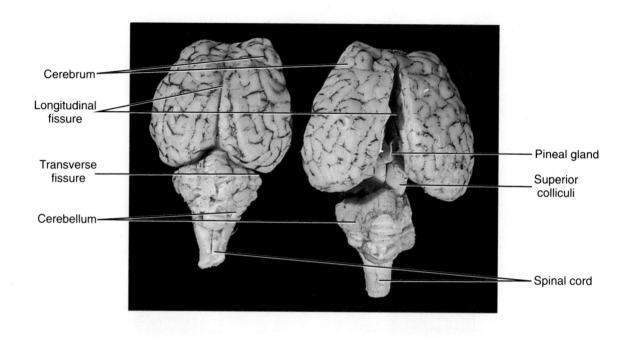

■ **FIGURE 11.6 The sheep brain: ventral view**

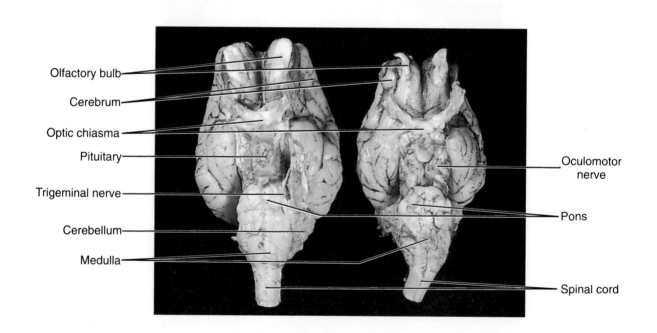

■ **FIGURE 11.7 The sheep brain: sagittal view**

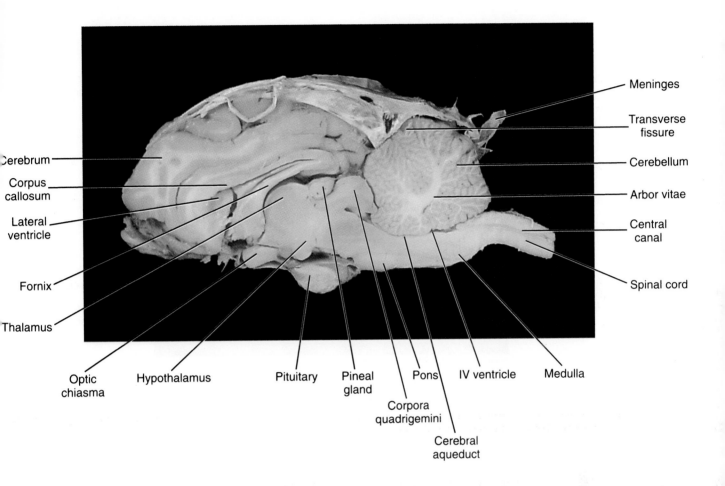

■ **FIGURE 13.4 The sheep eye: anterior view**

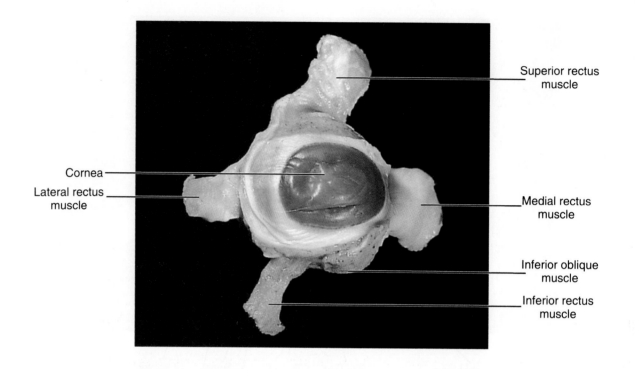

Superior rectus muscle

Cornea

Lateral rectus muscle

Medial rectus muscle

Inferior oblique muscle

Inferior rectus muscle

■ **FIGURE 13.5 The sheep eye: coronal section**

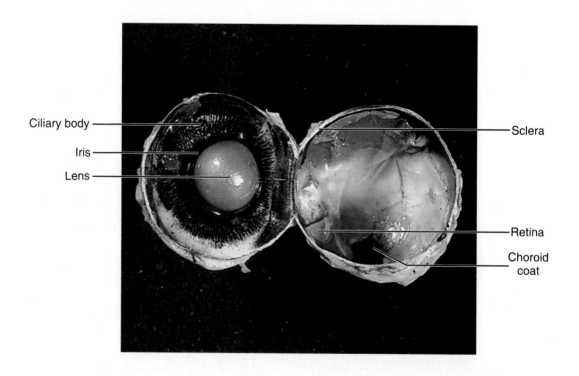

Ciliary body

Iris

Lens

Sclera

Retina

Choroid coat

■ **FIGURE 15.4 The sheep heart: exterior views of anterior portion**

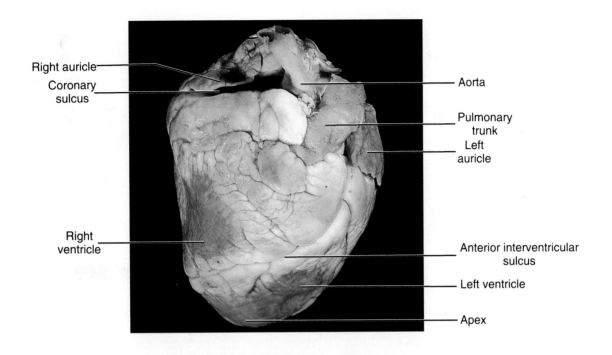

Right auricle

Coronary
sulcus

Aorta

Pulmonary
trunk

Left
auricle

Right
ventricle

Anterior interventricular
sulcus

Left ventricle

Apex

■ **FIGURE 15.5 The sheep heart: exterior views of posterior portion**

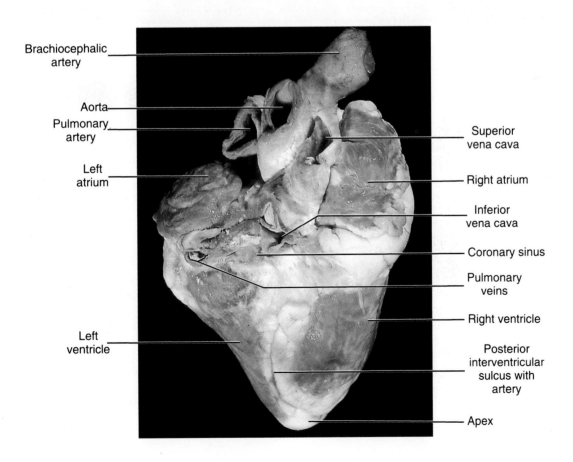

Brachiocephalic
artery

Aorta

Pulmonary
artery

Left
atrium

Superior
vena cava

Right atrium

Inferior
vena cava

Coronary sinus

Pulmonary
veins

Right ventricle

Left
ventricle

Posterior
interventricular
sulcus with
artery

Apex

■ **FIGURE 15.6 The sheep heart: interior view**

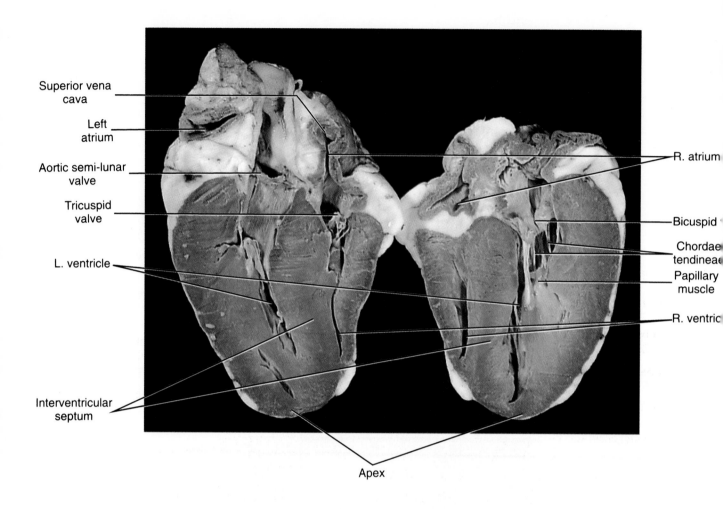

Superior vena cava

Left atrium

Aortic semi-lunar valve

Tricuspid valve

L. ventricle

Interventricular septum

R. atrium

Bicuspid

Chordae tendineae

Papillary muscle

R. ventric

Apex

■ **FIGURE 15.8 The thoracic cavity of the fetal pig**

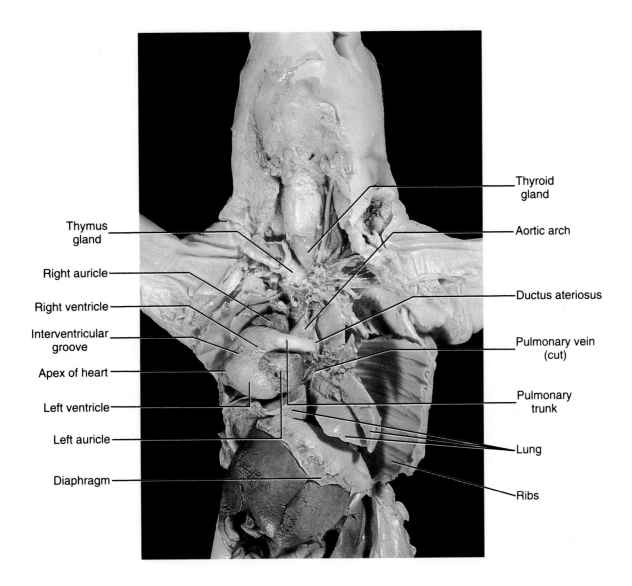

■ **FIGURE 17.4 Veins of the anterior region of the fetal pig**

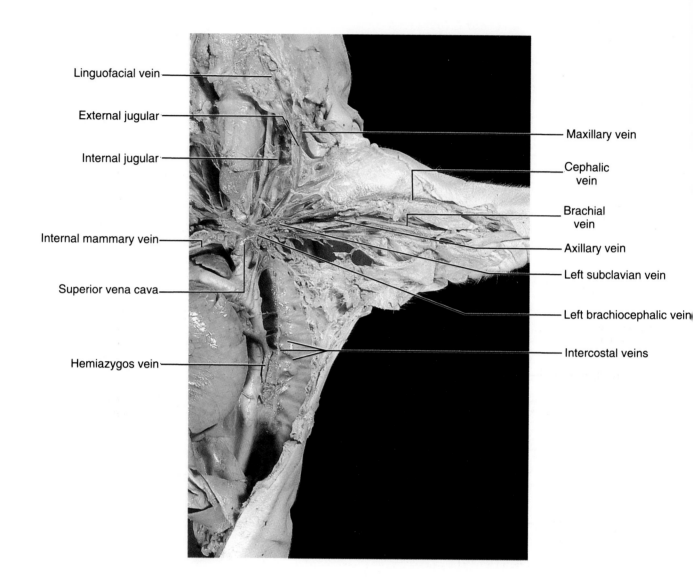

Linguofacial vein

External jugular

Internal jugular

Internal mammary vein

Superior vena cava

Hemiazygos vein

Maxillary vein

Cephalic vein

Brachial vein

Axillary vein

Left subclavian vein

Left brachiocephalic vein

Intercostal veins

■ FIGURE 17.5 Arteries of the anterior region of the fetal pig

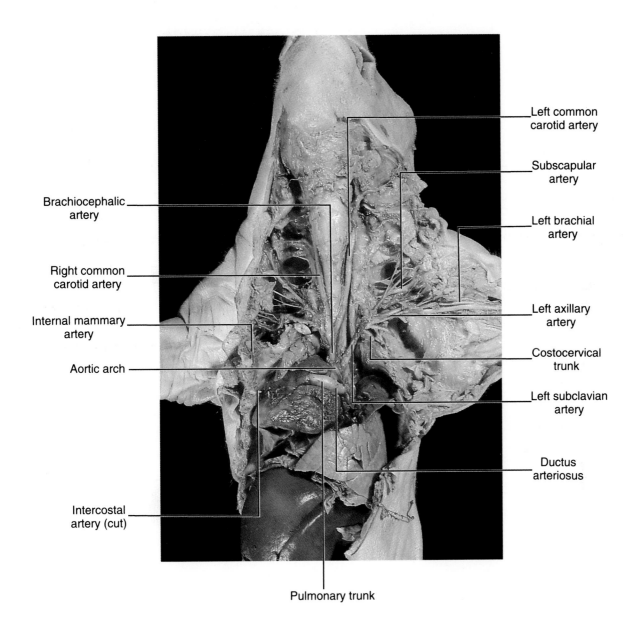

Left common
carotid artery

Subscapular
artery

Brachiocephalic
artery

Left brachial
artery

Right common
carotid artery

Internal mammary
artery

Left axillary
artery

Aortic arch

Costocervical
trunk

Left subclavian
artery

Ductus
arteriosus

Intercostal
artery (cut)

Pulmonary trunk

■ FIGURE 17.6 Selected veins of the posterior region of the fetal pig

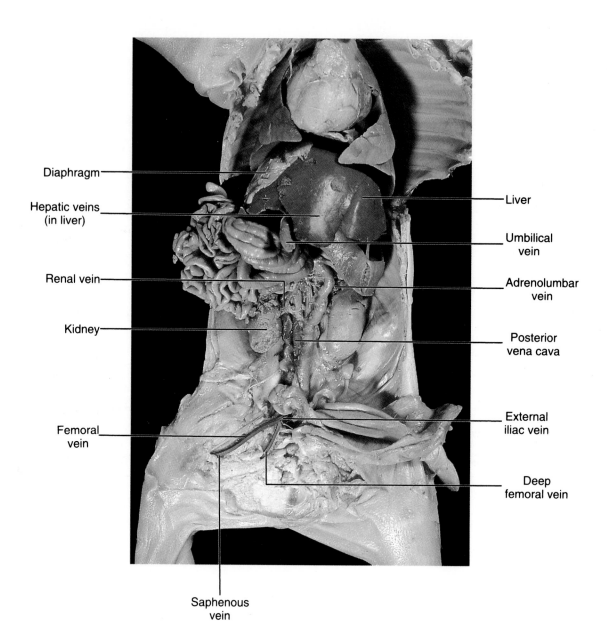

Diaphragm

Hepatic veins
(in liver)

Renal vein

Kidney

Femoral
vein

Liver

Umbilical
vein

Adrenolumbar
vein

Posterior
vena cava

External
iliac vein

Deep
femoral vein

Saphenous
vein

■ **FIGURE 17.7 Selected arteries of the posterior region of the fetal pig**

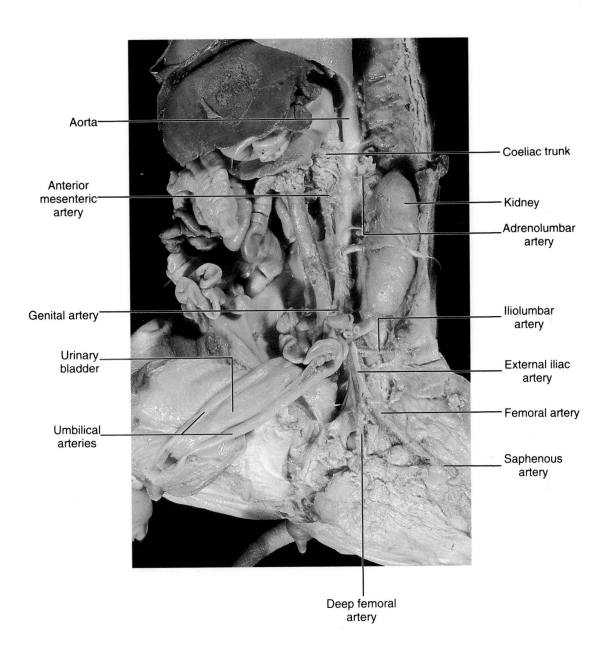

Aorta

Anterior
mesenteric
artery

Genital artery

Urinary
bladder

Umbilical
arteries

Coeliac trunk

Kidney

Adrenolumbar
artery

Iliolumbar
artery

External iliac
artery

Femoral artery

Saphenous
artery

Deep femoral
artery

■ **FIGURE 19.4a The upper respiratory system of the fetal pig**

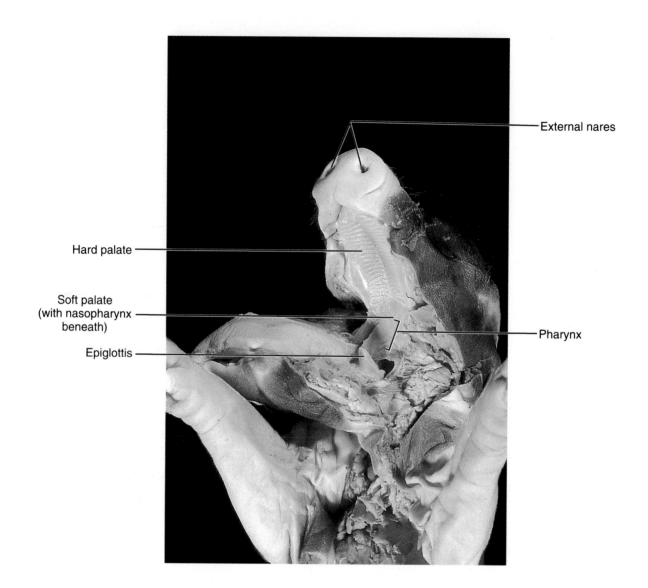

■ **FIGURE 19.4b The respiratory system of the fetal pig**

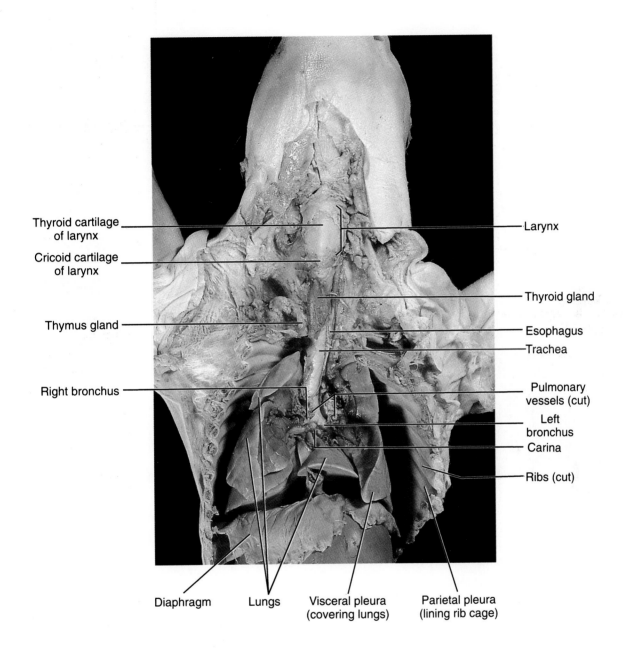

Thyroid cartilage of larynx

Cricoid cartilage of larynx

Thymus gland

Right bronchus

Larynx

Thyroid gland

Esophagus

Trachea

Pulmonary vessels (cut)

Left bronchus

Carina

Ribs (cut)

Diaphragm   Lungs   Visceral pleura (covering lungs)   Parietal pleura (lining rib cage)

■ **FIGURE 20.4 The salivary glands of the fetal pig**

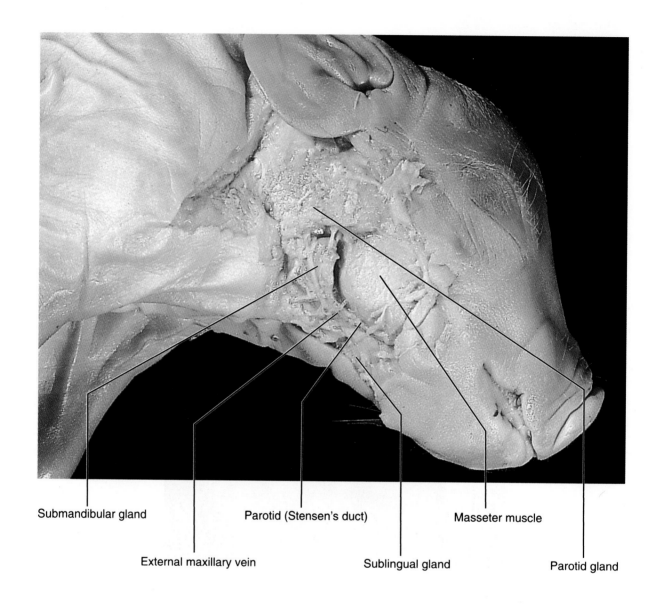

Submandibular gland

External maxillary vein

Parotid (Stensen's duct)

Sublingual gland

Masseter muscle

Parotid gland

■ **FIGURE 20.5 The oral cavity of the fetal pig**

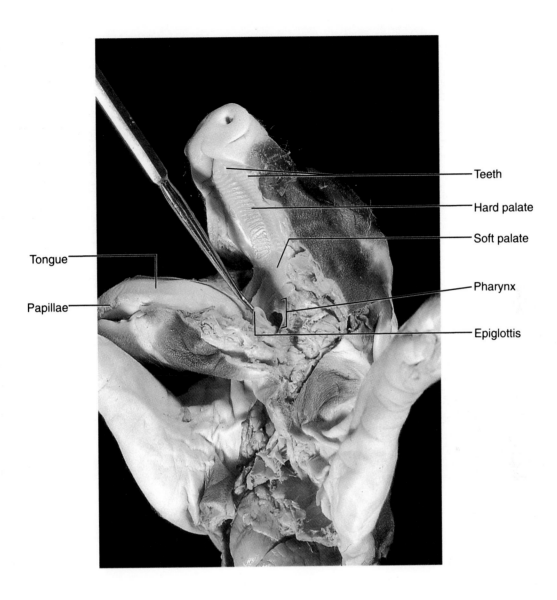

Teeth

Hard palate

Soft palate

Pharynx

Epiglottis

Tongue

Papillae

■ **FIGURE 20.6 Abdominal organs of the fetal pig**

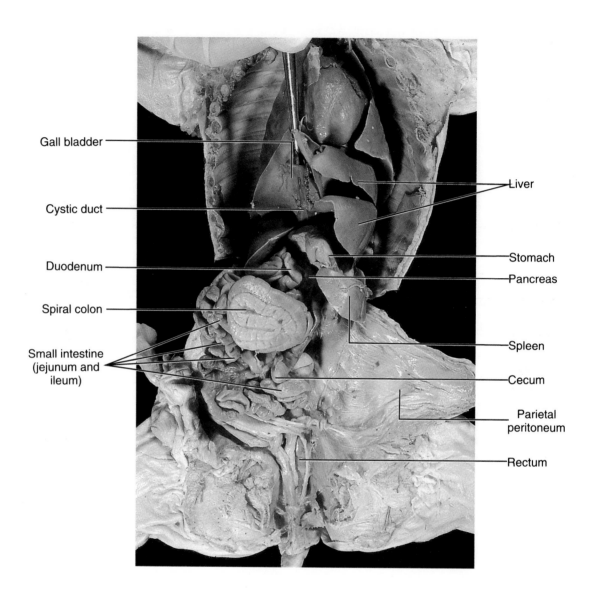

Gall bladder

Cystic duct

Duodenum

Spiral colon

Small intestine
(jejunum and
ileum)

Liver

Stomach

Pancreas

Spleen

Cecum

Parietal
peritoneum

Rectum

■ **FIGURE 21.4 The urogenital system of the male fetal pig**

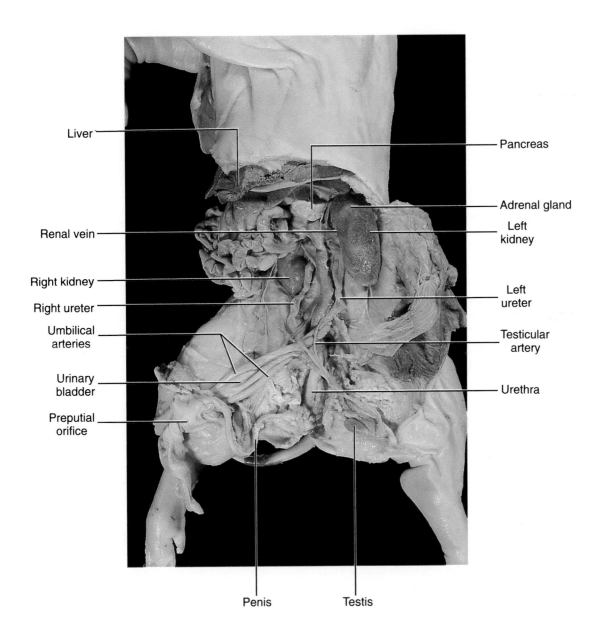

■ **FIGURE 21.5 The urogenital system of the female fetal pig**

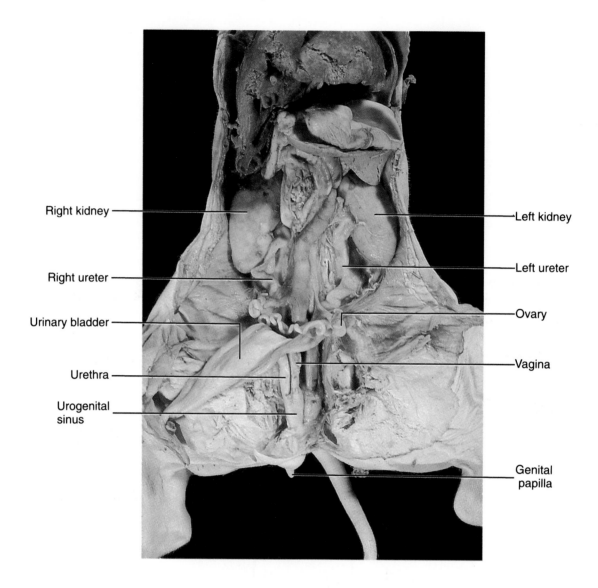

FIGURE 23.4 The reproductive system of the male fetal pig

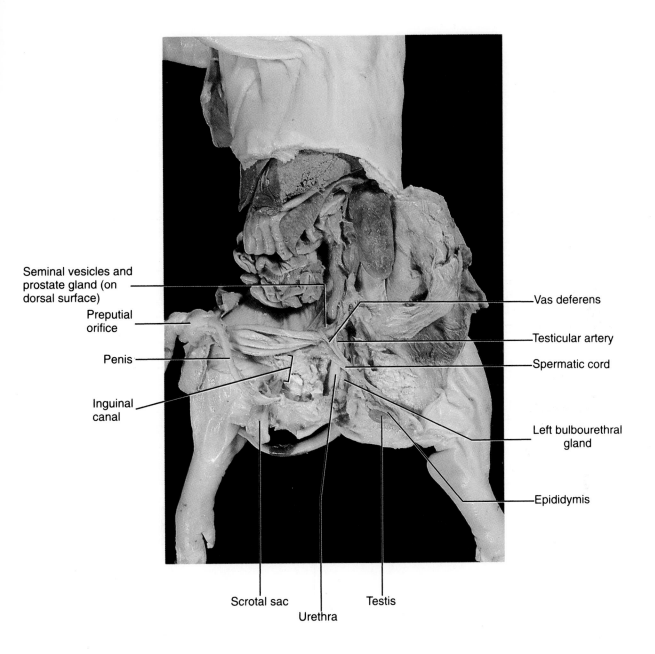

Seminal vesicles and prostate gland (on dorsal surface)

Preputial orifice

Penis

Inguinal canal

Vas deferens

Testicular artery

Spermatic cord

Left bulbourethral gland

Epididymis

Scrotal sac

Urethra

Testis

■ **FIGURE 23.6 The reproductive system of the female fetal pig**

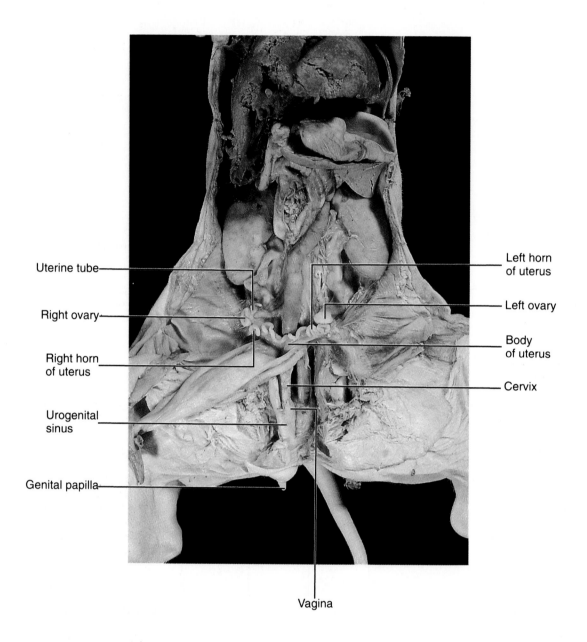

Index

NOTES

NOTES

NOTES

NOTES

NOTES

NOTES